COOPERATION AND COLLECTIVE ACTION

Cooperation & Collective Action

Archaeological Perspectives

EDITED BY DAVID M. CARBALLO

UNIVERSITY PRESS OF COLORADO
Boulder

Published by University Press of Colorado
5589 Arapahoe Avenue, Suite 206C
Boulder, Colorado 80303

Printed in the United States of America

The University Press of Colorado is a proud member of the Association of American University Presses.

The University Press of Colorado is a cooperative publishing enterprise supported, in part, by Adams State University, Colorado State University, Fort Lewis College, Metropolitan State University of Denver, Regis University, University of Colorado, University of Northern Colorado, Utah State University, and Western State Colorado University.

∞ This paper meets the requirements of the ANSI/NISO Z39.48-1992 (Permanence of Paper).

Library of Congress Cataloging-in-Publication Data

Cooperation and collective action : archaeological perspectives / edited by David M. Carballo.
pages cm
Includes bibliographical references and index.
ISBN 978-1-60732-197-2 (hardback) — ISBN 978-1-60732-208-5 (ebook) (print)
1. Commerce, Prehistoric. 2. Commerce, Prehistoric—Cross-cultural studies. 3. Economic anthropology. 4. Economic anthropology—Cross-cultural studies. I. Carballo, David M., editor of compilation.
GN799.C45C67 2012
306.3—dc23

2012038942

Design by Daniel Pratt

22 21 20 19 18 17 16 15 14 13   10 9 8 7 6 5 4 3 2 1

Contents

List of Figures | **vii**

List of Tables | **ix**

Part I: Theoretical Perspectives

1. Cultural and Evolutionary Dynamics of Cooperation in Archaeological Perspective | **3**
—*David M. Carballo*

2. The Emergence of Social Complexity: Why More than Population Size Matters | **35**
—*Gary M. Feinman*

3. War, Collective Action, and the "Evolution" of Human Polities | **57**
—*Paul Roscoe*

4. The Ritualized Economy and Cooperative Labor in Intermediate Societies | **83**
—*Charles Stanish*

5. Reconsidering Darwinian Anthropology: With Suggestions for a Revised Agenda for Cooperation Research | **93**
—*Richard E. Blanton and Lane F. Fargher*

6. Agency and Collective Action: Insights from North American Historical Archaeology | **129**
—Dean J. Saitta

Part II: Case Studies

7. Free-Riding, Cooperation, and Population Growth: The Evolution of Privatization and Leaders in Owens Valley, California | **151**
—Jelmer W. Eerkens

8. Cooperation and Competition among Late Woodland Households at Kolomoki, Georgia | **175**
—Thomas J. Pluckhahn

9. The Competitive Context of Cooperation in Pre-Hispanic Barinas, Venezuela: A Multilevel-Selection Approach | **197**
—Charles S. Spencer

10. Water Control and the Emergence of Polities in the Southern Maya Lowlands: Evolutionary, Economic, and Ecological Models | **223**
—Benjamin Chabot-Hanowell and Lisa J. Lucero

11. Labor Collectives and Group Cooperation in Pre-Hispanic Central Mexico | **243**
—David M. Carballo

12. Caste as a Cooperative Economic Entitlement Strategy in Complex Societies of the Indian Subcontinent and Sub-Saharan Africa | **275**
—Monica L. Smith

13. The Dynamics of Cooperation in Context: Concluding Thoughts | **299**
—Gary M. Feinman

List of Contributors | **309**

Index | **311**

Figures

1.1 Schematic representations of cooperation and other dimensions of group behavior | **8**

2.1 Relationship between number of potential interactions and community size | **39**

2.2 Relationship between number of types of political officials and the population of the largest organizational unit | **41**

2.3 Curvilinear relationship between scale and complexity | **43**

2.4 Scatterplot of use group population and high-level and low-level integrative facilities | **44**

2.5 Scatterplot of use group population and size of small integrative structures | **44**

2.6 Time from the establishment of the first sedentary agricultural communities to the presence of large villages | **45**

2.7 Two patterns of village growth from the establishment of the first sedentary agricultural communities to the presence of large villages | **46**

2.8 Relationship between population size and increasing complexity for collective and autocratic organization | **49**

5.1 Comparative cases discussed in text | **103**

6.1 Archaeological sites or nearby towns mentioned in text | **132**

7.1 Map of study area, centering on the Owens Valley and showing regional obsidian sources and surrounding geographic features | **156**

7.2 Density of seeds per liter of soil from house floor assemblages in the Owens Valley | **159**

7.3 Obsidian geochemical diversity in households over time in the southern Owens Valley | **161**

7.4 Distribution of microflakes of obsidian and carbon isotope ratios across two house floors | **163**

8.1 The Kolomoki site and the locations of Blocks A and D | **178**

8.2 Close-up of the pit structure in Block A | **181**

8.3 Close-up of the structure in Block D | **185**

9.1 Regional settlement patterns during the Late Gaván phase (AD 550–1000) | **203**

9.2 Map of El Gaván | **204**

9.3 Western Venezuelan llanos and adjacent Andes | **207**

9.4 Rafael Gassón's map of El Cedral | **208**

9.5 Area B excavation on the circumscribing oval earthwork at B12 | **213**

9.6 Drawing of the southwest profile of Mound A at B12 | **214**

10.1 Map of Maya area with sites mentioned in text | **224**

10.2 Flow chart of Smith and Choi patron-client simulation dynamics | **228**

11.1 Central Mexico with sites and regions discussed in text | **245**

11.2 Simplified map of central Teotihuacan | **254**

11.3 Ceremonial centers of three Late Formative sites | **256**

Tables

2.1 Three core dimensions of human social groups | **36**

2.2 Organizational thresholds of human groups | **40**

2.3 Effects of population range on population-complexity correlations | **40**

2.4 Relationship between maximal community size and scale of organizational complexity | **41**

2.5 Relationship between total population size and organizational complexity | **42**

2.6 Range of settlement population sizes in egalitarian societies | **42**

2.7 Relationship between number of administrative levels and maximal community size | **45**

3.1 Population distribution, community size, and the emergence of Big Men | **68**

5.1 The coded societies, indicating values for Collective Action Total | **106**

8.1 Relative frequencies of surface treatments in assemblages of identifiable Woodland pottery from Blocks A and D | **183**

8.2 Relative frequencies of vessel forms identified in MNV analysis of Blocks A and D | **183**

10.1 Per-period payoff structure for Smith and Choi's patron-client simulation | **227**

12.1 Caste occupations in the Indian subcontinent and in West Africa | **276**

13.1 Human behavior: basic principles | **302**

13.2 Perspectives on the preindustrial past | **303**

COOPERATION AND COLLECTIVE ACTION

PART I

Theoretical Perspectives

1

Cultural and Evolutionary Dynamics of Cooperation in Archaeological Perspective

David M. Carballo

Humans are excellent but strategically contingent cooperators. How we cooperate and the boundaries of our cooperative relations are two of the most important organizing principles for social groups. Not surprisingly, the cultural and evolutionary dynamics of cooperation represent a fertile topic of research in social and behavioral sciences such as anthropology, economics, political science, psychology, and sociology (Axelrod 1997; Bowles and Gintis 2011; Boyd and Richerson 1992, 2009; Dovido et al. 2006; Fehr and Schmidt 1999; Gintis et al. 2005; Gurven 2006; Hammerstein 2003; Henrich and Henrich 2007; Marshall 2010; Ostrom, Gardner, and Walker 2003; Patton 2009; Willer 2009). From a contemporary biological perspective, much of human uniqueness is said to rest in our abilities to cooperate at larger scales and in qualitatively different ways than all other animals, including nonhuman primates (Bingham 2000; Hill, Barton, and Hurtado 2009; Mitani 2009; Nowak 2006a, 2011; Sussman and Cloninger 2011; Tomasello 2009; Wilson, Timmel, and Miller 2004; cf. Kappeler and van Schaik 2006). Yet we can also be exceedingly competitive. These two sides of humanity are entwined, and may tragically converge in destructive forms of intergroup competition such as wars, which require high levels of intragroup cooperation and coordination. Disentangling the motivations and institutions that foster group cooperation among competitive individuals remains one of the few great conundrums within evolutionary theory. How, researchers ask, does cooperation evolve and thrive among individuals who strategically pursue self- or kin-interests despite all of the potential obstacles those interests present to group-oriented behaviors? What are the costs and benefits to individuals across the socioeconomic spectrum in participating in, or defecting from, cooperative endeavors? What suite of mechanisms for encouraging and maintaining cooperation exists within any particular society, and how does its composition evolve over time as a result of cumulative goal seeking by individuals and larger-scale environmental processes? Why does cooperation sometimes break down completely?

Archaeologists have been investigating the developmental trajectories of cooperation and competition in past societies for decades, but have tended to emphasize the latter in seeking to explain those processes underlying cultural evolution. As a result, bottom-up possibilities for group cooperation (or "self-organization") have been undertheorized in favor of political models stressing top-down leadership, often invoking compliance through coercion. In the meantime, evidence from a range of disciplines has demonstrated humans effectively sustain cooperative undertakings through a number of social norms and institutions that are applicable to archaeology on multiple analytical scales, including reciprocal exchanges, monitoring the reputation of others, and the retribution or rewarding of transgression or compliance. This important axis of variability in the dynamics of past human societies has received scant attention in archaeological theory, with notable exceptions discussed later in this chapter.

A focus on the interplay between cooperation and competition in past societies necessitates multiscalar approaches that consider the complete spectrum of human behavior, from the broad evolutionary processes instigated by aggregate individual actions, to the motivations for those actions at the level of households or individuals. Such approaches combine many of the strengths of existing theoretical paradigms in archaeology while offering productive means of reconciling entrenched divides between considerations of process and agency (compare Blanton and Fargher 2008; Boyd and Richerson 2008; Cowgill 2000; Feinman, Lightfoot, and Upham 2000; Flannery 1999; Pauketat 2001; Richerson and Boyd 1999; Shennan 2002; Spencer 1993). Contemporary models of cooperation are evolutionary, overlapping comfortably with traditional archaeological interests in elucidating the processes of diachronic social change. But they are also multi-actor, envisioning all individuals as pursuing goals that can be simultaneously individualistic/competitive and collective/cooperative in a manner consistent with approaches that emphasize human agency and strategic action. In turn, the diachronic breadth and material focus of archaeology provide a much-needed complement to existing research on cooperation and collective action, which thus far has relied largely on game-theoretic modeling, surveys of college students from affluent countries, brief ethnographic experiments, and limited historic cases. Archaeological perspectives draw on a comparative record of long cultural evolutionary sequences (Marcus 2008), containing the physical correlates of past cooperation and competition, including the particular resources that were utilized through collective action and the symbols people manipulated to define themselves as cooperative or antagonistic.

The contributions to this volume are not unified by a single paradigmatic approach to cooperation and collective action, yet the authors share the conviction that these issues should be foregrounded within contemporary archaeological discourse in order to better understand their dynamics in varied past and present contexts. Examples include non- or less coercive social mechanisms that operated in smaller-scale societies or in factions that primarily operated independently from the political institutions of larger ones, such as labor groups and social castes within early states and empires. Authors are interested in better defining the terms, appropriate units of analysis, and theoretical frameworks necessary for understanding group cooperation. We present diverse case studies with the aim of situating the diachronic and material foci of archaeology within the interdisciplinary dialogue on this issue of broad social concern. In this chapter I highlight some recent insights from research on cooperation across disciplines, use cross-cultural cases to suggest points of intersection with the archaeological record of cultural evolution, and outline the organization of the volume.

COOPERATION: DEFINITIONS AND APPROACHES

People cooperate within multiple, overlapping, and occasionally conflicting scales of social interaction, and they often do so in ways that are inconsistent with canonical models of rationality and self-interest. The structure of cooperative undertakings is segmentary, nested, and fluid, with the result that individuals who cooperate as groups in certain settings may be adversarial in others. This segmentary structure and the tensions inherent in reconciling individual and group interests pose dilemmas for sustaining cooperation, which has been analyzed within domains as diverse as treaties between sovereign nation-states, ethical codes established by enemies engaged in trench warfare, community-managed irrigation systems, and household recycling (e.g., Axelrod 1984: 73–87; Henrich and Henrich 2006; Ostrom 1990: 69–88; Wagner 1983). Individuals act within vastly different interpersonal parameters across the spectrum of potential cooperative undertakings, and cooperation is surely motivated and sustained by combinations of mechanisms depending on social context. Accordingly, classification of the types of undertakings that could be deemed cooperative and the sorts of mechanisms that promote them assists in comparative analysis and in evaluating the appropriateness of particular approaches to the archaeological record. Definitions of cooperation usually entail some calculation of cost or risk on the part of an individual so that another individual or group of individuals receives a benefit (e.g., Smith 2010; West, Griffin, and Gardner 2007). Yet authors envision the costs, benefits, and goals of cooperation differently, which is reflected in the approaches reviewed in this section, as well as in the subsequent chapters of this volume.

An initial distinction may be drawn between what could heuristically be termed ultimate and proximate causes of cooperation. Approaches focused on ultimate causes are more common in biology, evolutionary psychology, and human behavioral ecology. They seek to explain cooperation in terms of the evolved predispositions that humans are argued to possess that facilitate working in groups, and often discuss culture-gene coevolution (that human genetic evolution has been structured by life within cultural groups) and multilevel selection (that selective processes can operate at individual and group levels) in examining what evolved psychological mechanisms might be conducive to cooperation (e.g., Bingham 2000; Bowles 2006; Boyd and Richerson 1992, 2009; Fessler and Haley 2003; Fuentes, Wyczalkowski, and MacKinnon 2010; Gurven 2006; Henrich and Henrich 2006, 2007; Richerson, Boyd, and Henrich 2003; Sober and Wilson 1998; Traulsen and Nowak 2006; Wilson and Kniffin 1999; Wilson, Timmel, and Miller 2004). Although evolutionary themes focused on ultimate causality are central to the holistic study of cultural practices, the archaeological and historic cases discussed in this volume deal with the evolution of norms, institutions, and symbols that complex societies of the last ten thousand years created and reconfigured through time. Such cases are more aligned with developing proximate explanations regarding how particular cultural patterns either promoted or discouraged cooperation. Theories concerning the biological evolutionary bases of cooperation are incorporated into this volume and introductory chapter, but those relating to cultural evolution—the emergence of norms, institutions, and symbols through archaeological time—are emphasized for this reason.

Further classification of the relevant concepts for understanding cooperation forces us into semantic discussions of occasionally colorful terms, often derived from game-theoretic modeling, such as cheaters, defectors, free-riders, punishers, green beards, altruists, tit-for-tat, mutualism, common-pool resources, and the like. Several concise overviews of terminology have fortunately been compiled, though variability between and within disciplines is apparent (compare Dovido et al. 2006: 21–28; Henrich and Henrich 2006; Kapur and Kim-Chong 2002; Nowak 2006b; Tomasello, Kruger, and Ratner 1993; West, Griffin, and Gardner 2007). West and colleagues (2007: 416) provide an especially succinct glossary from a biological perspective. I draw on these recent works in discussing key terms, but should note that attention to such issues is not new; rather, it extends back centuries to earlier social theorists.

Conceptualizing Cooperation

The Enlightenment revival of classical scholarship saw many Western philosophers ponder why people participate in social groupings when their motivations tend to be selfish (see Baum 2004). For Hobbes (1958 [1651]: 142), a social contract based primarily on collective defense was the means by which self-interested individuals pursue collective goods through the "Leviathan" that is state governance: "The only way to erect such a common power as may be able to defend them from the invasion of foreigners and the injuries of one another, and thereby to secure them in such sort as that by their own industry and by the fruits of the earth they may nourish themselves and live contentedly, is to confer all their power and strength upon one man, or upon one assembly of men that may reduce all their wills, by plurality of voices, unto one will." Alternatively, in his *Discourse on the Origin and the Foundations of Inequality among Men*, Rousseau (1984 [1755]) considered how cooperation could shape social structure through the analogy of a stag hunt, in which individuals could choose to collectively hunt a larger stag or individually hunt smaller hares. Both scenarios suggest that mutual gains can be achieved through cooperation, but whereas the social backdrop for Hobbes was one of threat and competition, Rousseau's was one of more voluntary collaboration. The ideas of these two authors still resonate with contemporary social theorists, who have elaborated and refined them by approximating payoff matrices for collective or individual behavior drawing on game-theoretic modeling, human behavioral ecology, and related disciplines (Skyrms 2004). Rather than relying solely on the verbal logic of humanistic philosophy, contemporary behavioral-science approaches to decision making employ mathematical logic involving experimentally estimated costs and benefits, impacting archaeology in cases parallel to Rousseau's analogy, such as through the application of optimal foraging theory (e.g., Gremillion 2002).

Early anthropologists also took an interest in cooperation. Most notably, Mead's (1937c) edited volume *Cooperation and Competition among Primitive Peoples* assembled leading scholars of the time who presented ethnographic cases relating to the behavioral dynamics of these opposed tendencies within a number of societies that would be classified as "intermediate" by many contemporary archaeologists, because they are neither egalitarian nor do they possess institutionalized hierarchies. The modern equivalent of Mead's volume might well be *Foundations of Human Sociality: Economic Experiments and Ethnographic Evidence from Fifteen Small-Scale Societies* (Henrich et al. 2004), and a comparison of the two serves to develop a framework of terms and concepts that are applicable to archaeology. Soon after the Mead volume, Murdock (1945) included cooperative labor as one of the behaviors he deemed to be cultural universals (see also Brown 1991: 137–139). Later generations of anthropologists would object to the way that earlier approaches discussed cultural practices as habits, and their implicit assumptions of immutable group psychologies, as exemplifications of the normative model of culture (e.g., Fox 1991; Geertz 1973: 33–54). Indeed, an emphasis on individual action within anthropology began shortly after these publications (Hays 1958: 394–404). I return to this critique in outlining contemporary approaches to cooperation, which I argue dovetail with current archaeological interests in considering both process and agency in envisioning how individuals within groups strategize, moving us from the normative model to one of negotiated norms. But first I outline a few of the insights from Mead's volume and their relationship to contemporary models, such as those used in the volume by Henrich and colleagues.

In her introductory chapter, Mead (1937a) discussed cooperation and competition not as a singular axis, but rather as orthogonally balanced by individualistic

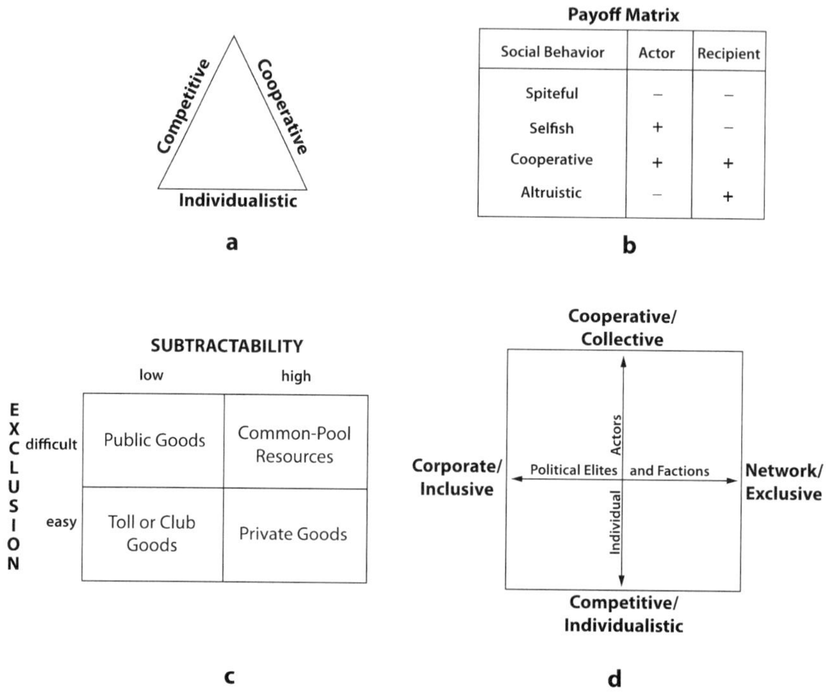

FIGURE 1.1 *Schematic representations of cooperation and other dimensions of group behavior: (a) Margaret Mead's triadic classification of intermediate societies (modified from* Mead 1937b: *461); (b) idealized payoff matrix for social behaviors, which following biological frameworks classify impact in terms of reproductive success (e.g.,* West, Griffin, and Gardner 2007*); (c) classification of types of resource problems by Elinor Ostrom and colleagues (modified from* Ostrom, *Gardner, and Walker* 1994: *Figure 1.1); (d) comparative axes drawing on the work of* Blanton et al. (1996) *and* Feinman, Lightfoot, and Upham (2000: *Figure 1).*

behaviors (Figure 1.1a). These three terms were then distinguished as (1) competitive: individuals striving to gain what another is simultaneously striving to gain; (2) cooperative: individuals working together to one end; and (3) individualistic: individuals striving toward goals without reference to others (Mead 1937a: 8, 16). In concluding the volume, Mead (1937b: 461) classified the societies discussed along a triangular plot with these three terms as midpoints. Mead's definitions for competitive and individualistic behaviors are straightforward, while her definition of cooperation deserves some unpacking for our purposes because it implies mutualism (Tomasello 2009: 41) or mutual benefit (West, Griffin, and Gardner 2007: 416), as used by contemporary scholars. In the sense of ultimate causality, Tomasello (2009) argues that mutualism among early human ancestors was the primary selective pressure affecting the evolution of our faculties promoting cooperation. He defines collaboration for mutual benefits as a phenomenon of cooperation, of which altruism is another.

Definitions of cooperation involving mutualism find wide support within many disciplines, but definitions of altruism vary greatly, with some researchers questioning whether "pure" altruism, involving a cost with no payoff, actually exists. West and colleagues (2007: 419–423) outline several uses of the term *altruism* and critique its application to cases when the cost-benefit matrix for an individual actor is anything other than a net decrease in fitness relative to another actor or group of actors, measured over the lifelong effects of that behavior on all parties (Figure 1.1b). For instance, reciprocal altruism has cost-benefit matrices that offset or are mutually beneficial ("a win-win situation"), and is therefore not pure altruism according to these authors. Other behaviors potentially defined as altruistic may carry more cost than benefit for an individual actor, but those effects could be mediated by genetic relatedness, such as in calculations of inclusive fitness, whereby actions are considered by their impact on the aggregate fitness of all the relatives of that individual who may potentially be impacted, weighted by genetic relatedness (Grafen 1984; Hamilton 1964, 1972; cf. Nowak, Tarnita, and Wilson 2010). This definition of altruism is significantly more restrictive than one from social psychology, in which an act may be considered altruistic if it is performed with no expectation of rewards or benefits to one's self (Dovido et al. 2006: 26). Mead (1937a: 17) also considered the necessity of distinguishing between cooperative behaviors and help. In cooperation, shared goals or mutual benefit keep individuals working in a coordinated manner, while for help the immediate goal only benefits a single individual, but the relationship between helper and helped is shared. Mead's distinction of these terms is similar to the one drawn in contemporary approaches between public goods or common-pool resource problems and forms of direct reciprocity that operate at smaller scales, such as is common for planting or harvesting between households in agrarian societies.

Public Goods and Common-Pool Resource Problems

The vast, multidisciplinary corpus of literature on public goods and common-pool resources is of significant value to archaeology. Some of the central tenets of public goods are encapsulated by the familiar parable of the tragedy of the commons (Hardin 1968, 1998), which stipulates that people face certain cooperative dilemmas in which it is in everybody's individual interest to pursue one strategy (in this parable, grazing one's animals as much as possible on communally owned pasture) that is at odds with the collective interests of the community as a whole (in this case, to avoid overgrazing).

In his classic work on collective action, Olson (1965: 28) noted that for something to truly be a public good, its benefits must be available to all and cannot be easily excluded (see also Hardin 1982: 17–20). Contemporary theorists continue to distinguish between excludable and nonexcludable goods problems (e.g., Boyd and Mathew 2007; Ostrom and Walker 1997). Excludable goods are those in which the benefits of some undertaking can be excluded from a noncontributing segment of the population to a significant degree (Figure 1.1c). Nonexcludable goods confer important benefits on the members of a population irrespective of an individual's or faction's level of participation. This distinction is used to separate private or toll/club goods (where exclusion is easy) from public goods or common-pool resources (where exclusion is difficult), and is of greatest relevance for defining cooperative dilemmas, as individuals can only free-ride if it is difficult to exclude them from benefits (Ostrom, Gardner, and Walker 1994: 6–8). Subtractability presents another important axis of variability in defining the degree to which the exploitation of a resource precludes others from doing the same. Common-pool resources and private goods are by definition unevenly distributed among a population, and have high subtractability, while public and toll/club goods are theoretically available to anyone, but cultural institutions may regulate the means and frequency with which they are exploited (see also Acheson 2011).

As examples, economic collectivities such as guilds, in which individuals cooperate in different aspects of the production and/or distribution of craft items, can easily exclude nonparticipators from the goods or the profits derived from them; the end products are largely private or club goods, depending on the subtractability of the materials involved. However, the building of a palisade around a community engaged in violent conflict with neighbors confers a public-goods benefit to all residing within that palisade irrespective of whether they contributed to its construction. Zero-sum resources, which can be overexploited to the point of no or very low regeneration (high subtractability) and where exclusion is possible, represent classic common-pool resource problems and include illuminating cases of self-organization in undertakings such as forest and fishery maintenance (e.g., Ostrom 1990). Between these more clear-cut examples lie many subtler ones, with culturally or historically contingent dimensions that could result in their being excludable or nonexcludable goods problems. For instance, the benefits derived from the construction and maintenance of an irrigation system may be more excludable if diversionary canals can be managed (i.e., open/closed) or landholding is more private, but they become nonexcludable in situations of low management or more collective landholding—making irrigated farmland in such cases a common-pool resource. Likewise, cooperative craft production is less excludable and becomes more of a public good if the products are intended for a redistributive economic system, in which the households within a community will all receive a share (e.g., Stanish 2004). The resource extraction involved in craft production could also create cooperation problems in cases where scarcity affects net yields for crafters, whereby high subtractability results in a common-pool resource problem (e.g., Ostrom and Gardner 1993: 93).

Nonexcludable goods problems have been of significant interest to theorists who model cooperation because they represent the strongest cases of free-rider dilemmas: where nonparticipators can reap the benefits of collective action. Nevertheless, problems involving relatively more excludable goods characterize a wide array of cultural phenomena that researchers are interested in explaining, including archaeologists. Common-pool resource problems include cases of collective ownership with some possibility of exclusion, while their high degree of subtractability makes overexploitation a possible cause for social crises or outright collapse. Ostrom's work on such problems draws on a number of multigenerational and cross-cultural cases that should be of great interest to archaeologists, including the local management of irrigation systems and other ecological resources (1990, 1992; Ostrom, Gardner, and Walker 1994).

Just as the free-rider dilemma affects the expected social dynamics between excludable and nonexcludable goods problems, it may also be used to distinguish between cooperation and coordination. Coordination problems may be defined as those in which an individual's level of participation is a function of how many other individuals participate (Chwe 2001). For example, a critical mass of people must arrive in costume for an event to qualify as a successful "costume party," rather than a party with a few oddballs in costume, and individuals are more likely to dress unconventionally if there are assurances that others will as well (e.g., from previous experience or communication with other partygoers). Within group coordination problems, conformity is usually beneficial to all individuals, while defecting from the norm may even incur greater costs to an individual defector. As an example of the latter, cultures may decide to drive on the right side of the road or the left side of the road, and individuals benefit from conforming to the system around them, whereas they incur costs by driving on the left side in the United States or the right side in the United Kingdom (Henrich and Henrich 2006: 242). Free-riding in such situations is not an issue, therefore, and while the particular dynamics of how groups coordinate their actions remains an essential issue in the social sciences, coordination problems do not pose the same evolutionary conundrum as do cooperation problems in terms of reconciling individual and group interests.

Mechanisms Promoting Cooperation: The Four Rs

Many contemporary evolutionary approaches emphasize the following four mechanisms in promoting cooperation, conveniently all beginning with the letter *r*: (1) reciprocity, (2) reputation, (3) retribution, and (4) rewards (compare Baumard 2010; Boyd and Richerson 1992, 2009; Boyd et al. 2003; Fehr and Gächter 2000; Fehr and Schmidt 1999; Henrich and Henrich 2006, 2007; Milinski, Semmann, and Krambeck 2002; Nowak 2006b; Ostrom and Walker 1997; Richerson, Boyd, and Henrich 2003). Reciprocity is perhaps the best known within anthropological archaeology, following classic ethnographic cases such as the Trobriand *kula* exchange and Pacific Northwest potlatch, and has already been incorporated into archaeological theory. Game-theoretic models such as tit-for-tat (i.e., "You scratch my back, I scratch yours") center on reciprocity and have been influential in theorizing contemporary international politics (e.g., Axelrod 1997). Tit-for-tat models have many iterations depending on how forgiving actors are considered to be in cases of deviation from reciprocal relations. Some researchers critique these models on the grounds that they only work well when groups are small and there is little restraint in the transmission of information on cooperative intent among actors, which is usually not how humans interact in real-world settings (e.g., Henrich and Henrich 2007: 51). They argue that such mechanisms must be bolstered by other, stronger forms of affiliation such as kinship and ethnic marking as strategies for determining with whom one should reciprocate (e.g., Dawkins 1976: 89; Gil White 2001; Hamilton 1964, 1972).

Reputation is directly tied to reciprocity because, aside from kinship or other corporate-group ties, individuals should make decisions concerning the choice of partners based on favorable or unfavorable information (Baumard 2010). Indeed, reputation is often referred to as *indirect reciprocity* in the cooperation literature, and negative reputations—which may be justly earned based on previous behavior, or unjustly earned through malicious gossip—could be used by third parties as a light form of retribution (Nowak and Sigmund 2005). Reputation effects within communities may act as strong leveling mechanisms in certain instances or, alternatively, could be effectively manipulated by individuals in order to aggrandize themselves through strategic displays of largesse. The aggrandizing form relates to behavioral ecology models of costly signals within cooperative undertakings—meaning, signals that reliably convey an individual's likelihood of cooperating and are not worth the cost for a potential free-rider to attempt to fake (see Gintis, Smith, and Bowles 2001; Smith 2003; Smith and Bleige Bird 2005). Theorists focused on more proximate explanations for cooperation more frequently refer to mutual monitoring and the generation of common knowledge (e.g., Chwe 2001; Ostrom 1990; Ostrom, Gardner, and Walker 1994; Ostrom and Walker 1997), with the same understanding that individuals who have been witnessed transgressing suffer poor reputations. The process of mutual monitoring is of direct relevance to archaeology due to the spatial component of how actions are monitored by members of communities, and the topic is explored further in the next section.

Retribution may be better known under the more frequently used terms *punishment* or *sanctioning*. In either case, it is a central feature of many proposed frameworks for the evolution of cooperation (e.g., Boyd et al. 2003, 2010; Henrich and Boyd 2001; Henrich et al. 2006; O'Gorman, Henrich, and Van Vugt 2009; Ostrom and Walker 1997; Richerson, Boyd, and Henrich 2003). A number of mathematical models suggest that retribution against defectors and free-riders—along with retribution of individuals who do not punish such transgressions (i.e., second-order free-rider problems or third-party sanctions)—can effectively stabilize norms of cooperation within a population. It should be noted that models of the formalization of retribution suggest that it can serve to stabilize virtually any norm within a population (Boyd and Richerson 1992), but those populations that cooperate are hypothesized to possess group-selection advantages over those that do not.

Overcoming the second-order free-rider problem through the development of norms of retribution against individuals who do not punish is termed *strong reciprocity*. In terms of ultimate causality, Bowles and Gintis (2004) model how strong reciprocity could proliferate in even heterogeneous populations based on traits that are unique to humans, particularly language and remote punishment. They acknowledge that Hobbes and earlier classical philosophers emphasized social punishment in the maintenance of cooperative human behavior (Bingham 2000: 49; Bowles and Gintis 2002: 419), yet contemporary approaches have benefited from centuries of historical and ethnographic cases to reflect on, as well as from a range of continually more refined cost-benefit calculations derived from mathematical models and simulated experimental cases. For instance, Bingham (2000) explores ultimate evolutionary explanations in proposing that the selective roots of cooperation through retribution lie in the reduced costs of punishing that developed within our early *Homo* ancestors as a result of their adoption of group-coordinated remote killing using projectile technologies (i.e., thrown rocks or spears). Given the individual and group benefits of cooperation and the reduced costs of enforcement, Bingham outlines a "coalitional enforcement hypothesis" for human uniqueness. Unlike other animals, humans can physically punish individual transgressors collectively, but transgression need not be culturally defined exclusively as cheating in cooperative endeavors—as many rationales for death by stoning make (painfully) clear. In terms of more proximate causality, Dubreuil (2008) notes that any evolved tendencies for strong reciprocity do not explain the scales of cooperation seen in complex human societies, and the cultural evolution of larger social groups required divisions of labor associated with retribution. Further, Baumard (2010) suggests that the ethnographic record of small-scale societies does not support a central role for retribution in group cooperation.

To our list of mechanisms promoting cooperation following these first "three Rs" we could add a fourth based on more recent studies: rewards. Several classic works in sociology list rewards as a natural counterpart to sanctions, or the latter as being defined either positively or negatively (e.g., Giddens 1979; Parsons and Smelser 1956). Contemporary modeling efforts and experimental studies may suggest that rewarding cooperative behaviors, or withholding reciprocity from noncooperators, encourages cooperation more effectively than does retribution (Ohtsuki, Iwasa, and Nowak 2009; Rand et al. 2009; Rand, Ohtsuki, and Nowak 2009). This work is more recent within the evolutionary literature, and is currently supported by fewer models and experiments, but it is certainly a line of investigation worth following as it carries important implications for understanding the dynamics of cooperation.

Cooperation and Collective Action in Cultural Evolution

The diverse disciplines represented above reflect the broad concern for the issues discussed in this volume, and the truly multidisciplinary nature of researching cooperation and collective action. It is only natural that lively debates exist within and between these strands of research, including over terminology and even the very terms *cooperation* and *collective action* themselves. Some tensions relate to differing goals, as with seeking ultimate versus proximate causes; others have to do with scalar issues, such as population size, or differing views on the importance of culture or history (Is an ant colony *really* like a complex human society?); while yet others are methodological debates regarding the relative merits of mathematical modeling, experimental games, ethnographic observation, historical or sociological survey, or other social- and behavioral-science methods that operate at variable levels of abstraction, rigor, and verifiability.

In general, the researchers cited above discussing the "evolution of cooperation" tend to look for more ultimate causes; think in terms of smaller scales (at least for human groups, but not for cells, social insects, and other biota); and draw primarily on game-theoretic modeling of costs and benefits with the understanding, following natural selection, that if some action does not confer a greater benefit than cost, immediately or somewhere down the line of reproductive success, then it should not proliferate in a population. Researchers discussing "collective action" tend to emphasize more proximate, historically contingent causes, rather than cost-benefit matrices; think in terms of large human groups with marked differences in power, wealth, and hierarchy; and favor the compilation of case-based surveys as an analytical method. These differences are apparent in the contributions to this volume, but so too are their important points of intersection.

Regularization of the relevant concepts is useful for drawing comparisons across disciplines and culture regions. Applying the terminology outlined above to describe something familiar, such as contemporary recycling practices, serves as an illustration. The goals of recycling programs (minimizing pollution and the unnecessary exploitation of nonrenewable resources) represent nonexcludable goods problems (public goods or common-pool resources) involving costs and benefits that may be assessed at the level of individuals, communities, or the entire planet. Most recycling programs are voluntary, and the reputation impacts of compliance vary greatly by community and the context and visibility of associated actions; for instance, there are relatively high reputation impacts on many college campuses. However, some municipalities have begun to levy fines on individuals for not recycling, employing retribution, punishment, or negative sanctioning for compliance. The payoff matrices for these cases are thereby different. In states with redemption values for cans or bottles, organized groups of individuals (often households) collect them for mutual gains that are excludable from other individuals who are not participating in the enterprise, making it classifiable as a toll good. Taken together, we see that a single issue involves a range of potential cultural dynamics that implicate differing forms of cooperation.

Some readers may question the utility of reductive logic such as mathematical modeling and experimental games, used by researchers who favor *both* the terms *cooperation* and *collective action* to characterize what they study (e.g., Henrich et al. 2004; Ostrom, Gardner, and Walker 1994, 2003). A critique might be that simplified exercises cannot account for the complex webs of action and meaning that all humans perpetuate and act within, studied by more traditionally allied social-science disciplines, like in the works of Mead, Giddens, and Parsons cited above. Yet the simplification of models and experiments does not purport to capture all of the intricacies of cultural interaction; rather, they provide the opportunity to reduce the spectrum of potential variables in order to examine whether certain premises are logically sound (Maynard Smith 1982; McElreath and Boyd 2007). These premises need to then be evaluated with data, which might include ethnographic observation, sociological or historical survey, archaeological materials analysis, or other methods of cataloging and quantifying human behavior (see Gurven and Winking 2008). Modeling assists in developing multiscalar perspectives on cultural evolution that simultaneously consider broad processes, group action, and individual decision making (e.g., Kohler and Gumerman 2000; Kohler and van der Leeuw 2007).

Research on cooperation and collective action may be evaluated in light of the archaeological record of cultural evolution, including issues such as why humans formed larger and more internally differentiated groups through time, and the importance of material symbols to group identification and affiliation. The above literature makes two important points regarding the application of these concepts to archaeology: (1) helping and altruism are behavioral phenomena that are best understood at the level of the individual, while cooperation is best understood at the level of the group (Dovido et al. 2006: 269); and (2) cooperation should be viewed as a process that can be mediated by any number of social institutions, rather than as a discrete event (Boyd and Richerson 2008; Ostrom 1990; Ostrom and Walker 1997). These points relate to archaeology in that, while archaeologists should be aware of work in other disciplines on individual motivations and interactions, archaeological data are usually better suited to understanding cooperation within and among groups: households, corporate factions, communities. Further, the unique archaeological perspective on the processes by which institutions originate and develop through time is well suited to addressing cooperative dilemmas and adding a more diachronic perspective on how human groups looked to solve them. The approaches reviewed in the rest of this chapter are among those that are examining and refining the logic of models and experiments through the study of diachronic change in material culture that defines archaeology.

DEVELOPING ARCHAEOLOGICAL PERSPECTIVES ON COOPERATION

The archaeology of complex societies has progressed significantly beyond the identification of indices of complexity (asking: *Is* a society complex?) toward a better appreciation of the manner in which myriad possibilities for complex human action intersect with one another to create, sustain, and dissolve social institutions (asking: *How* are societies complex?). The behavioral axis of cooperation and competition is a major determinant of how groups organize themselves, and is therefore fundamental to archaeological explanations that consider the broadest possible array of social institutions. Institutions may be defined as "sets of working rules that are used to determine who is eligible to make decisions in some arena, what actions are allowed or constrained, what aggregation rules will be used, what procedures must be followed, what information must or must not be provided, and what payoffs will be assigned to individuals dependent on their actions" (Ostrom 1990: 51). We know that some level of cooperation takes place in all societies, but how do individuals and groups cooperate in particular cases, and how do these relations define the organization and stability of their institutions?

Archaeological investigations drawing on cooperation and collective action theory have increased over the last two decades, but in many cases scholars attempt to reinvent the wheel by ignoring the large multidisciplinary dialogue on these issues. In this section I highlight three ways in which that literature is relevant to archaeology: (1) the consideration of public goods or common-pool resource problems within the natural and social environments of past peoples; (2) greater attention to self-organization in the emergence of institutions, differential political strategies on the part of actors and factions among and within institutions, and bottom-up collective action as checks on, and resistance to, top-down power; and (3) clearer elucidation of the material correlates of mutual monitoring, signaling of affiliation, and social obligations associated with community membership. Any single archaeological approach may incorporate many of these dimensions, but the discussion is divided thematically in order to differentiate the possible conflicts of interest people faced, the institutions they developed to deal with them, and the symbols and built environments that served to mediate such behavioral dynamics.

Resource Problems

Public goods and common-pool resource problems have long been incorporated into archaeological perspectives on the organization of human societies, with many of these problems proposed in earlier literature as "prime movers" that catalyzed, sustained, or diverged particular cultural evolutionary trajectories. Far fewer investigations have incorporated the logic of cooperation or collective action theory. We will begin by considering land and water, those essential elements for human subsistence, and how they may be conceptualized in terms of their associated resource problems for groups of conditionally cooperative actors. Next we consider resource problems involving warfare and economic specialization.

The potential for past systems of land tenure to have created resource problems consistent with the tragedy of the commons has been explored by Kohler (1992), who combines ethnohistoric information on Puebloan collective land holding and use rights with archaeological indices of population increase and resource competition in explaining the appearance of architectural units designated as field houses in the American Southwest. Kohler argues that families attempted to establish more private claims to agricultural plots by building field houses following the overexploitation of land in a system that was previously more loosely regulated. Bayman and Sullivan (2008) build on Kohler's logic in examining two other regions of the Southwest. They argue that agricultural terraces in the Trincheras region and public mounds in the Hohokam region may be indicative of shifts from systems in which land was treated as a common-pool resource to ones in which it became more privatized. Both of these studies succeed in demonstrating how common-pool resource theory might be applied to archaeological cases. They also make judicious use of ethnohistoric data in proposing what social norms and institutions may have regulated one of the more elusive dimensions of past economies lacking textual evidence (land tenure), and the possible archaeological indices of the operation or absence of those norms and institutions.

Eerkens (1999) has explored similar issues but among forager populations in arid portions of Southeastern California. He draws on the work of Ostrom (1990) and ethnographic data from neighboring regions to suggest that foraging territories were managed as common-pool resources, with disputes having been resolved by face-to-face interaction and symbolic signaling of affiliation during the first to mid second millennium. Eerkens (2004) documents a shift approximately six hundred years ago to more intensive seed collecting, which he attributes to increased privatization of resources and decreased norms of food sharing with a rise in regional population. Eerkens builds on his earlier work in chapter 7. Kohler and van West's (1996) study of households in the Mesa Verde region focuses on food sharing as well. These authors argue that the cooperative pooling of food coincides with village nucleation creating social circumscription, integrative rituals within kivas, and relatively high agricultural yields, rather than the low population and low yield assumptions for when food pooling might serve as a buffer against risk.

The control and management of water resources has been debated as factor in the development of complex societies for decades, particularly associated with the works of Steward (1955) and Wittfogel (1957). Debates include whether sophisticated irrigation systems preceded or followed complex social intuitions, and how these systems could be developed and maintained without centralized organization such as state governance or temple ritual (Hunt 1988; Mitchell 1973). Scarborough (2003) provides a recent overview in which he proposes that past societies differed in whether they addressed water systems through greater division of labor, more advanced technologies, or more diversified and decentralized strategies (see also Janusek and Kolata 2004). Contemporary water systems and historically documented cases that span centuries, such as those that Ostrom (1990, 1992) has discussed, are directly relevant to archaeological models. In analyzing these cases as common-pool resource problems, Ostrom (1990: 27) emphasizes the importance of solving problems of individual commitment and the threats posed by free-riding through mutual monitoring and the maintenance of legitimate institutions with clearly defined rules. As her definition cited at the beginning of this section makes clear, institutions may be more centralized or more decentralized within archaeological cases like those reviewed by Scarborough. Chabot-Hanowell and Lucero bring a new perspective to these issues in chapter 10 (see also Lucero and Fash 2006).

Spencer's (1993) model for the evolution of institutionalized leadership features common-pool resources and public goods prominently, though not explicitly using these terms, and suggests a relationship between such issues and the development of formal, heritable leadership through a process that Boyd and Richerson (1985) termed *indirect bias* in cultural transmission, but now is more descriptively termed *prestige bias* (Henrich and Henrich 2006). One of the common-pool resources Spencer (1993: 48–58) discusses is the irrigation system built in Mexico's arid Tehuacán Valley during the first millennium BC. He argues that the coordination and monitoring of participation in the construction of a large dam and in the maintenance of canals may have permitted a leader with achieved status, within a more egalitarian setting, to persuade community members to accept a "leadership package" that included rank differences, or ascribed status, for members of their family. Early institutionalized leadership may therefore have evolved in certain cases through cooperation dilemmas that were confronted by communities through accepting institutionalized retribution coordinated by a single or limited number of individuals (O'Gorman, Henrich, and Van Vugt 2009; Richerson and Boyd 1999, 2001; Van Vugt, Hogan, and Kaiser 2008). These models demonstrate how cost-benefit considerations may be assessed at the level of individuals and the group, or among agents and their aggregate actions that create communities and drive cultural change. They suggest that an important avenue by which individuals or groups gain power in intermediate societies is success in coordinating and sustaining large-scale cooperation that benefits a community or a large segment of one.

The model developed by Spencer is also applicable to raised fields and warfare in the context of competing villages in western Venezuela during the first millennium AD (Spencer 1993: 58–69). While raised fields may have constituted another common-pool resource system (Spencer, Redmond, and Rinaldi 1994), defensive works, like those of the largest town Spencer discusses, may be considered a public good because the benefits of attacks from raiding neighbors cannot be excluded from individuals residing within the defensive structure. In this case, safety and/or group coercive abilities are considered a resource and, just like with land tenure or water systems, social hierarchy may have resulted as an unintended consequence of groups attempting to mediate free-riding dilemmas. Roscoe (2009) discusses similar dynamics relating to warfare within the ethnographic and historic record of New Guinea. He suggests that social signaling in small-scale societies is an institutional response that addresses conflicts of interest in cooperation through rewarding individuals who contribute more to group viability in violent conflict, serving to catalyze incipient hierarchies in the process (see also Shennan 2002: 239–261; Turchin and Gavrilets 2009). Roscoe (chapter 3) and Spencer (chapter 9) elaborate on many of these points in their contributions to this volume.

A final set of resource problems revolve around economic goods such as crafts. Evolutionary models suggest that economies of scale change the potential payoffs associated with cooperative dilemmas (Boyd and Mathew 2007; Kaplan, Hooper, and Gurven 2009; Matheau and Boyd 2009). As discussed above, the division of labor involved in creating economies of scale is generally an excludable goods issue, unless those goods are intended for a redistributive system. The fact that members of some societies relinquished their productive autonomy for the Faustian bargain of greater efficiency combined with greater social inequality is one of the central topics in the cultural evolution of complex societies (Henrich and Boyd 2008). Stanish (2004; Stanish and Haley 2005) has explored this issue by drawing on contemporary cooperation theory and develops his arguments in chapter 4 (see also Shennan 2002: 165–168).

Some of the potential resource problems just noted are environmentally specific (such as irrigation networks or fisheries), while others are widespread (such as forest or soil maintenance), or have the potential to be present anywhere there are people (exchange networks or warfare/defense). Studies of cultural evolution should consider the relevance of any number of problems to a particular study region and assess what social institutions and strategies mediated such problems.

Institutions and Strategies

Archaeologists interested in the evolution of social institutions must address the interrelatedness of individual and group strategies and the tensions in analyzing variable scales of action. Over the last few decades, frameworks often termed *aggrandizer models* have focused on how individual actors compete for power within a milieu of aggregate actions or processes (e.g., Clark and Blake 1994; Flannery 1999; Hayden 1995). Also important are those frameworks that focus on institutional variability related to group and individual strategies, often termed the *corporate/network* or *inclusive/exclusive spectrum* (e.g., Blanton et al. 1996; Feinman 1995, 2010; Feinman, Lightfoot, and Upham 2000; Renfrew 1974). Such approaches have resonated within anthropology for some time. Within the competitive-cooperative-individualistic triad and the scalar distinctions between collective and individual activities proposed by Mead (1937b) are many of the elements of the corporate/network spectrum of political strategies. These axes of variability have heuristic value for understanding the goal-seeking behaviors of individual actors across the socioeconomic spectrum and those of political elites and factions, respectively (Figure 1.1d).

In building from these frameworks and drawing more explicitly from cooperation and collective action theory, archaeologists have come to question the utility of models for achieving group cohesion based primarily on coercion (e.g., Blanton and Fargher 2008, 2009; Kohler, VanBuskirk, and Ruscavage-Barz 2004; Shennan 2002: 206–238; Stanish 2004). Coercive theories must be tempered by considering self-organization and resistance while still allowing for the significant social and institutional inequalities in influence and power that exist in complex societies. In the evolutionary conundrum of why individuals might accept inequality, contemporary cooperation theory considers how despotic motives and group demand for rulership may be entangled (Kaplan, Hooper, and Gurven 2009), as was explored by Spencer (1993). Blanton and Fargher (2008) present a detailed treatment of collective action in premodern states drawing especially from Levi's (1988) work on mutual accommodation between principals (rulers) and taxpayers (commoners). They demonstrate a range of potential collective enterprises within state societies and a variety of factional strategies based on factors of class, economy, cultural norms, and identity. Saitta (2007) and Galle (2010) both incorporate historical texts and archaeological remains in showing how collective action among socioeconomically disenfranchised groups constituted active forms of resistance against coercive authority in two distinct chapters of US history: industrialization and antebellum slavery, respectively. Feinman (chapter 2), Blanton and Fargher (chapter 5), and Saitta (chapter 6) elaborate more on these issues in subsequent chapters.

In considering the relationship between strategies and institutions, models of cooperation offer suggestions for evaluating interpersonal motivations and dyadic interactions, but these are infrequently visible in the archaeological record. Patterning in archaeological data is more often the result of cumulative group behaviors, particularly households, but also corporate kin and economic groups, political factions, communities, polities, and ethnic groups. Developing archaeological approaches to cooperation therefore places one of the discipline's primary units of analysis, the household, at the forefront in attempts to consider both process and agency. Multiactor (*n*-person) models of public goods problems are more appropriate than dyadic games for this level of analysis (Henrich et al. 2004). At broader scales of analysis, cooperation models developed for interactions between limited parties, such as between sovereign polities, are also applicable. Such is the approach followed by Bonhage-Freund and Kurland (1994) in their analysis of the Hodenosaunee, or League of the Iroquois, using tit-for-tat models outlined by researchers such as Axelrod (1984, 1997). The models serve for analyzing peer-polity interactions and diplomacy in the past just as they do for contemporary international politics.

Material Cultures of Cooperation

Moving from the historical cases of cooperation and collective action discussed by many of the authors cited above to prehistoric cases in the archaeological record requires identification of the material correlates of cooperation through reasoned analogy or, if appropriate, the direct-historical approach. The tangible components of cooperation should be discernible in the built environments that past peoples created to facilitate mutual monitoring and group identification, as well as the material symbols that people manipulated to affirm or negate bonds of affiliation and their concomitant social obligations. Theorists have repeatedly emphasized the importance of culturally constructed notions of fairness and the upholding of legitimate institutions through monitoring and retribution (Bowles and Gintis 2002; Fehr and Gächter 2000; Fehr and Schmidt 1999; Gächter and Herrmann 2009; Ostrom, Gardner, and Walker 2003). Bowles and Gintis (2002) note that for cooperation to thrive within communities of self-interested actors, individuals need to perceive that tangible benefits are a result of their participation in cooperative undertakings. They argue that cooperation can be sustained if, "opportunities for mutual monitoring and punishment of noncooperators are built into the structure of social interactions" (Bowles and Gintis 2002: 430).

Along these lines, Stanish and Haley (2005) explore the role of ceremonial architecture in fostering social integration and sustaining cooperative action by social retribution (such as scorn or ostracism) and reward (such as redistributive feasts or earned status recognition) within the context of public rituals in which social roles and responsibilities are mediated transparently, in full view of the community (see also Blanton and Fargher 2008: 22; Ostrom 1990: 73–75; Richerson and Boyd 1999: 271–272). Stanish's work dovetails with longstanding archaeological interests in how ceremonial architecture simultaneously integrated and differentiated members of past societies (e.g., Adler and Wilshusen 1990; Lindauer and Blitz 1997; Lucero 2003; Pluckhahn 2003). The large open plazas, temple platforms, and spacious enclosed communal structures of early complex societies may therefore be productively investigated as venues for community self-organization of cooperation involving mutual monitoring within ritualized settings. I explore these issues further in chapter 11.

Style in the multiple artifact classes that archaeologists study could also be considered as communicating mutual bonds of trust and obligation related to cooperative undertakings. Signals of affiliation as part of group cooperation have been proposed for Southwestern pottery decorations (Kohler et al. 2004) and in the formalization of representations of deities in central Mexico (Carballo 2007). In such cases, belief and ideology are primary motivations for the materialization of supernatural concepts, but the processes by which local traditions converge over large areas also involve group cooperation/integration or prestige competition/emulation, and discerning which processes are more likely for a given area requires combining multiple lines of evidence. Symbols of affiliation and cooperative intent are likely to converge within built environments involving public ritual and mutual monitoring. An excellent example of such convergence is pilgrimage networks, which involve honest and transparent signals of participation and affiliation (Kantner and Vaughn 2012; Vaughn and Van Gijseghem 2007). The material correlates of prestige bias conferred on incipient leaders who coordinate cooperation and punish defectors may include the cultural elaboration of particular material goods that served as signals of prestige prior to more formal distinctions of social rank, as outlined by Plourde (2008, 2009).

The research reviewed above demonstrates how theories of cooperation and collective action are moving archaeological inquiries of topics such as human ecology, social complexity, and group identity in new directions. The broader themes are not new, but the logic differs, with implications for understanding past dynamics such as resource usage, institutional development and collapse, and symbolic behavior. Contributors to this volume explore these dynamics from a range of theoretical and geographic perspectives that we believe will be useful for reconciling the multiscalar tensions in archaeological analysis and for improving understanding of human cooperation through the diachronic and materially focused lens of archaeology.

ORGANIZATION OF THE VOLUME

The chapters that follow are divided into two parts: the first composed of studies focused more broadly on how cooperation relates to cultural evolutionary theory, and the second directed toward elucidating the dynamics of cooperation in particular cases from across the globe. Gary Feinman (chapter 2) continues Part I by considering the relationship between socioeconomic complexity and cooperation, focusing particularly on questions of social scales (i.e., population) and integration (i.e., cooperation). Using comparative data, Feinman notes that while a positive correlation exists between the size of political formations and cooperative undertakings, heterogeneity is apparent among the cooperative institutions of societies with similar population densities. He focuses on potential reasons for divergent historical pathways of social integration, including more group-based (corporate) versus individual-based (network) strategies, and their relationship to the mutualistic and hierarchical motives of individual actors.

Jim (Paul) Roscoe (chapter 3) also discusses scalar issues and their relationship to cooperation. Focusing especially on warfare, he proposes that political organizations may be conceived as possessing a vertical dimension (leaders, control) and a horizontal one (mutualism, particularly in defense). The interrelatedness of these dimensions in what researchers of political evolution term *polities* creates unnecessary divisions between more voluntaristic and more coercive models, and Roscoe cogently argues for their reconciliation based on better definition and analysis of human interests.

In chapter 4 Charles Stanish elaborates further on his (Stanish 2004) model for collective labor and ritualized economy in intermediate societies. He identifies cooperative relations as the most important social variable in what changes through processes of cultural evolution. Drawing on classic ethnographic cases, Stanish argues that ritual is essential for regulating the political economy of labor and redistribution in societies without developed markets and bureaucratic institutions of governance. In short, public ritual permits the predictability necessary to sustain the economic cooperation characteristic of complex societies.

Richard Blanton and Lane Fargher critique facets of the evolutionary cooperation literature reviewed above in chapter 5. They question the validity of evolved models of cognition, central to many evolutionary psychology approaches, and insights derived from experimental games and mathematical modeling, rather than historically or ethnographically based analyses. To bolster reciprocity, reputation, and rewards, Blanton and Fargher propose that social memory, theory of mind, analysis of intentions, and representational understanding (STAR) provides a better set of ultimate, evolved mechanisms for investigating cooperation within human societies. They then incorporate rational-choice theory (e.g., Levi 1988; Lichbach 1996; Ostrom 2007) in evaluating the dynamics of cooperation in the formation of premodern states documented in historical records, arguing for an inverse correlation between divinely sanctified leadership and indices of collectivity within historically documented civilizations.

Rounding out the more conceptually based first part, Dean Saitta reviews cases of collective action from historical archaeology in chapter 6, focusing particularly on the dimensions of race, class, and gender, and their tensions within structures of institutionalized power. He suggests that evolutionary and historical perspectives on cooperation offer different goals and analytical strengths, but can be reconcilable and mutually informative through reflexive archaeological practices.

Part II of the volume contains essays that are more case specific. Contributions by Jelmer Eerkens and Thomas Pluckhahn present cases from North America that focus especially on households, that fundamental unit of analysis which typically represents the finest scale that archaeologists use in evaluating strategic decision making. The analyses of household assemblages from the Owens Valley of California led Eerkens (chapter 7) to suggest a shift in cooperative behavior in which households became more heterogeneous and insular in their cooperative networks, while intravillage cooperation declined yet extravillage cooperation, measured through exchange, may have increased. Eerkens connects these shifts in cooperative practices to foraging strategies, in which households pursued diversified foraging ranges and negotiated reciprocal access with households in other clines that were less likely to free-ride because of potential mutualistic gains.

In chapter 8 Pluckhahn presents a somewhat parallel case based on detailed evidence from two residential areas of Kolomoki, Georgia. These areas suggest a growth in household size associated with more autonomy in food storage and consumption, decreased public ritual, and the introduction of the bow and arrow between the Middle to Late Woodland periods. Pluckhahn concludes that increased subsistence independence through bow hunting may have served to sever certain bonds of community cooperation, but that the changes are more nuanced than a simple binary distinction and instead represent a spectrum of relative competitive and cooperative emphases.

Charles Spencer (chapter 9) applies archaeological data from Barinas, Venezuela, to mathematical models of multilevel selection and suggests that payoff matrices involving intercommunity cooperation and conflict are higher than for those focused at the intracommunity level. Though the backdrop of his case is one of intense competition and warfare, and competition of some sort is implicit in all contemporary evolutionary models of cooperation, Spencer's models provide an archaeologically relevant example of how polity formation can be a noncoercive process.

Benjamin Chabot-Hanowell and Lisa Lucero (chapter 10) use bargaining models as a framework for understanding early Maya settlement in the lowlands and the origins of patron-client relations, as well as their dissolution in the Classic period interconnected with societal collapse. Incorporating a predictive model, they focus particularly on agricultural land and water as critical resource problems for colonizing populations and the densely settled urban landscape on the eve of the political decentralization.

In chapter 11 I draw on many of the frameworks for studying cooperation discussed in this chapter in examining how institutions of collective labor known ethnographically and historically from central Mexico might have operated millennia earlier, during the region's initial period of urbanization. I outline several resource problems that inhabitants of the period would have faced, and their potential mediation involving the formalization of public ritual space and religious symbols, the growth of corporate-kin groups, and the, later, political mobilization of collective labor as tax.

Rounding out Part II is Monica Smith's (chapter 12) comparative case of caste systems in the Indian subcontinent and West Africa. Smith argues that in both instances caste systems arose during periods of dramatic reorganization and social crisis that were, nevertheless, not followed by polity collapse. In such cases, strictly defined cooperative labor groups presented benefits to their members, who may have accepted lower status in exchange for economic stability—another example of how consideration of strategies and trade-offs across the socioeconomic spectrum provides a more nuanced explanation for the emergence of social institutions. Finally, in chapter 13 Gary Feinman concludes the volume by contextualizing the value of research on cooperation and collective action within the trajectory of archaeological theory over the last few decades.

As the study of how individuals act within groups, cooperation research offers a productive means of bridging considerations of process and agency in past societies, and allows archaeologists to consider issues such as resource problems, social institutions, economic production, public rituals, and material symbols in a new light. The archaeological record provides a critical means of evaluating and refining multidisciplinary models of cooperation derived from experimental games and other abstracted exercises. Contributions in this volume attest to the fact that culture-evolutionary theory as a whole is enhanced by archaeological perspectives on the material dynamics of collective behaviors, and their long-term social transformations within diverse regions of the world.

ACKNOWLEDGMENTS

Sincere thanks go to Darrin Pratt, Jessica d'Arbonne, Laura Furney, and the members of the University Press of Colorado for their assistance in preparing this volume. Its genesis is in a session organized for the Society of American Archaeology Annual Meeting of 2009 and a second organized for the American

Association of the Advancement of Science of 2010. I thank the participants in these sessions—including those from the first not represented here: Elizabeth Chilton, Kevin Vaughn, and our discussant, Tim Kohler—for sharing their thoughts on certain themes in this paper, as well as all the contributors to the volume for continuing the dialog through our correspondence. Jeff Brantingham and Jeff Kurland stimulated my early thinking on cooperation and pointed me to relevant works, for which I am grateful. The comments of two anonymous reviewers were very valuable in clarifying portions of this chapter and the volume as a whole.

REFERENCES

Acheson, James M. 2011. "Ostrom for Anthropologists." *International Journal of the Commons* 5 (2): 319–39.

Adler, Michael A., and Richard H. Wilshusen. 1990. "Large-Scale Integrative Facilities in Tribal Societies: Cross-Cultural and Southwestern US Examples." *World Archaeology* 22 (2): 133–46. http://dx.doi.org/10.1080/00438243.1990.9980136.

Axelrod, Robert. 1984. *The Evolution of Cooperation.* New York: Basic Books.

Axelrod, Robert. 1997. *The Complexity of Cooperation: Agent-Based Models of Competition and Collaboration.* Princeton, NJ: Princeton University Press.

Baum, Richard. 2004. "The Origin of Polities: A Preliminary Inquiry into the Evolution of Command and Control in Human Collectivities." *Social Evolution & History* 3 (2): 55–92.

Baumard, Nicolas. 2010. "Has Punishment Played a Role in the Evolution of Cooperation? A Critical Review." *Mind and Society* 9 (2): 171–92. http://dx.doi.org/10.1007/s11299-010-0079-9.

Bayman, James M., and Alan P. Sullivan, III. 2008. "Property, Identity, and Macroeconomy in the Prehispanic Southwest." *American Anthropologist* 110 (1): 6–20. http://dx.doi.org/10.1111/j.1548-1433.2008.00004.x.

Bingham, Paul M. 2000. "Human Evolution and Human History: A Complete Theory." *Evolutionary Anthropology* 9 (6): 248–57. http://dx.doi.org/10.1002/1520-6505(2000)9:6<248::AID-EVAN1003>3.0.CO;2-X.

Blanton, Richard, and Lane Fargher. 2008. *Collective Action in the Formation of Pre-Modern States.* New York: Springer.

Blanton, Richard, and Lane Fargher. 2009. "Collective Action in the Evolution of Pre-Modern States." *Social Evolution & History* 8 (2): 133–66.

Blanton, Richard E., Gary M. Feinman, Stephen A. Kowalewski, and Peter N. Peregrine. 1996. "A Dual-Processual Theory for the Evolution of Mesoamerican Civilization." *Current Anthropology* 37 (1): 1–14. http://dx.doi.org/10.1086/204471.

Bonhage-Freund, Mary Theresa, and Jeffrey A. Kurland. 1994. "Tit-For-Tat among the Iroquois: A Game Theoretic Perspective on Inter-Tribal Political Organization." *Journal of Anthropological Archaeology* 13 (3): 278–305. http://dx.doi.org/10.1006/jaar.1994.1016.

Bowles, Samuel. 2006. "Group Competition, Reproductive Leveling, and the Evolution of Human Altruism." *Science* 314 (5805): 1569–72. http://dx.doi.org/10.1126/science.1134829.

Bowles, Samuel, and Herbert Gintis. 2002. "Social Capital and Community Governance." *Economic Journal* 112 (483): 419–36. http://dx.doi.org/10.1111/1468-0297.00077.

Bowles, Samuel, and Herbert Gintis. 2004. "The Evolution of Strong Reciprocity: Cooperation in Heterogeneous Populations." *Theoretical Population Biology* 65 (1): 17–28. http://dx.doi.org/10.1016/j.tpb.2003.07.001.

Bowles, Samuel, and Herbert Gintis. 2011. *A Cooperative Species: Human Reciprocity and Its Evolution*. Princeton, NJ: Princeton University Press.

Boyd, Robert, Herbert Gintis, and Samuel Bowles. 2010. "Coordinated Punishment of Defectors Sustains Cooperation and Can Proliferate When Rare." *Science* 328 (5978): 617–20. http://dx.doi.org/10.1126/science.1183665.

Boyd, Robert, Herbert Gintis, Samuel Bowles, and Peter J. Richerson. 2003. "The Evolution of Altruistic Punishment." *Proceedings of the National Academy of Sciences of the United States of America* 100 (6): 3531–35. http://dx.doi.org/10.1073/pnas.0630443100.

Boyd, Robert, and Sarah Mathew. 2007. "A Narrow Road to Cooperation." *Science* 316 (5833): 1858–59. http://dx.doi.org/10.1126/science.1144339.

Boyd, Robert, and Peter J. Richerson. 1985. *Culture and the Evolutionary Process*. Chicago: University of Chicago Press.

Boyd, Robert, and Peter J. Richerson. 1992. "Punishment Allows the Evolution of Cooperation (or Anything Else) in Sizable Groups." *Ethology and Sociobiology* 13 (3): 171–95. http://dx.doi.org/10.1016/0162-3095(92)90032-Y.

Boyd, Robert, and Peter J. Richerson. 2008. "Gene-Culture Coevolution and the Evolution of Social Institutions." In *Better than Conscious? Decision Making, the Human Mind, and Implications for Institutions*, ed. Christoph Engel and Wolf Singer, 305–23. Cambridge, MA: MIT Press.

Boyd, Robert, and Peter J. Richerson. 2009. "Culture and the Evolution of Human Cooperation." *Philosophical Transactions of the Royal Society of London. Series B, Biological Sciences* 364 (1533): 3281–88. http://dx.doi.org/10.1098/rstb.2009.0134.

Brown, Donald E. 1991. *Human Universals*. New York: McGraw-Hill, Inc.

Carballo, David M. 2007. "Effigy Vessels, Religious Integration, and the Origins of the Central Mexican Pantheon." *Ancient Mesoamerica* 18 (1): 53–67. http://dx.doi.org/10.1017/S0956536107000028.

Chwe, Michael Suk-Young. 2001. *Rational Ritual: Culture, Coordination, and Common Knowledge*. Princeton: Princeton University Press.

Clark, John E., and Michael Blake. 1994. "The Power of Prestige: Competitive Generosity and the Emergence of Rank Societies in Lowland Mesoamerica." In *Factional Competition and Political Development in the New World*, ed. Elizabeth Brumfiel and John Fox, 17–30. Cambridge: Cambridge University Press. http://dx.doi.org/10.1017/CBO9780511598401.003.

Cowgill, George L. 2000. "'Rationality' and Contexts in Agency Theory." In *Agency in Archaeology*, ed. Marcia-Ann Dobres and John Robb, 51–60. London: Routledge.

Dawkins, Richard. 1976. *The Selfish Gene*. Oxford: Oxford University Press.

Dovido, John F., Jane Allyn Piliavin, David A. Schroeder, and Louis A. Penner. 2006. *The Social Psychology of Prosocial Behavior*. Mahwah, NJ: Lawrence Erlbaum Associates.

Dubreuil, Benoît. 2008. "Strong Reciprocity and the Emergence of Large-Scale Societies." *Philosophy of the Social Sciences* 38 (2): 192–210. http://dx.doi.org/10.1177/0048393108315509.

Eerkens, Jelmer W. 1999. "Common Pool Resources, Buffer Zones, and Jointly Owned Territories: Hunter-Gatherer Land and Resource Tenure in Fort Irwin, Southeastern California." *Human Ecology* 27 (2): 297–318. http://dx.doi.org/10.1023/A:1018777311943.

Eerkens, Jelmer W. 2004. "Privatization, Small-Seed Intensification, and the Origins of Pottery in the Western Great Basin." *American Antiquity* 69 (4): 653–70. http://dx.doi.org/10.2307/4128442.

Fehr, Ernst, and Simon Gächter. 2000. "Fairness and Retaliation: The Economics of Reciprocity." *Journal of Economic Perspectives* 14 (3): 159–81. http://dx.doi.org/10.1257/jep.14.3.159.

Fehr, Ernst, and Klaus M. Schmidt. 1999. "A Theory of Fairness, Competition, and Cooperation." *Quarterly Journal of Economics* 114 (3): 817–68. http://dx.doi.org/10.1162/003355399556151.

Feinman, Gary M. 1995. "The Emergence of Inequality: A Focus on Strategies and Processes." In *Foundations of Social Inequality*, ed. T. Douglas Price and Gary M. Feinman, 255–279. New York: Plenum Press.

Feinman, Gary M. 2010. "A Dual-Processual Perspective on the Power and Inequality in the Contemporary United States: Framing Political Economy for the Present and the Past." In *Pathways to Power: New Perspectives on the Emergence of Social Inequality*, ed. T. Douglas Price and Gary M. Feinman, 255–288. New York: Springer. http://dx.doi.org/10.1007/978-1-4419-6300-0_9

Feinman, Gary M., Kent G. Lightfoot, and Steadman Upham. 2000. "Political Hierarchies and Organizational Strategies in the Puebloan Southwest." *American Antiquity* 65 (3): 449–70. http://dx.doi.org/10.2307/2694530.

Fessler, Daniel M.T., and Kevin J. Haley. 2003. "The Strategy of Affect: Emotions in Human Cooperation." In *The Genetic and Cultural Evolution of Cooperation*, ed. Peter Hammerstein, 7–36. Cambridge, Berlin: MIT Press and Preie Universität.

Flannery, Kent V. 1999. "Process and Agency in Early State Formation." *Cambridge Archaeological Journal* 9 (1): 3–21. http://dx.doi.org/10.1017/S0959774300015183.

Fox, Richard G. 1991. "For a Nearly New Culture History." In *Recapturing Anthropology: Working in the Present*, ed. Richard G. Fox, 93–113. Santa Fe: School of American Research Press.

Fuentes, Agustín, Matthew A. Wyczalkowski, and Katherine C. MacKinnon. 2010. "Niche Construction through Cooperation: A Nonlinear Dynamics Contribution to Modeling Facets of the Evolutionary History in the Genus Homo." *Current Anthropology* 51 (3): 435–44. http://dx.doi.org/10.1086/651221.

Gächter, Simon, and Benedikt Herrmann. 2009. "Reciprocity, Culture, and Human Cooperation: Previous Insights and a New Cross-Cultural Experiment." *Philosophical Transactions of the Royal Society B—Biological Sciences* 364 (1518): 791–806. http://dx.doi.org/10.1098/rstb.2008.0275.

Galle, Jillian E. 2010. "Costly Signaling and Gendered Social Strategies among Slaves in the Eighteenth-Century Chesapeake: An Archaeological Perspective." *American Antiquity* 75 (1): 19–43.

Geertz, Clifford. 1973. *The Interpretation of Cultures*. New York: Basic Books.

Giddens, Anthony. 1979. *Central Problems in Social Theory: Action, Structure, and Contradiction in Social Analysis*. Berkeley, Los Angeles: University of California.

Gil-White, Francisco. 2001. "Are Ethnic Groups Biological 'Species' to the Human Brain? Essentialism in our Cognition of some Social Categories." *Current Anthropology* 42 (4): 515–54. http://dx.doi.org/10.1086/321802.

Gintis, Herbert, Samuel Bowles, Robert Boyd, and Ernst Fehr, eds. 2005. *Moral Sentiments and Material Interests: The Foundations of Cooperation in Economic Life*. Cambridge: The MIT Press.

Gintis, Herbert, Eric Alden Smith, and Samuel Bowles. 2001. "Costly Signaling and Cooperation." *Journal of Theoretical Biology* 213 (1): 103–19. http://dx.doi.org/10.1006/jtbi.2001.2406.

Grafen, A. 1984. "Natural Selection, Kin Selection and Group Selection." In *Behavioural Ecology: An Evolutionary Approach*, ed. J. R. Krebs and N. B. Davies, 62–84. Oxford: Blackwell Scientific Publications.

Gremillion, Kristen J. 2002. "Foraging Theory and Hypothesis Testing in Archaeology: An Exploration of Methodological Problems and Solutions." *Journal of Anthropological Archaeology* 21 (2): 142–64. http://dx.doi.org/10.1006/jaar.2001.0391.

Gurven, Michael. 2006. "The Evolution of Contingent Cooperation." *Current Anthropology* 47 (1): 185–92. http://dx.doi.org/10.1086/499552.

Gurven, Michael, and Jeffrey Winking. 2008. "Collective Action in Action: Pro-Social Behavior In and Out of the Laboratory." *American Anthropologist* 110 (2): 179–90. http://dx.doi.org/10.1111/j.1548-1433.2008.00024.x.

Hammerstein, Peter, ed. 2003. *Genetic and Cultural Evolution of Cooperation*. Cambridge, Berlin: MIT Press and Preie Universität.

Hamilton, William D. 1964. "The Genetical Evolution of Social Behaviour." *Journal of Theoretical Biology* 7 (1): 1–16. http://dx.doi.org/10.1016/0022-5193(64)90038-4.

Hamilton, William D. 1972. "Altruism and Related Phenomena, Mainly in the Social Insects." *Annual Review of Ecology and Systematics* 3 (1): 193–232. http://dx.doi.org/10.1146/annurev.es.03.110172.001205.

Hardin, Garrett. 1968. "The Tragedy of the Commons." *Science* 162 (3859): 1243–48. http://dx.doi.org/10.1126/science.162.3859.1243.

Hardin, Garrett. 1998. "Extensions of "The Tragedy of the Commons"." *Science* 280 (5364): 682–3. http://dx.doi.org/10.1126/science.280.5364.682.

Hardin, Russell. 1982. *Collective Action*. Baltimore: The Johns Hopkins University Press.

Hayden, Brian. 1995. "Pathways to Power: Principles for Creating Socioeconomic Inequalities." In *Foundations of Social Inequality*, ed. T. Douglas Price and Gary M. Feinman, 15–86. New York: Plenum Press.

Hays, H. R. 1958. *From Ape to Angel: An Informal History of Social Anthropology*. New York: Alfred A. Knopf.

Henrich, Joseph, and Robert Boyd. 2001. "Why People Punish Defectors. Weak Conformist Transmission can Stabilize the Costly Enforcement of Norms in Cooperative Dilemmas." *Journal of Theoretical Biology* 208 (1): 79–89. http://dx.doi.org/10.1006/jtbi.2000.2202.

Henrich, Joseph, and Robert Boyd. 2008. "Division of Labor, Economic Specialization, and the Evolution of Social Stratification." *Current Anthropology* 49 (4): 715–24. http://dx.doi.org/10.1086/587889.

Henrich, Joseph, Robert Boyd, Samuel Bowles, Colin Camerer, Ernst Fehr, and Herbert Gintis, eds. 2004. *Foundations of Human Sociality: Economic Experiments and Ethnographic Evidence from Fifteen Small-Scale Societies*. Oxford: Oxford University Press.

Henrich, Joseph, and Natalie Henrich. 2006. "Culture, Evolution and the Puzzle of Human Cooperation." *Cognitive Systems Research* 7 (2-3): 220–45. http://dx.doi.org/10.1016/j.cogsys.2005.11.010.

Henrich, Natalie, and Joseph Henrich. 2007. *Why Humans Cooperate: A Cultural and Evolutionary Explanation*. Oxford: Oxford University.

Henrich, Joseph, Richard McElreath, Abigail Barr, Jean Ensminger, Clark Barrett, Alexander Bolyanatz, Juan Camilo Cardenas, Michael Gurven, Edwins Gwako,

Natalie Henrich, et al. 2006. "Costly Punishment Across Human Societies." *Science* 312 (5781): 1767–70. http://dx.doi.org/10.1126/science.1127333.

Hill, Kim, Michael Barton, and A. Magdalena Hurtado. 2009. "The Emergence of Human Uniqueness: Characters Underlying Behavioral Modernity." *Evolutionary Anthropology* 18 (5): 187–200. http://dx.doi.org/10.1002/evan.20224.

Hobbes, Thomas. 1958 [1651]. *Leviathan, Parts One and Two.* New York: The Liberal Arts Press.

Hunt, Robert C. 1988. "Size and Structure of Authority in Canal Irrigation Systems." *Journal of Anthropological Research* 44: 325–55.

Janusek, John Wayne, and Alan L. Kolata. 2004. "Top-Down or Bottom-Up: Rural Settlement and Raised Field Agriculture in the Lake Titicaca Basin, Bolivia." *Journal of Anthropological Archaeology* 23 (4): 404–30. http://dx.doi.org/10.1016/j.jaa.2004.08.001.

Kantner, John, and Kevin Vaughn. 2012. "Pilgrimage as Costly Signal: Religiously Motivated Cooperation in Chaco and Nasca." *Journal of Anthropological Archaeology* 33: 66–82.

Kaplan, Hillard S., Paul L. Hooper, and Michael Gurven. 2009. "The Evolutionary and Ecological Roots of Human Social Organization." *Philosophical Transactions of the Royal Society of London. Series B, Biological Sciences* 364 (1533): 3289–99. http://dx.doi.org/10.1098/rstb.2009.0115.

Kappeler, Peter M., and Carel P. van Schaik, eds. 2006. *Cooperation in Primates and Humans: Mechanisms and Evolution.* Heidelberg: Springer. http://dx.doi.org/10.1007/3-540-28277-7.

Kapur, Basant K., and Kim-Chong Chong, eds. 2002. *Altruistic Reveries: Perspectives from the Humanities and Social Sciences.* Boston: Kluwer Academic Publications. http://dx.doi.org/10.1007/978-1-4615-0809-0.

Kohler, Timothy A. 1992. "Field Houses, Villages, and the Tragedy of the Commons in the Early Northern Anasazi Southwest." *American Antiquity* 57 (4): 617–35. http://dx.doi.org/10.2307/280826.

Kohler, Timothy A., and George J. Gumerman, eds. 2000. *Dynamics in Human and Primate Societies: Agent-Based Modeling of Social and Spatial Processes.* New York: Oxford University.

Kohler, Timothy A, Stephanie VanBuskirk, and Samantha Ruscavage-Barz. 2004. "Vessels and Villages: Evidence for Conformist Transmission in Early Village Aggregations on the Pajarito Plateau, New Mexico." *Journal of Anthropological Archaeology* 23 (1): 100–18. http://dx.doi.org/10.1016/j.jaa.2003.12.003.

Kohler, Timothy A., and Sander van der Leeuw, eds. 2007. *The Model-Based Archaeology of Socionatural Systems.* Santa Fe: School for Advanced Research Press.

Kohler, Timothy A., and Carla van West. 1996. "The Calculus of Self-Interest in the Development of Cooperation: Sociopolitical Development and Risk among the Northern Anasazi." In *Evolving Complexity and Environment: Risk in the Prehistoric Southwest,* ed. Joseph A. Tainter and Bonnie Bagley Tainter, 169–96. Reading, MA: Addison-Wesley Press.

Levi, Margaret. 1988. *Of Rule and Revenue.* Berkeley: University of California Press.

Lichbach, Mark I. 1996. *The Cooperator's Dilemma.* Ann Arbor: University of Michigan Press.

Lindauer, Owen Lindauer, and John H. Blitz. 1997. "Higher Ground: The Archaeology of North American Platform Mounds." *Journal of Archaeological Research* 5 (2): 169–207. http://dx.doi.org/10.1007/BF02229110.

Lucero, Lisa J. 2003. "The Politics of Ritual: The Emergence of Classic Maya Rulers." *Current Anthropology* 44 (4): 523–58. http://dx.doi.org/10.1086/375870.
Lucero, Lisa J., and Barbara W. Fash, eds. 2006. *Precolumbian Water Management: Ideology, Ritual, and Power.* Tucson: University of Arizona Press.
Marcus, Joyce. 2008. "The Archaeological Evidence for Social Evolution." *Annual Review of Anthropology* 37 (1): 251–66. http://dx.doi.org/10.1146/annurev.anthro.37.081407.085246.
Marshall, Robert C., ed. 2010. *Cooperation in Economy and Society.* Lanham, MD: Alta Mira Press.
Mathew, Sarah, and Robert Boyd. 2009. "When does Optional Participation Allow the Evolution of Cooperation?" *Proceedings of the Royal Society* 276 (1659): 1167–74. http://dx.doi.org/10.1098/rspb.2008.1623.
Maynard Smith, John. 1982. *Evolution and the Theory of Games.* Cambridge: Cambridge University Press.
McElreath, Richard, and Robert Boyd. 2007. *Mathematical Models of Social Evolution: A Guide for the Perplexed.* Chicago: University of Chicago Press.
Mead, Margaret. 1937a. "Introduction." In *Cooperation and Competition among Primitive Peoples,* ed. Margaret Mead, 1–19. New York: McGraw-Hill Book Company, Inc.
Mead, Margaret. 1937b. "Interpretative Statement." In *Cooperation and Competition among Primitive Peoples,* ed. Margaret Mead, 458–515. New York: McGraw-Hill Book Company, Inc.
Mead, Margaret, ed. 1937c. *Cooperation and Competition among Primitive Peoples.* New York: McGraw-Hill Book Company, Inc.
Milinski, Manfred, Dirk Semmann, and Hans-Jürgen Krambeck. 2002. "Reputation Helps Solve the 'Tragedy of the Commons.'" *Nature* 415 (6870): 424–6. http://dx.doi.org/10.1038/415424a.
Mitani, John C. 2009. "Cooperation and Competition in Chimpanzees: Current Understanding and Future Challenges." *Evolutionary Anthropology* 18 (5): 215–27. http://dx.doi.org/10.1002/evan.20229.
Mitchell, William P. 1973. "The Hydraulic Hypothesis: A Reappraisal." *Current Anthropology* 14 (5): 532–4. http://dx.doi.org/10.1086/201379.
Murdock, George Peter. 1945. "The Common Denominator of Cultures." In *The Science of Man in the World Crisis,* ed. Ralph Linton, 123–42. New York: Columbia University Press.
Nowak, Martin A. 2006a "Evolutionary Dynamics of Cooperation." In *Proceedings of the International Congress of Mathematicians* 1523–40.
Nowak, Martin A. 2006b. "Five Rules for the Evolution of Cooperation." *Science* 314 (5805): 1560–3. http://dx.doi.org/10.1126/science.1133755.
Nowak, Martin A, with Roger Highfield. 2011. *Supercooperators: Altruism, Evolution, and Why We Need Each Other to Succeed.* New York: Free Press.
Nowak, Martin A., and Karl Sigmund. 2005. "Evolution of Indirect Reciprocity." *Nature* 437 (7063): 1291–98. http://dx.doi.org/10.1038/nature04131.
Nowak, Martin A., Corina E. Tarnita, and Edward O. Wilson. 2010. "The Evolution of Eusociality." *Nature* 466 (7310): 1057–62. http://dx.doi.org/10.1038/nature09205.
O'Gorman, Rick, Joseph Henrich, and Mark Van Vugt. 2009. "Constraining Free Riding in Public Goods Games: Designated Solitary Punishers can Sustain Human Cooperation." *Proceedings. Biological Sciences* 276 (1655): 323–9. http://dx.doi.org/10.1098/rspb.2008.1082.

Ohtsuki, Hisashi, Yoh Iwasa, and Martin A. Nowak. 2009. "Indirect Reciprocity Provides Only a Narrow Margin of Efficiency for Costly Punishment." *Nature* 457 (7225): 79–82. http://dx.doi.org/10.1038/nature07601.

Olson, Mancur. 1965. *The Logic of Collective Action: Public Goods and the Theory of Groups.* Cambridge: Harvard University Press.

Ostrom, Elinor. 1990. *Governing the Commons: The Evolution of Institutions for Collective Action.* Cambridge: Cambridge University Press.

Ostrom, Elinor. 1992. *Crafting Institutions for Self-Governing Irrigation Systems.* San Francisco: Institute for Contemporary Studies Press.

Ostrom, Elinor. 2007. "Collective Action Theory." In *The Oxford Handbook of Comparative Politics*, ed. Carles Boix and Susan C. Stokes, 186–208. Oxford: Oxford University Press.

Ostrom, Elinor, and Roy Gardner. 1993. "Coping with Asymmetries in the Commons: Self-Governing Irrigation Systems Can Work." *Journal of Economic Perspectives* 7: 93–112.

Ostrom, Elinor, Roy Gardner, and James M. Walker. 1994. *Rules, Games, and Common-Pool Resources.* Ann Arbor: University of Michigan Press.

Ostrom, Elinor, Roy Gardner, and James M. Walker, eds. 2003. *Trust and Reciprocity: Interdisciplinary Lessons from Experimental Research.* New York: Russell Sage Foundation.

Ostrom, Elinor, and James M. Walker. 1997. "Neither Markets nor States: Linking Transformation Processes in Collective Action Arenas." In *Perspectives on Public Choice: A Handbook*, ed. Dennis C. Mueller, 35–72. Cambridge: Cambridge University Press. http://dx.doi.org/10.1017/CBO9780511664458.004.

Parsons, Talcott, and Neil J. Smelser. 1956. *Economy and Society: A Study in the Integration of Economic and Social Theory.* London: Routledge and Kegan Paul.

Patton, John Q. 2009. Evolutionary Studies of Cooperation. Special Edition of *Human Nature* 20 (4): 351–446. http://dx.doi.org/10.1007/s12110-009-9077-1.

Pauketat, Timothy R. 2001. "Practice and History in Archaeology: An Emerging Paradigm." *Anthropological Theory* 1 (1): 73–98.

Plourde, Aimée M. 2008. "The Origins of Prestige Goods as Honest Signals of Skill and Knowledge." *Human Nature (Hawthorne, N.Y.)* 19 (4): 374–88. http://dx.doi.org/10.1007/s12110-008-9050-4.

Plourde, Aimée M. 2009. "Prestige Goods and the Formation of Political Hierarchy: A Costly Signaling Model." In *Pattern and Process in Cultural Evolution*, ed. Stephen Shennan, 265–75. Berkeley: University of California Press.

Pluckhahn, Thomas J. 2003. *Kolomoki: Settlement, Ceremony, and Status in the Deep South, A.D. 350 to 750.* Tuscaloosa: University of Alabama Press.

Rand, David G., Anna Dreber, Tore Ellingsen, Drew Fudenberg, and Martin A. Nowak. 2009. "Positive Interactions Promote Public Cooperation." *Science* 325 (5945): 1272–75. http://dx.doi.org/10.1126/science.1177418.

Rand, David G., Hisashi Ohtsuki, and Martin A. Nowak. 2009. "Direct Reciprocity with Costly Punishment: Generous Tit-for-Tat Prevails." *Journal of Theoretical Biology* 256 (1): 45–57. http://dx.doi.org/10.1016/j.jtbi.2008.09.015.

Renfrew, Colin. 1974. "Beyond a Subsistence Economy: The Evolution of Social Organization in Prehistoric Europe." In *Reconstructing Complex Societies: An Archaeological Colloquium*, ed. C. B. Moore, 69–95. Bulletin of the American Schools of Oriental Research 20.

Richerson, Peter J., and Robert Boyd. 1999. "Complex Societies: The Evolutionary Origins of a Crude Superorganism." *Human Nature (Hawthorne, N.Y.)* 10 (3): 253–89. http://dx.doi.org/10.1007/s12110-999-1004-y.

Richerson, Peter J., and Robert Boyd. 2001. "Institutional Evolution in the Holocene: The Rise of Complex Societies." In *The Origins of Human Social Institutions*, ed. W. G. Runciman, 197–234. Oxford: Oxford University Press.

Richerson, Peter J., Robert T. Boyd, and Joseph Henrich. 2003. "Cultural Evolution of Human Cooperation." In *Genetic and Cultural Evolution of Cooperation*, ed. Peter Hammerstein, 357–88. Cambridge, Berlin: MIT Press and Preie Universität.

Roscoe, Paul. 2009. "Social Signaling and the Organization of Small-Scale Society: The Case of Contact-Era New Guinea." *Journal of Archaeological Method and Theory* 16 (2): 69–116. http://dx.doi.org/10.1007/s10816-009-9062-3.

Rousseau, Jean-Jacques. 1984 [1755]. *A Discourse on Inequality*. Trans. Maurice Cranston. New York: Penguin Books.

Saitta, Dean J. 2007. *The Archaeology of Collective Action*. Gainesville: University of Florida Press.

Scarborough, Vernon L. 2003. *The Flow of Power: Ancient Water Systems and Landscapes*. Santa Fe: School of American Research Press.

Shennan, Stephen. 2002. *Genes, Memes and Human History: Darwinian Archaeology and Cultural Evolution*. London: Thames and Hudson.

Skyrms, Brian. 2004. *The Stag Hunt and the Evolution of Social Structure*. Cambridge: Cambridge University Press.

Smith, Eric A. 2003. "Human Cooperation: Perspectives from Behavioral Ecology." In *Genetic and Cultural Evolution of Cooperation*, ed. Peter Hammerstein, 401–27. Cambridge, Berlin: MIT Press and Preie Universität.

Smith, Eric A. 2010. "Communication and Collective Action: Language and the Evolution of Human Cooperation." *Evolution and Human Behavior* 31 (4): 231–45. http://dx.doi.org/10.1016/j.evolhumbehav.2010.03.001.

Smith, Eric A., and Rebecca Bliege Bird. 2005. "Costly Signaling and Cooperative Behavior." In *Moral Sentiments and Material Interests: The Foundations of Cooperation in Economic Life*, ed. Herbert Gintis, Samuel Bowles, Robert Boyd, and Ernst Fehr, 115–48. Cambridge, MA: MIT Press.

Sober, Elliott, and David Sloan Wilson. 1998. *Unto Others: The Evolution and Psychology of Unselfish Behavior.* Cambridge: Harvard University Press.

Spencer, Charles S. 1993. "Human Agency, Biased Transmission, and the Cultural Evolution of Chiefly Authority." *Journal of Anthropological Archaeology* 12 (1): 41–74. http://dx.doi.org/10.1006/jaar.1993.1002.

Spencer, Charles S., Elsa M. Redmond, and Milagro Rinaldi. 1994. "Drained Fields at La Tigra, Venezuela Llanos: A Regional Perspective." *Latin American Antiquity* 5 (2): 119–43. http://dx.doi.org/10.2307/971559.

Stanish, Charles. 2004. "The Evolution of Chiefdoms: An Economic Anthropological Model." In *Archaeological Perspectives on Political Economies*, ed. Gary M. Feinman and Linda M. Nicholas, 7–24. Salt Lake City: University of Utah.

Stanish, Charles, and Kevin J. Haley. 2005. "Power, Fairness, and Architecture: Modeling Early Chiefdom Development in the Central Andes." In *Foundations of Power in the Prehispanic Andes*, ed. Kevin J. Vaughn, Dennis Ogburn, and Christina A. Conlee, 53–70. Archaeological Papers of the American Anthropological Association, No. 14. Arlington, VA: American Anthropological Association.

Steward, Julian H., ed. 1955. *Irrigation Civilizations: A Comparative Study*. Washington, DC: Pan American Union.

Sussman, R. W., and R. C. Cloninger, eds. 2011. *Origins of Altruism and Cooperation*. New York: Springer.

Tomasello, Michael. 2009. *Why We Cooperate*. Cambridge: The MIT Press.
Tomasello, M., A. Kruger, and H. Ratner. 1993. "Cultural Learning." *Behavioral and Brain Sciences* 16 (3): 495–511. http://dx.doi.org/10.1017/S0140525X0003123X.
Traulsen, Arne, and Martin A. Nowak. 2006. "Evolution of Cooperation by Multilevel Selection." *Proceedings of the National Academy of Sciences of the United States of America* 103 (29): 10952–55. http://dx.doi.org/10.1073/pnas.0602530103.
Turchin, Peter, and Sergey Gavrilets. 2009. "Evolution of Complex Hierarchical Societies." *Social Evolution & History* 8 (2): 167–98.
Van Vugt, Mark, Robert Hogan, and Robert B. Kaiser. 2008. "Leadership, Followership, and Evolution: Some Lessons from the Past." *American Psychologist* 63 (3): 182–96. http://dx.doi.org/10.1037/0003-066X.63.3.182.
Vaughn, Kevin J., and Hendrik Van Gijseghem. 2007. "A Compositional Perspective on the Origins of the "Nasca Cult" at Cahuachi." *Journal of Archaeological Science* 34 (5): 814–22. http://dx.doi.org/10.1016/j.jas.2006.08.008.
Wagner, R. Harrison. 1983. "The Theory of Games and Problem of International Cooperation." *American Political Science Review* 77 (2): 330–46. http://dx.doi.org/10.2307/1958919.
West, S. A., A. S. Griffin, and A. Gardner. 2007. "Social Semantics: Altruism, Cooperation, Mutualism, Strong Reciprocity and Group Selection." *Journal of Evolutionary Biology* 20 (2): 415–32. http://dx.doi.org/10.1111/j.1420-9101.2006.01258.x.
Willer, Robb. 2009. "Groups Reward Individual Sacrifice: The Status Solution to the Collective Action Problem." *American Sociological Review* 74 (1): 23–43. http://dx.doi.org/10.1177/000312240907400102.
Wilson, David Sloan, and Kevin M. Kniffin. 1999. "Multilevel Selection and the Social Transmission of Behavior." *Human Nature (Hawthorne, N.Y.)* 10 (3): 291–310. http://dx.doi.org/10.1007/s12110-999-1005-x.
Wilson, David Sloan, John J. Timmel, and Ralph R. Miller. 2004. "Cognitive Cooperation: When the Going Gets Tough, Think as a Group." *Human Nature (Hawthorne, N.Y.)* 15 (3): 225–50. http://dx.doi.org/10.1007/s12110-004-1007-7.
Wittfogel, Karl A. 1957. *Oriental Despotism: A Comparative Study of Total Power.* New Haven: Yale University Press.

2

The Emergence of Social Complexity

Why More than Population Size Matters

Gary M. Feinman

The emergence and manifestation of socioeconomic complexity in human societies is one of those research questions that is big, intrinsically complicated, and important. It is a topic that has intrigued scholars from a broad range of disciplines for centuries, if not longer, and yet those of us interested in this issue remain far from an answer. In fact, there is no consensus concerning the most fruitful paradigms to employ in order to address this topic (e.g., Clutton-Brock et al. 2009; Fuentes 2004; Price and Feinman 1995, 2010; Spencer 1997; West, Griffin, and Gardner 2007; Wilson and Wilson 2007).

In this discussion, socioeconomic complexity is considered within the broader comparative context of human cooperation and socioeconomic networks. Here cooperation is defined broadly, following Mead (1937: 8), as "the act of working together to one end" (see Carballo, chapter 1). Mead's definition corresponds with Melis and Semmann (2010: 2663), who see cooperation as "behaviours which provide a benefit to another individual or are beneficial to both the actor and the recipient." In both cases, cooperation clearly is foundational to the formation and persistence of human groups.

In this analysis, elements of both a structural and an agency approach are interwoven (Wiessner 2002: 234) to examine the dynamic between scale (size), complexity, and integration (sensu Blanton et al. 1993: 14–17) (Table 2.1). The empirical synthesis and discussion draws most heavily on comparative studies from archaeology and anthropology, but my aim is to build an intellectual foundation

TABLE 2.1 Three core dimensions of human social groups

Complexity	The extent of functional differentiation among social units; may be vertical or horizontal. Vertical complexity is hierarchical governance with a degree of concentration in decision making and power. Horizontal complexity is the differentiation of a population into various roles or subgroups.
Scale	Size, overall population, maximal community size, geographic extent.
Integration	The various means by which social units and their members are interconnected. The nature and degree of interdependence and self-sufficiency.

(Blanton, Kowalewski, Feinman, and Finsten 1993)

that broadens the potential for dialogue with other disciplines. The overarching message is that while much extant research has probed the link between scale or size and differentiation or complexity in human social formations, that relationship (and our understanding of the emergence of social complexity) will remain somewhat fuzzy until variation in the ways that groups are integrated (cooperate) also is given more systematic consideration. By understanding the interplay of scale, complexity, and integration in different human groups, we strive to understand some of the tendencies, parameters, and constraints that underpin human cooperative behaviors and help account for key axes of diversity in human cooperative networks and groupings.

HUMAN GROUPS ARE DISTINCTIVE

Key aspects of human sociality, such as kin-based relations and reciprocity, status competition, and dominance, have a foundation in our primate biology that still affect our behavior today (Mazur 1985; Silk 2009; Zink et al. 2008). And yet, for at least the past tens of thousands of years, human networks of cooperation have been both far more diverse and often much larger in scale relative to those of other higher primates (e.g., Boyd and Richerson 2009; Dunbar 1993; Henrich 2006; Melis and Semmann 2010). Unlike any other animal, humans form large-scale cooperative groups that are made up of genetically distant and unrelated individuals (Dunbar 1993; Fehr and Fischbacher 2004). Neither kin selection nor pair-wise reciprocity among non-kin can account for the scale of human cooperation (Boyd and Richerson 2009: 3283; Richerson and Boyd 1999; Stone 2008). Based on the size of the human brain and its neocortex, extrapolations from other primates predict human group sizes of less than 200, but of course our groups are orders of magnitude larger, and they have been for thousands of years (Dunbar 1992, 1998; Fehr and Fischbacher 2004; Hill and Dunbar 2003). Clearly, compared to other primates, humans have an advanced capacity for social cognition and learning (Dunbar 1998; Richerson, Boyd, and Henrich 2003; Tomasello 1999) and a highly plastic neural structure (e.g., Gibson 2005) that allows them to initiate and sustain new forms and scales of social behavior and cooperation.

EGALITARIANISM

Given our species' primate legacy toward dominance relations, human egalitarianism (social cooperative networks that lack institutionalized inequalities beyond those based on age or sex) is more productively viewed as a later Pleistocene social adaptation or innovation rather than some kind of human "blank slate" (Boehm 1993; Wiessner 2002). Egalitarianism in small-scale human societies likely is a facet of our tendency toward "groupishness" (or our species' advanced capacities for and reliance on social learning) as it helped check social differences and competition, allowing for larger and more resilient groups (Boehm 1993, 1997; Gavrilets, Duenez-Guzman, and Vose 2008; Knauft 1991; Ridley 1996; Svensson 2009; Wiessner 2002). If egalitarianism is viewed as a "reverse dominance hierarchy" or a way to damp down hierarchical expression (sensu Boehm 1993), then it is reasonable that relatively minor degrees of inequality have been found in select Upper Paleolithic/Late Pleistocene societies (sensu Hayden 1995). Yet at the same time, these differences in inequality were never sustained for long periods, nor were they formally institutionalized.

Although institutionalized social complexity (the emergence of societies marked by hierarchical leadership and socioeconomic inequalities) is post-Pleistocene in date, this discussion briefly delves back into earlier times because of new research, which provides a somewhat different vantage on human cooperation coming out of the later Paleolithic than the perspective that predominated a decade ago. It may seem semantic, but if Upper Paleolithic egalitarian social formations were not merely a "blank slate" on which exogenous forces acted to trigger change, then our basic understanding of human sociality and its deep history must be reconsidered. Extended human sociality may have set the stage for an "egalitarian revolution" (Gavrilets et al. 2008), but it was itself underpinned by our earlier primate heritage (Dunbar and Shultz 2007).

It now becomes clearer that human social formations were (and are) marked by a fundamental tension or dynamic between relations based on dominance, hierarchy, and kin altruism (part of our primate heritage) and newer capacities for social cognition, cultural learning, alliance building, and cooperation (Stone 2008: 79). The recognition of these tensions between the interests and aims of individuals and groups, and between the different relational building blocks or "work-arounds" (Richerson and Boyd 1999: 267–268; 2001a, 2001b; Richerson, Boyd, and Henrich 2003) that humans use to form groups, is important as it provides key insights into the diversity of later human social formations, variation in the integrative means and mechanisms that interlink them, and the alternative historical pathways that these groups ultimately have followed. Here it is important to stress that the concept of work-arounds is used rather more broadly than in previous studies (e.g., Richerson, Boyd, and Henrich 2003) not solely as the outgrowth of psychological propensities or discrete behavioral modules, but rather as the expansion of certain sets of interpersonal ties, behaviors, and meanings that foster cooperation through the enhancement of trust and constraint of free-riding albeit in different ways (see Blanton and Fargher, chapter 5).

To put it succinctly: "Our evolutionary heritage has hardwired us to be boundedly self-seeking at the same time that we are capable of learning heuristics and norms, such as reciprocity, that help achieve successful collective action" (Ostrom 1998: 2). Yet since these norms and social contracts that are the basis of our relationships and groups are neither uniform nor static, and since humans are imbued with the capacity to make rational choices regarding social actions, it becomes necessary to probe the variation in (and implications of) these integrative social ties (e.g., Dunbar and Shultz 2007: 1346).

THE NEOLITHIC PACKAGE, POPULATION, AND COMPLEXITY

Archaeologists have long recognized that human social complexity and more sustained socioeconomic inequalities were linked to the post-Pleistocene advent of the "Neolithic package"—domesticated plants and animals, larger and more enduring settlements, and population growth—in many regions of the globe (Price 2000). But an understanding of both the causal underpinnings and the different historical pathways taken from region to region has proven elusive, perhaps in part because the specific contents and synchrony of the elements in this package varied from case to case across diverse global settings.

Repeatedly, social scientists from several disciplines (e.g., Blau 1968, 1970; Bodley 2003; Carneiro 1967; Dubreuil 2010: 164–166; Ember 1963; Fletcher 1995; Johnson 1982; Kosse 1990, 2000; Naroll 1956; Scott 1975) have recognized the broad-brush and cross-cultural relationship between the size of human groups and their organizational complexity. Based on the principle of interaction, these studies argue that the larger the size of groups, the greater the number of person-to-person contacts (in a geometric progression) (Figure 2.1), hence the larger chance of disputes and the greater likelihood that leaders/administrators would arise to mediate, keep order, and prevent fission. Variation in community size (Carneiro 1967; Ember 1963; Lekson 1985; Naroll 1956) repeatedly has been found to be particularly sensitive to this correlation, since the intensity of interpersonal contacts would be magnified within a single settlement as compared to a society, nation, or other more dispersed cooperative network (Carneiro 1967: 238; Fletcher 1995: 71; Mayhew and Levinger 1976).

Yet despite the repeated correlations found between population and complexity, demographic determinism (either population growth or population pressure on resources) has not proven a convincing explanation for Holocene increases in societal complexity. For one thing, archaeological studies have found different rates of population change from one area to the next—not a good attribute for a supposedly independent variable or prime mover (Hassan 1981). In addition, population increases do not necessarily translate into larger social units because of the ever-present options of fission and migration. Population pressure arguments are even less convincing, as key Neolithic organizational transitions often occurred at low demographic densities, well below resource constraints (Blanton et al. 1993; Cowgill 1975; Feinman 1991, 1995; Feinman and Neitzel

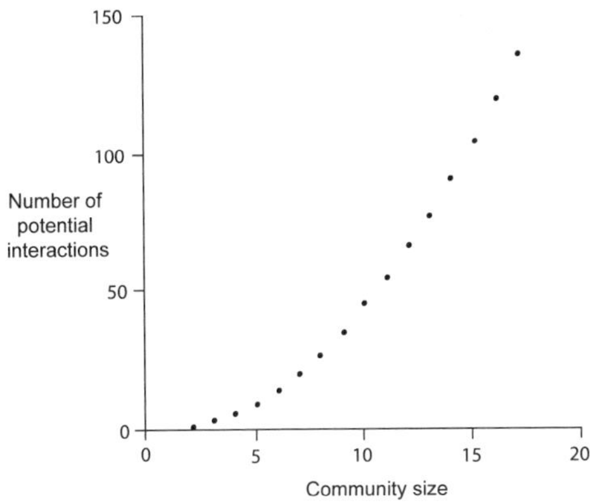

FIGURE 2.1 *Relationship between number of potential interactions and community size (redrawn from Fletcher 1995: fig. 4.1).*

1984; Hayden 1986; Kosse 1994). At the same time, the repeated recognition of certain demographic stress points in the size-complexity relationship that occur in vastly different environmental-resource contexts (and even when sustenance is not an issue) would seem to imply that this relationship has more to do with human information-processing/cognitive constraints (Dunbar 1993, 1998; Johnson 1982; Kosse 1994) and efficiencies in the sanctioning of free-riders (Dubreuil 2010: 166–170; O'Gorman, Henrich, and Van Vugt 2009) than with resource availabilities.

ZOOMING DOWN ON SIZE-COMPLEXITY

When examined over the wide range of human societies, both total population and maximal community size correspond closely with organizational complexity (hierarchical organization) (e.g., Carneiro 1967; Ember 1963; Feinman and Neitzel 1984; Kosse 1990; Johnson 1982; Lekson 1985; Naroll 1956). This relationship has been referred to as the "size-complexity rule," with analogical extrapolations made beyond human social formations (Bonner 2004; see also Naroll and von Bertalanffy 1956). For example, reaffirming earlier findings from a study by Mel Ember (1963), Gregory Johnson (1982: 389–390) (Figure 2.2) found a correlation of .828 (r) between the size of the largest territorial unit and the number of types of political officials in a sample of 23 societies (see also Kosse 1990: 287–288). If all 24 cases in Ember's original sample are included, then the correlation increases to .905 (r). In these synchronic cross-cultural analyses, groups over 400–500 generally have suprahousehold integrative institutions of some kind, while populations numbering 2,000–3,000 are almost always organized hierarchically (Feinman 1995, 1998) (Table 2.2).

Nevertheless, the population size/organizational complexity relationship is more complicated than it first appears. When Johnson (1982: 391) focused down on narrower population ranges, the correlation between size and complexity weakened considerably (Table 2.3). This pattern of variation has been repeated in

TABLE 2.2 Organizational thresholds of human groups

Source	*150–200*	*Source*	*2,000–3,000*
Hill and Dunbar (2003)	150	Carneiro (1967)	2,000
Forge (1972)	150	Forge (1972)	2,000
Adler and Wilshusen (1990)	~200	Sinha (1978)	2,000–3,000
		Kosse (1990)	2,000–3,000
		Johnson (1982)	2,400
		Bernard and Killworth (1973)	2,460
		Brown and Podolefsky (1976)	2,500
		Lekson (1985)	2,500

TABLE 2.3 Effects of population range on population-complexity correlations (variables in logarithmic transformation)

Population range	*N*	*r*	R
50–500	10	–.019	.000
50–1,000	12	.612	.375
50–5,000	17	.626	.392
50–10,000	20	.738	.545
50–75,000	23	.828	.686
50–20,000,000	24	.905	.819

(Johnson 1982: Table 21.1; drawing on data in Ember 1963)

several subsequent comparative studies, which clearly illustrate that there are no simple thresholds or magic numbers beyond which the organizations of human groups always change in uniform ways (see MacSweeney 2004). Rather, larger group sizes or social fields generally entail organizational shifts, but such changes take different forms and at somewhat variable rates.

For example, Lekson (1985) (Table 2.4) compared the size of the largest settlements in certain societies with their organizational complexity, and while the relationship is strong, there also is a good deal of variation, particularly for societies with maximal settlements that range between 1,000 and 5,000. In a study of middle-range societies (Feinman and Neitzel 1984), similar findings were reached in comparisons of organizational complexity with both maximal community size and total population (Table 2.5). Likewise, in a sample of cross-cultural data on 339 hunting and gathering populations compiled by Lewis Binford (2001: Tables 5.01, 8.01), comparisons of total population and the size of maximal aggregations with his measures of leadership and political centralization produce similar patterns (general relationships with notable variation).

Based on these results, the size of human cooperation networks or social fields is clearly, but not rigidly, related to the hierarchical complexity of their political institutions. Yet equivalent increments in size do not have uniform cross-

FIGURE 2.2 *Relationship between number of types of political officials and the population of the largest organizational unit (redrawn from Johnson 1982: fig. 21.1).*

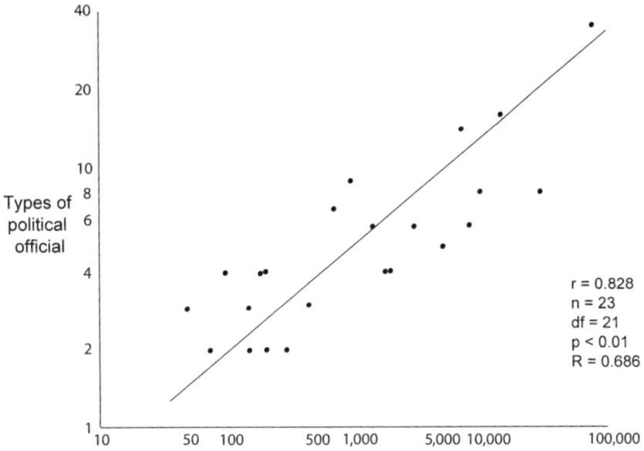

TABLE 2.4 Relationship between maximal community size and scale of organizational complexity

Maximal community size	*Stateless*	*Petty chief*	*Paramount*	*State*
0–500	12	6	2	0
501–1,000	1	4	1	0
1,001–5,000	2	5	2	1
5,001–10,000	0	0	2	1
10,001–75,000	0	0	1	2

(adapted from Lekson 1985: Table 1a)

cultural effects on organization. Human groups above a certain scale are almost always hierarchically complex. At the same time, aggregations of equivalent size may be organized in various ways, and a degree of leadership and hierarchical complexity can be found even in small groups. These patterns are illustrated (Table 2.6) in the synthesized results of George Murdock's (1967) compilation of settlement sizes in societies from the *Ethnographic Atlas* that he judged as egalitarian or lacking hierarchical complexity.

So what accounts for the lack of a tighter relationship or "messiness" (e.g., Binford 2001: 317–318) when you zoom down on the size-complexity relationship? Johnson (1982: 415, 1983) hypothesized that larger domestic group or basal unit size might mitigate the scalar stress associated with larger social groupings, thereby accounting for the lack of a closer fit between total population and the degree of societal complexity. Yet his suggestion cannot explain all of the organizational variation in these comparative cases (e.g., Binford 2001: 316–344).

BRINGING IN INTEGRATION

Nevertheless, Johnson's hypothesis that the size and organizational role of households can alter the societal relationship between scale and complexity highlights integration as a meaningful third factor to consider. Forty years ago, in a

TABLE 2.5 Relationship between total population size and organizational complexity

	Number of administrative levels		
Total population	*1–1.5*	*2–2.5*	*3*
0–4,000	11	5	0
4,000–13,000	3	7	1
13,000+	0	5	2
Number of cases	14	17	3

(adapted from Feinman and Neitzel 1984: Table 2.18)

TABLE 2.6 Range of settlement population sizes in egalitarian societies

Settlement population size range	*Number of Societies*	*Percentage of Societies*
< 50	34	21.9
50–99	37	23.9
100–199	30	19.4
200–399	29	18.7
400–1,000	18	11.6
> 1,000, no urban aggregations	1	3.9
5,000–50,000	1	0.6
> 50,000	0	0.0
Total	155	100.0

(sample derived from Murdock 1967)

comparative study of governmental employment agencies, the sociologist Peter Blau (1970) noted that as the size of a work unit increased, so did its organizational complexity (differentiation). But, in this sample, complexity grew at a slower pace so that larger groups have a smaller proportion of administrators and wider span of control. Johnson (1982: 413–414) (Figure 2.3) found a similar result in his reanalysis of Ember's (1963) earlier cross-cultural sample of societies. These largely structural analyses hint that the intensity or degree of coordination and integration is likely diminished as these social groups increase in size (see also Scott 1975: 8–15), although the matter was formally interpreted largely in structural terms as being due to the economies of larger size (Blau 1970; cf. Kasarda 1974).

Subsequent sociological analyses questioned the uniform applicability of the economies of scale argument (Clarke 1983; Kasarda 1974; Nolan 1979), while also evaluating why organizational scale and complexity/differentiation regularly correlated, but not in an entirely consistent way. Most notable, in a comparison of US state governments from the 1960 census, James Noell (1974: 556) found that, in addition to population size, a series of socioeconomic factors such as a state's relative wealth and the nature of the governance that was expected and provided were correlated with its bureaucratic complexity.

FIGURE 2.3 *Curvilinear relationship between scale and complexity (shown on logarithmic scale) (redrawn from Johnson 1982: fig. 21.5).*

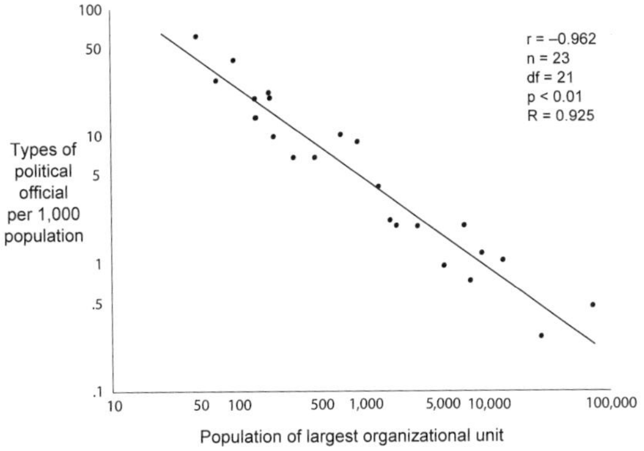

DIFFERENT HISTORICAL PATHWAYS

These analyses illustrate that the scale and complexity relationship is mediated or affected by the nature and degree of connectivity or integration in large sociopolitical entities. At the same time, other studies, building on Johnson's (1982, 1983) discussion of scalar stress, point to the importance of integrative variation in the organization and histories of smaller human groups. For example, Michael Adler and Richard Wilshusen (Adler 1989; Adler and Wilshusen 1990) studied 28 sedentary "tribal" or village-level societies, examining the nature and size of integrative architecture. Such nonresidential facilities are found in most (22 of 28, or 79 percent) of the groups. Importantly, in some cases, these structures were small (or what they termed "low level"), while in others they were larger (referred to as "high level"). Societies with low-level facilities tended to have more than one of these buildings, and these served small segments or components of a society (e.g., individual communities) or parts of/factions within a community. The high-level facilities often served the largest (or more than one) settlement (Figure 2.4). In other words, the high-level facilities were used by the entire or large segments of the society, while the low-level facilities had more constricted spatial ranges for participation.

Significantly, high-level facilities tended to be found in communities with larger populations (Adler and Wilshusen 1990: 135), but once again there was no simple threshold or cutoff that divided the sample. In those societies with small integrative features, the floor space of these facilities corresponded with the group's total population, and the space was generally adequate to accommodate most or all adults (Figure 2.5). In groups with larger integrative facilities, there is actually less floor space per person so that only a smaller subset of adults could participate in the activities held in those structures, indicating a lower degree of participation, representation, and voice (Adler and Wilshusen 1990: 136).

Several observations can be drawn from this research. First, as noted earlier from other comparative studies, when several hundred humans interact closely for an extended period, new integrative practices or institutions tend to emerge.

FIGURE 2.4 *Scatterplot of use group population and high-level and low-level integrative facilities (redrawn from Adler 1989: fig. 5; regression line from Adler and Wilshusen 1990: fig. 1).*

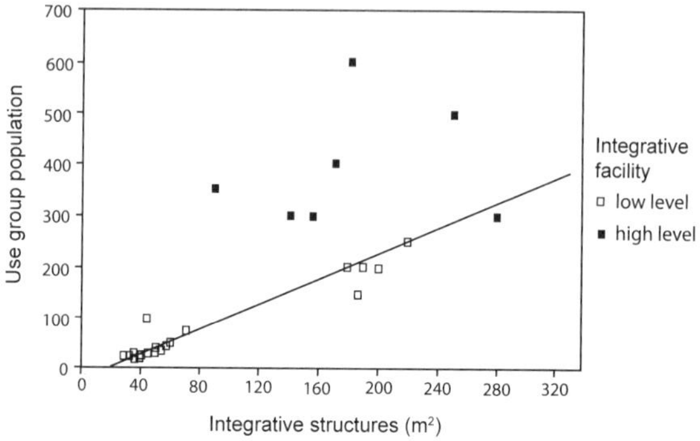

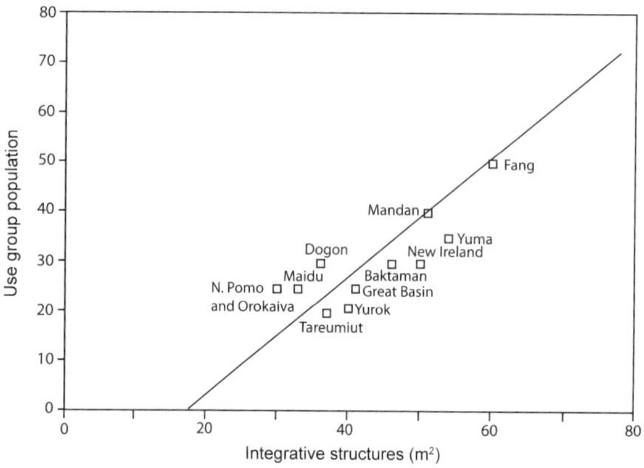

FIGURE 2.5 *Scatterplot of use group population and size of small integrative structures (all groups have more than one small integrative structure) (redrawn from Adler 1989: Fig. 4b).*

These institutions take different forms, but some (as described here) stress broad face-to-face participation and group solidarity. Nevertheless, as group size increases, elements of unequal access, differential participation, and, perhaps, inequities of power arise even in comparatively egalitarian/tribal contexts.

What makes these findings even more interesting is the contrast that the cases in the Adler-Wilshusen study illustrate with aggrandizer/accumulator models of leadership (e.g., Clark and Blake 1994; Hayden and Gargett 1990) that tend to view societal cooperation as more directly related to the individual networks of emerging power brokers. In other words, in societies of roughly comparable scale (e.g., Feinman and Neitzel 1984: 69) (Table 2.7), we see not only varying degrees of political complexity but different integrative modes of fostering cooperation.

Corresponding variation is evident in several recent comparative analyses of the archaeological record as well (Bandy 2008; see also Drennan and Peterson 2006, 2008; Lesure 2008). Matthew Bandy's synthetic efforts, which rely heavily on settlement pattern studies, measured the time lag from the beginnings of the first agricultural villages to the emergence of large villages (more than 300

TABLE 2.7 Relationship between number of administrative levels and maximal community size

	Number of administrative levels		
Max. community size	*1–1.5*	*2–2.5*	*3*
100–400	Obispano Chumash	Timucua	
500–1,000	Hopi	Costanoan	Cherokee
	Nisenan	Aruacay	
		Creek	
		Iroquois	
1,200 +	Barbareno Chumash	Huron	Cuna
			Hispaniola Arawak

(Feinman and Neitzel 1984: Table 2.16)

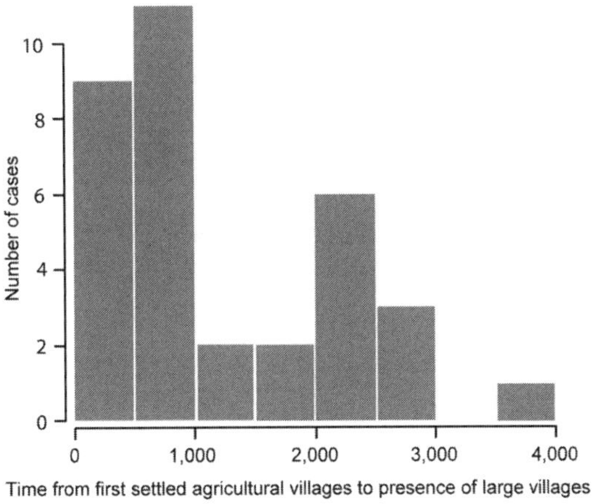

FIGURE 2.6 *Time (in years) from the establishment of the first sedentary agricultural communities to the presence of large villages (redrawn from Bandy 2008: fig. 2).*

people/3 hectares) in 36 regions of the world. He noted that the timing from sedentary communities to large communities is highly variable (Figure 2.6), ranging from almost immediately to up to 4,000 years later, thereby weakening an argument for the complete independence of demographic processes. Yet also key is the finding that the trajectories of community growth follow two distinct types or modes. Sometimes a series of villages of roughly comparable size are established in a region. That pattern then endures for centuries, even millennia, until one or more settlements eventually take off, becoming larger than others. Alternatively, in other global areas, as soon as one large community arises it dwarfs all other villages (Figure 2.7), and a more centralized system of settlement is established at least for a time. It may seem counterintuitive, but more cases of primary state formation ultimately arise from the former circumstances (the initial absence of a dominating center), which likely were founded on consensus building, intercommunity alliances, shared power, and broader webs of interconnection.

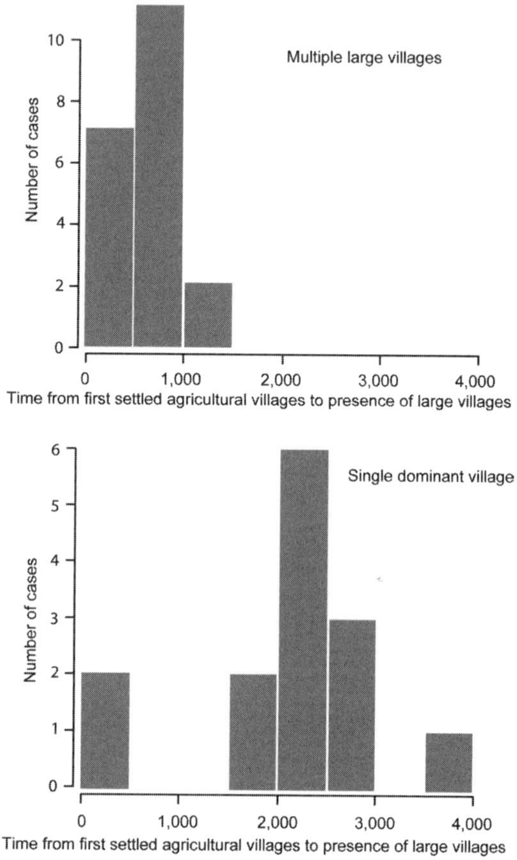

FIGURE 2.7 *Two patterns of village growth from the establishment of the first sedentary agricultural communities to the presence of large villages (redrawn from Bandy 2008: fig. 3).*

VARIATION IN INTEGRATIVE STRATEGIES: AN EXPLANATORY FRAMEWORK

In the remainder of this discussion, I briefly highlight different means of cooperation, integration, and leadership found cross-culturally. I draw heavily on my previous writings with Richard Blanton, Stephen Kowalewski, and Peter Peregrine (Blanton et al. 1996; Feinman 1995) that defined a contrast between corporate and network or exclusionary modes of interconnection between leaders and followers. Our original formulations were stimulated by Colin Renfrew's (1974; see also D'Altroy and Earle 1985; Lehman 1969) definition of "group-oriented" and "individualizing" chiefdoms. We outlined contrasts between the corporate mode, characterized by distributed power, muffled wealth disparities, corporate codes, consensus building, and an economic reliance on basic production, and the exclusionary arrangements, associated with highly centralized individual rule, a reliance on personal networks, more ostentatious expressions of inequalities and wealth, and an economy underpinned by long-distance networks and flows of free-floating resources.

This contrastive axis of integrative modes has been implemented in diverse global contexts (e,g., Coon 2009; Earle 1997; Mills 2000; Schachner 2010; Thurston 2001), where they have been found to shift over time within a given region or cultural tradition, and so are not bound immutably to specific societies or groups. What is striking about these different integrative modes is that they are not entirely mutually exclusive, and dialectical tension exists in each specific historical context. In fact, returning to the discussion at the outset of this essay, these different integrative modes—one based on dominance, hierarchy, and individualized ties (through kin, marriage, pairwise or reciprocal relations, etc.), and the other more reliant on consensus building, shared power, group cohesion, and broadly held values—may have their roots in the much earlier interpersonal relations and stresses that were integral to Late Pleistocene group formation (Richerson and Boyd 1999).

The central argument is that human groups or networks of cooperation vary in the ways that they are interconnected. Yet group size alone is not a necessary and sufficient predicator of the specific means of group integration or organization; mechanisms of cooperation vary based on more than group size alone (Gintis 2000; Gintis et al. 2008; Henrich et al. 2010). Reciprocity, reputation, retribution, and rewards are part of the relational means that keep both corporate and exclusionary groups interconnected (Carballo, chapter 1), but the diverse ways these practices are/were implemented and their relative importance in specific cultural contexts is/was variable, potentially shifting in particular societal contexts over time.

CONCLUDING THOUGHTS AND FUTURE DIRECTIONS

In human groupings, scale is related to societal complexity, but not in a strictly finite or mechanical way. The more one focuses down on a narrower range of group sizes, the messier the relationship is. Different modes and means of integration/agency can seemingly modify the relationship between scale and hierarchical complexity. Since large human groupings are at least in part sustained through trust, reciprocity, cooperation, and sanctioning (e.g., Boyd and Richerson 2009; Dubreuil 2008; Henrich 2006; Richerson and Boyd 2001a), it is relatively easy to see how the specific nature of the relational links between individuals could ultimately have a major effect on group cohesiveness and size (Dunbar 2008; Norenzayan and Shariff 2008; Roberts et al. 2009; Sosis and Ruffle 2004; Stiller and Dunbar 2007).

Consequently, to understand these divergent historical pathways through which organizational complexity emerged, we must give more formal focus to different modes of integration/connectivity, not solely as unique, culturally specific practices (e.g., Henrich, Heine, and Norenzayan 2010) but also as more general means of group interconnection that foster cooperation on a large scale. To a degree, these modes of interpersonal connectivity have their legacy in early human tensions between coercive dominance and kin-based relations and more consensual, cooperative means that prompted alliance building and larger social group cohesion (Boyd and Richerson 2009: 3287; Richerson, Boyd, and Henrich 2003; Stone 2008). Both of these countervailing proclivities, which enable the building of large social networks, have deep roots in the human career, and their countervailing tensions still are evident today (e.g., Feinman 2010). Yet this long-standing dialectic alone is not adequate to account for the diversity of human cooperative arrangements.

Lastly, to look forward, recent writings by Richard Blanton and Lane Fargher (2008, 2009, chapter 5; Fargher and Blanton 2007) build on rational choice models (Levi 1988, 2006; Olson 1965) to address some of the lingering questions concerning the different integrative and organizational mechanisms that hold large complex societies together despite self-interest. Furthermore, they query why the prevalence of different integrative modes varies through time and across space in large preindustrial societies. Through this work, they anchor the anthropological research that contrasts corporate-network modes of interconnection within a more encompassing social science theoretical tradition (see also Acemoglu and Robinson 2006, 2012; Willer 2009) that recognizes a relationship in large social networks and groups between the economic foundations of power and the nature of political integration.

More specifically, Blanton and Fargher (2008) argue, based on a sizable comparative sample, that collective or corporately integrated polities are found where the leaders are more dependent on local populations for their economic support, whereas exclusionary rule tends to occur where leaders exact less from the populace (dependent instead on spot resources, the control of trade, or other comparable means) and so are freer to yield them less voice. In other words, in larger polities, the more rulers depend directly on their immediate sustaining/local population for their resource support, the more agency and voice that population is likely to assert, whereas the more unfettered rulers are from that compact, the more autocratic they are able and apt to be. It is potentially significant that collaborative/consensual forms of decision making may forestall hierarchical development in smaller human groups with high amounts of face-to-face contact (e.g., Johnson 1982), while greater degrees of collective action and democratic practices may require more rapid increments of administrative complexity in large-scale social formations (Blanton and Fargher 2008; Levin 2010: 17) (Figure 2.8). In other words, the quantitative calculus between demographic scale and organizational complexity may vary even in groups with distinct modes and means of leadership, connectivity, and cooperation. Although these ideas require wider consideration and scrutiny, the recognition of the countervailing tensions that are fundamental to large human social formations serves to affirm that these complex groupings are not always or exclusively girded or premised on dominance, greed, costly signaling, personal power, and autocracy, as is all too often presumed in theoretical perspectives that tend to privilege unduly the

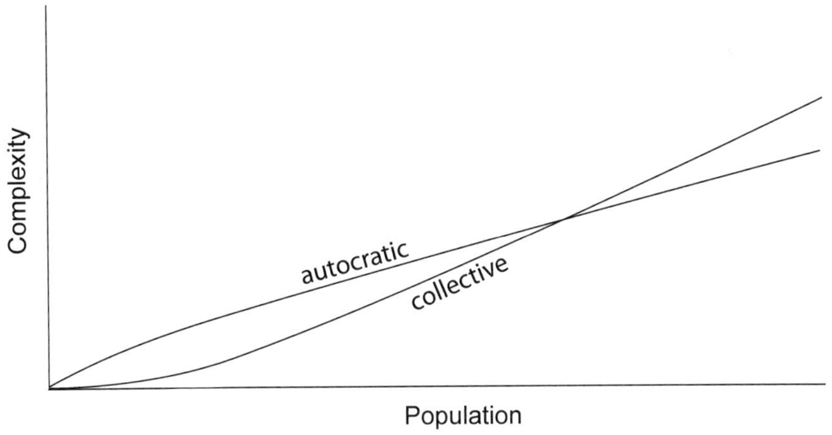

FIGURE 2.8. *Relationship between population size and increasing complexity for collective and autocratic organization.*

agency and actions of the powerful or reproductively successful in models of social change and human organization.

For example, two recent studies (Dubreuil 2010; O'Gorman, Henrich, and Van Vugt 2009) link human cognitive constraints with the sanctioning of free-riders to construct arguments concerning the general relationship between social group size and complexity. Each study builds on the tenet that the sanctioning of unfamiliar free-riders presents a challenge for the persistence and coherence of social groups that could be addressed through the designation of positions of authority. Yet while in one instance the authors (O'Gorman, Henrich, and Van Vugt 2009) simply assume that the authorized punisher is a single individual, Dubreuil (2010: 157–186) argues that such sanctions could be meted out efficiently and effectively by different kinds of punishers or modes of leadership, including some in which power and authority are less centralized or more broadly shared (see also Iannaccone 1992).

In sum, sociality and cooperation are key parts of what it means to be human, and only further systematic and comparative focus on (and understanding of) these integrative practices will give us a fuller explanation of the diversity of human social arrangements and why demographic factors alone can never fully account for the organizational diversity of such human formations. The social and biological legacies of human history also are critical for understanding the foundations and key parameters of human cooperation and sociality. Yet more complete understandings will emerge only when the plasticity of human behavior likewise is more fully recognized. Human cooperation in its variable forms and manifest at diverse scales ultimately is grounded in the rational choices made, negotiated, and constrained in the forging of the social relations and networks that we as a species have constructed across time and continue to make over geographic space.

ACKNOWLEDGMENTS

An earlier version of this essay was presented at the 2009 Annual Meeting of the American Association for the Advancement of Science. I would like to thank David Carballo both for asking me to contribute to this collection and for his dedicated efforts to coordinate and edit this volume. Richard E. Blanton, David Carballo, Robert D. Martin, Linda M. Nicholas, and two anonymous reviewers provided insightful suggestions on previous drafts of this work, for which I am very grateful. Linda M. Nicholas prepared the tables and figures, and I thank her for this critically important assistance.

REFERENCES

Acemoglu, Daron, and James A. Robinson. 2006. *Economic Origins of Dictatorship and Democracy*. Cambridge: Cambridge University Press.

Acemoglu, Daron, and James A. Robinson. 2012. *Why Nations Fail: The Origins of Power, Prosperity, and Poverty*. New York: Crown Business.

Adler, Michael A. 1989. "Ritual Facilities and Social Integration in Nonranked Societies." In *The Architecture of Social Integration in Prehistoric Pueblos*, ed. W. D. Lipe and Michelle Hegmon, 35–52. Cortez, CO: Crow Canyon Archaeological Center.

Adler, Michael A., and Richard H. Wilshusen. 1990. "Large-Scale Integrative Facilities in Tribal Societies: Cross-Cultural and Southwestern U.S. Examples." *World Archaeology* 22 (2): 133–46. http://dx.doi.org/10.1080/00438243.1990.9980136.

Bandy, Matthew. 2008. "Global Patterns of Early Village Development." In *The Neolithic Demographic Transition and Its Consequences*, ed. Jean-Pierre Bocquet-Appel and Ofer Bar-Yosef, 333–57. Berlin: Springer. http://dx.doi.org/10.1007/978-1-4020-8539-0_13.

Bernard, H. Russell, and Peter D. Killworth. 1973. "On the Social Structure of an Ocean-Going Research Vessel and Other Important Things." *Social Science Research* 2 (2): 145–84. http://dx.doi.org/10.1016/0049-089X(73)90016-1.

Binford, Lewis R. 2001. *Constructing Frames of Reference: An Analytical Method for Archaeological Theory Building Using Ethnographic and Environmental Data Sets*. Berkeley: University of California Press.

Blanton, Richard E., and Lane F. Fargher. 2008. *Collective Action in the Formation of Pre-Modern States*. New York: Springer.

Blanton, Richard E., and Lane F. Fargher. 2009. "Collective Action in the Evolution of Pre-Modern States." *Social Evolution and History* 8:133–66.

Blanton, Richard E., Gary M. Feinman, Stephen A. Kowalewski, and Peter N. Peregrine. 1996. "A Dual-Processual Theory for the Evolution of Mesoamerican Civilization." *Current Anthropology* 37 (1): 1–14. http://dx.doi.org/10.1086/204471.

Blanton, Richard E., Stephen A. Kowalewski, Gary M. Feinman, and Laura Finsten. 1993. *Ancient Mesoamerica: A Comparison of Change in Three Regions*. 2nd ed. Cambridge: Cambridge University Press.

Blau, Peter. 1968. "The Hierarchy of Authority in Organizations." *American Journal of Sociology* 73 (4): 453–67. http://dx.doi.org/10.1086/224506.

Blau, Peter. 1970. "A Formal Theory of Differentiation in Organizations." *American Sociological Review* 35 (2): 201–18. http://dx.doi.org/10.2307/2093199.

Bodley, John H. 2003. *The Power of Scale: A Global History Approach*. Armonk, NY: M. E. Sharpe.

Boehm, Christopher. 1993. "Egalitarian Society and Reverse Dominance Hierarchy." *Current Anthropology* 34 (3): 227–54. http://dx.doi.org/10.1086/204166.

Boehm, Christopher. 1997. "Impact of the Human Egalitarian Syndrome on Darwinian Selection Mechanics." *American Naturalist* 150 (s1 supplement): 100–21. http://dx.doi.org/10.1086/286052.

Bonner, John T. 2004. "Perspective: The Size-Complexity Rule." *Evolution; International Journal of Organic Evolution* 58:1883–90.

Boyd, Robert, and Peter J. Richerson. 2009. "Culture and the Evolution of Human Cooperation." *Philosophical Transactions of the Royal Society of London. Series B, Biological Sciences* 364 (1533): 3281–88. http://dx.doi.org/10.1098/rstb.2009.0134.

Brown, Paula, and Aaron Podolefsky. 1976. "Population Density, Agricultural Intensity, Land Tenure and Group Size in the New Guinea Highlands." *Ethnology* 15 (3): 211–38. http://dx.doi.org/10.2307/3773132.

Carneiro, Robert L. 1967. "On the Relationship between Size of Population and Complexity of Social Organization." *Southwestern Journal of Anthropology* 23: 234–43.

Clark, John E., and Michael Blake. 1994. "The Power of Prestige: Competitive Generosity and the Emergence of Rank in Lowland Mesoamerica." In *Factional Competition and Political Development in the New World*, ed. Elizabeth M. Brumfiel and John W. Fox, 17–30. Cambridge: Cambridge University Press. http://dx.doi.org/10.1017/CBO9780511598401.003.

Clarke, Clifford J. 1983. "Size and Administrative Intensity in Macro-Level Systems: A Reconcilliation." *Social Forces* 61 (3): 886–92.

Clutton-Brock, Tim, Stuart West, Francis Ratnieks, and Robert Foley. 2009. "The Evolution of Society." *Philosophical Transactions of the Royal Society of London. Series B, Biological Sciences* 364 (1533): 3127–33. http://dx.doi.org/10.1098/rstb.2009.0207.

Coon, Matthew S. 2009. "Variation in Ohio Hopewell Political Economies." *American Antiquity* 74:49–76.

Cowgill, George L. 1975. "On Causes and Consequences of Ancient and Modern Population Changes." *American Anthropologist* 77 (3): 505–25. http://dx.doi.org/10.1525/aa.1975.77.3.02a00030.

D'Altroy, Terence, and Timothy Earle. 1985. "Staple Finance, Wealth Finance, and Storage in the Inka Political Economy." *Current Anthropology* 26 (2): 187–206. http://dx.doi.org/10.1086/203249.

Drennan, Robert D., and Christian E. Peterson. 2006. "Patterned Variation in Prehistoric Chiefdoms." *Proceedings of the National Academy of Sciences of the United States of America* 103 (11): 3960–7. http://dx.doi.org/10.1073/pnas.0510862103.

Drennan, Robert D., and Christian E. Peterson. 2008. "Centralized Communities, Population, and Social Complexity after Sedentarization." In *The Neolithic Demographic Transition and Its Consequences*, ed. Jean-Pierre Bocquet-Appel and Ofer Bar-Yosef, 359–86. Berlin: Springer. http://dx.doi.org/10.1007/978-1-4020-8539-0_14.

Dubreuil, Benoît. 2008. "Strong Reciprocity and the Emergence of Large-Scale Societies." *Philosophy of the Social Sciences* 38 (2): 192–210. http://dx.doi.org/10.1177/0048393108315509.

Dubreuil, Benoît. 2010. *Human Evolution and the Origins of Hierarchies: The State of Nature.* Cambridge: Cambridge University Press. http://dx.doi.org/10.1017/CBO9780511780035

Dunbar, Robin I.M. 1992. "Neocortex Size as a Constraint on Group Size in Primates." *Journal of Human Evolution* 22 (6): 469–93. http://dx.doi.org/10.1016/0047-2484(92)90081-J.

Dunbar, Robin I. M. 1993. "Co-evolution of Neocortex Size, Group Size and Language in Humans." *Behavioral and Brain Sciences* 16 (04): 681–735. http://dx.doi.org/10.1017/S0140525X00032325.

Dunbar, Robin I. M. 1998. "The Social Brain Hypothesis." *Evolutionary Anthropology* 6 (5): 178–90. http://dx.doi.org/10.1002/(SICI)1520-6505(1998)6:5<178::AID-EVAN5>3.0.CO;2-8.

Dunbar, Robin I. M. 2008. "Cognitive Constraints on the Structure and Dynamics of Social Networks." *Group Dynamics* 12 (1): 7–16. http://dx.doi.org/10.1037/1089-2699.12.1.7.

Dunbar, Robin I. M., and Suzanne Shultz. 2007. "Evolution in the Social Brain." *Science* 317 (5843): 1344–47. http://dx.doi.org/10.1126/science.1145463.

Earle, Timothy. 1997. *How Chiefs Come to Power: The Political Economy in Prehistory*. Stanford, CA: Stanford University Press.

Ember, Melvin. 1963. "The Relationship between Economic and Political Development in Nonindustrialized Societies." *Ethnology* 2 (2): 228–48. http://dx.doi.org/10.2307/3772821.

Fargher, Lane F., and Richard E. Blanton. 2007. "Revenue, Voice, and Public Goods in Three Pre-Modern States." *Comparative Studies in Society and History* 49 (04): 848–82. http://dx.doi.org/10.1017/S0010417507000783.

Fehr, Ernst, and Urs Fischbacher. 2004. "Social Norms and Human Cooperation." *Trends in Cognitive Sciences* 8 (4): 185–90. http://dx.doi.org/10.1016/j.tics.2004.02.007.

Feinman, Gary M. 1991. "Demography, Surplus, and Inequality: Early Political Formations in Highland Mesoamerica." In *Chiefdoms: Power, Economy, and Ideology*, ed. Timothy Earle, 229–62. Cambridge: Cambridge University Press.

Feinman, Gary M. 1995. "The Emergence of Inequality: A Focus on Strategies and Processes." In *Foundations of Social Inequality*, ed. T. Douglas Price and Gary M. Feinman, 255–79. New York: Plenum Press.

Feinman, Gary M. 1998. "Scale and Social Organization: Perspectives on the Archaic State." In *Archaic States*, ed. Gary M. Feinman and Joyce Marcus, 95–133. Santa Fe, NM: School of American Research Press.

Feinman, Gary M. 2010. "A Dual-Processual Perspective on the Power and Inequality in the Contemporary United States: Framing Political Economy for the Present and the Past." In Path*ways to Power: New Perspectives on the Emergence of Social Inequality*, ed. T. Douglas Price and Gary M. Feinman, 255–88. New York: Springer. http://dx.doi.org/10.1007/978-1-4419-6300-0_9.

Feinman, Gary M., and Jill Neitzel. 1984. "Too Many Types: An Overview of Sedentary Prestate Societies in the Americas." *Advances in Archaeological Method and Theory* 7: 39–102.

Fletcher, Roland. 1995. *The Limits of Settlement Growth: A Theoretical Outline*. Cambridge: Cambridge University Press.

Forge, Anthony. 1972. "Normative Factors in Settlement Size of Neolithic Cultivators (New Guinea)." In *Man, Settlement, and Urbanism*, ed. Peter J. Ucko, Ruth Tringham, and George W. Dimbleby, 363–76. London: Duckworth.

Fuentes, Agustin. 2004. "It's Not All Sex and Violence: Integrated Anthropology and the Role of Cooperation and Social Complexity in Human Evolution." *American Anthropologist* 106 (4): 710–18. http://dx.doi.org/10.1525/aa.2004.106.4.710.

Gavrilets, Sergey, Edgar A. Duenez-Guzman, and Michael D. Vose. 2008. "Dynamics of Alliance Formation and the Egalitarian Revolution." *PLoS ONE* 3 (10, e3279): 1–9. http://dx.doi.org/10.1371/journal.pone.0003293.

Gibson, Kathleen R. 2005. "Epigenesis, Brain Plasticity, and Behavioral Versatility: Alternatives to Standard Evolutionary Psychology Models." In *Complexities: Beyond Nature & Nurture*, ed. Susan McKinnon and Sydel Silverman, 23–42. Chicago: University of Chicago Press.

Gintis, Herbert. 2000. "Strong Reciprocity and Human Sociality." *Journal of Theoretical Biology* 206 (2): 169–79. http://dx.doi.org/10.1006/jtbi.2000.2111.

Gintis, Herbert, Joseph Henrich, Samuel Bowles, Robert Boyd, and Ernst Fehr. 2008. "Strong Reciprocity and the Roots of Human Morality." *Social Justice Research* 21 (2): 241–53. http://dx.doi.org/10.1007/s11211-008-0067-y.

Hassan, Fekri A. 1981. *Demographic Archaeology*. New York: Academic Press.

Hayden, Brian. 1986. "Resources, Rivalry, and Reproduction: The Influence of Basic Resource Characteristics on Reproductive Behavior." In *Culture and Reproduction*, ed. Penn Handwerker, 176–95. Boulder, CO: Westview Press.

Hayden, Brian. 1995. Pathways to Power: Principles for Creating Socioeconomic Inequalities. In *Foundations of Social Inequality*, ed. T. Douglas Price and Gary M. Feinman, 15–86. New York: Plenum Press.

Hayden, Brian, and Rob Gargett. 1990. "Big Man, Big Heart? A Mesoamerican View of the Emergence of Complex Society." *Ancient Mesoamerica* 1 (01): 3–20. http://dx.doi.org/10.1017/S0956536100000043.

Henrich, Joseph. 2006. "Cooperation, Punishment, and the Evolution of Human Institutions." *Science* 312 (5770): 60–61. http://dx.doi.org/10.1126/science.1126398.

Henrich, Joseph, Jean Ensminger, Richard McElreath, Abigail Barr, Clark Barrett, Alexander Bolyanatz, Juan Camilo Cardenas, Michael Gurven, Edwins Gwako, Natalie Henrich, et al. 2010. "Markets, Religion, Community Size, and the Evolution of Fairness and Punishment." *Science* 327 (5972): 1480–84. http://dx.doi.org/10.1126/science.1182238.

Henrich, Joseph, Steven J. Heine, and Ara Norenzayan. 2010. "The Weirdest People in the World?" *Behavioral and Brain Sciences* 33 (2-3): 61–83. http://dx.doi.org/10.1017/S0140525X0999152X.

Hill, R. A., and Robin I. M. Dunbar. 2003. "Social Network Size in Humans." *Human Nature (Hawthorne, N.Y.)* 14 (1): 53–72. http://dx.doi.org/10.1007/s12110-003-1016-y.

Iannaccone, Laurence R. 1992. "Sacrifice and Stigma: Reducing Free-Riding in Cults, Communes, and Other Collectives." *Journal of Political Economy* 100 (2): 271–91. http://dx.doi.org/10.1086/261818.

Johnson, Gregory A. 1982. "Organizational Structure and Scalar Stress." In *Theory and Explanation in Archaeology*, ed. Colin Renfrew, Michael J. Rowlands, and Barbara A. Seagraves, 389–412. New York: Academic Press.

Johnson, Gregory A. 1983. "Decision-Making Organization and Pastoral Nomad Camp Size." *Human Ecology* 11 (2): 175–99. http://dx.doi.org/10.1007/BF00891742.

Kasarda, John D. 1974. "The Structural Implications of Social System Size: A Three-Level Analysis." *American Sociological Review* 39 (1): 19–28. http://dx.doi.org/10.2307/2094273.

Knauft, Bruce M. 1991. "Violence and Sociality in Human Evolution." *Current Anthropology* 32 (4): 391–428. http://dx.doi.org/10.1086/203975.

Kosse, Krisztina. 1990. "Group Size and Societal Complexity: Thresholds in the Long-Term Memory." *Journal of Anthropological Archaeology* 9 (3): 275–303. http://dx.doi.org/10.1016/0278-4165(90)90009-3.

Kosse, Krisztina. 1994. "The Evolution of Large, Complex Groups: A Hypothesis." *Journal of Anthropological Archaeology* 13 (1): 35–50. http://dx.doi.org/10.1006/jaar.1994.1003.

Kosse, Krisztina. 2000. "Some Regularities in Human Group Formation and the Evolution of Societal Complexity." *Complexity* 6 (1): 60–64. http://dx.doi.org/10.1002/1099-0526(200009/10)6:1<60::AID-CPLX1010>3.0.CO;2-I.

Lehman, Edward H. 1969. "Toward a Macrosociology of Power." *American Sociological Review* 34 (4): 453–65. http://dx.doi.org/10.2307/2091956.

Lekson, Stephen. 1985. "Largest Settlement Size and the Interpretation of Socio-political Complexity at Chaco Canyon, New Mexico." *Haliksa'i: UNM Contributions to Anthropology* 4: 68–75.

Lesure, Richard G. 2008. "The Neolithic Demographic Transition in Mesoamerica? Larger Implications of the Strategy of Relative Chronology." In *The Neolithic Demographic Transition and Its Consequences*, ed. Jean-Pierre Bocquet-Appel and Ofer Bar-Yosef, 107–138. Berlin: Springer. http://dx.doi.org/10.1007/978-1-4020-8539-0_6.

Levi, Margaret. 1988. *Of Rule and Revenue*. Berkeley: University of California Press.

Levi, Margaret. 2006. "Why We Need a New Theory of Government." *Perspectives on Politics* 4 (01): 5–19. http://dx.doi.org/10.1017/S1537592706060038.

Levin, Simon A. 2010. "Crossing Scales, Crossing Disciplines: Collective Motion and Collective Action in the Global Commons." *Philosophical Transactions of the Royal Society of London. Series B, Biological Sciences* 365 (1537): 13–18. http://dx.doi.org/10.1098/rstb.2009.0197.

MacSweeney, Naoise. 2004. "Social Complexity and Population: A Study in the Early Bronze Age Aegean." *Papers from the Institute of Archaeology* 15: 52–65.

Mayhew, Bruce H., and Roger L. Levinger. 1976. "Size and the Density of Interaction in Human Aggregates." *American Journal of Sociology* 82 (1): 86–110. http://dx.doi.org/10.1086/226271.

Mazur, Allan. 1985. "A Biosocial Model of Status in Face-to-Face Groups." *Social Forces* 64: 377–402.

Mead, Margaret. 1937. "Introduction." In *Cooperation and Competition among Primitive Peoples*, ed. Margaret Mead, 1–19. New York: McGraw-Hill Book Company, Inc.

Melis, Alicia P., and Dirk Semmann. 2010. "How is Human Cooperation Different?" *Philosophical Transactions of the Royal Society of London. Series B, Biological Sciences* 365 (1553): 2663–74. http://dx.doi.org/10.1098/rstb.2010.0157.

Mills, Barbara, ed. 2000. *Alternative Leadership Strategies in the Prehispanic Southwest*. Tucson: University of Arizona Press.

Murdock, George P. 1967. *Ethnographic Atlas*. Pittsburgh, PA: University of Pittsburgh Press.

Naroll, Raoul. 1956. "A Preliminary Index of Social Development." *American Anthropologist* 58 (4): 687–715. http://dx.doi.org/10.1525/aa.1956.58.4.02a00080.

Naroll, Raoul, and Ludwig von Bertalanffy. 1956. "The Principle of Allometry in Biology and the Social Sciences." *General Systems* 1: 76–89.

Noell, James J. 1974. "On the Administrative Sector of Social Systems: An Analysis of the Size and Complexity of Government Bureaucracies in the American States." *Social Forces* 52 (4): 549–58.

Nolan, Patrick D. 1979. "Size and Administrative Intensity in Nations." *American Sociological Review* 44 (1): 110–25. http://dx.doi.org/10.2307/2094821.

Norenzayan, Ara, and Azim F. Shariff. 2008. "The Origin and Evolution of Religious Prosociality." *Science* 322 (5898): 58–62. http://dx.doi.org/10.1126/science.1158757.

O'Gorman, Rick, Joseph Henrich, and Mark Van Vugt. 2009. "Constraining Free Riding in Public Goods Games: Designated Solitary Punishers Can Sustain Human Cooperation." *Proceedings. Biological Sciences* 276 (1655): 323–9. http://dx.doi.org/10.1098/rspb.2008.1082.

Olson, Mancur. 1965. *The Logic of Collective Action: Public Goods and the Theory of Groups.* Cambridge, MA: Harvard University Press.

Ostrom, Elinor. 1998. "A Behavioral Approach to the Rational Choice Theory of Collective Action." *American Political Science Review* 92 (1): 1–22. http://dx.doi.org/10.2307/2585925.

Price, T. Douglas. 2000. Europe's First Farmers: An Introduction. In *Europe's First Farmers*, ed. T. Douglas Price, 1–18. Cambridge: Cambridge University Press. http://dx.doi.org/10.1017/CBO9780511607851.002.

Price, T. Douglas, and Gary M. Feinman, eds. 1995. *Foundations of Social Inequality.* New York: Plenum Press.

Price, T. Douglas, and Gary M. Feinman, eds. 2010. *Pathways to Power: New Perspectives on the Emergence of Social Inequality.* New York: Springer.

Renfrew, Colin. 1974. "Beyond a Subsistence Economy: The Evolution of Social Organization in Prehistoric Europe." In *Reconstructing Complex Societies: An Archaeological Colloquium*, ed. Charlotte B. Moore, 69–85. Cambridge, MA: American Schools of Oriental Research.

Richerson, Peter J., and Robert Boyd. 1999. "Complex Societies: The Evolutionary Origins of a Crude Superorganism." *Human Nature (Hawthorne, N.Y.)* 10 (3): 253–89. http://dx.doi.org/10.1007/s12110-999-1004-y.

Richerson, Peter J., and Robert Boyd. 2001a. "Institutional Evolution in the Holocene: The Rise of Complex Societies." In *The Origins of Human Social Institutions*, ed. W. G. Runciman, 197–234. Oxford: Oxford University Press.

Richerson, Peter J., and Robert Boyd. 2001b. "The Evolution of Subjective Commitments to Groups: A Tribal Instincts Hypothesis." In *Evolution and The Capacity for Commitment*, ed. Randolph M. Nesse, 186–220. New York: Russell Sage Foundation.

Richerson, Peter J., Robert T. Boyd, and Joseph Henrich. 2003. "Cultural Evolution of Human Cooperation." In *Genetic and Cultural Evolution of Cooperation*, ed. Peter Hammerstein, 357–88. Cambridge, MA: MIT Press.

Ridley, Matt. 1996. *The Origins of Virtue: Human Instincts and the Evolution of Cooperation.* London: Penquin Books.

Roberts, Sam G. B., Robin I. M. Dunbar, Thomas V. Pollet, and Toon Kuppens. 2009. "Exploring Variation in Active Network Size: Constraints and Ego Characteristics." *Social Networks* 31 (2): 138–46. http://dx.doi.org/10.1016/j.socnet.2008.12.002.

Schachner, Gregson. 2010. "Corporate Group Formation and Differentiation in Early Puebloan Villages of the American Southwest." *American Antiquity* 75: 473–96.

Scott, Richard W. 1975. "Organizational Structure." *Annual Review of Sociology* 1 (1): 1–20. http://dx.doi.org/10.1146/annurev.so.01.080175.000245.

Silk, Joan B. 2009. "Nepotistic Cooperation in Non-human Primate Groups." *Philosophical Transactions of the Royal Society of London. Series B, Biological Sciences* 364 (1533): 3243–54. http://dx.doi.org/10.1098/rstb.2009.0118.

Sinha, Surajit. 1978. "Coexistence of Multiple Scales and Networks of a Civilization: India." In *Scale and Social Organization*, ed. Fredrik Barth, 122–32. Oslo: Universitetsforlaget.

Sosis, Richard, and Bradley J. Ruffle. 2004. "Ideology, Religion, and the Evolution of Cooperation: Field Experiments on Israeli Kibbutzim." *Research in Economic Anthropology* 23: 89–117. http://dx.doi.org/10.1016/S0190-1281(04)23004-9.

Spencer, Charles S. 1997. "Evolutionary Approaches in Archaeology." *Journal of Archaeological Research* 5 (3): 209–64. http://dx.doi.org/10.1007/BF02229153.

Stiller, James, and Robin I. M. Dunbar. 2007. "Perspective-taking and Memory Capacity Predict Social Network Size." *Social Networks* 29 (1): 93–104. http://dx.doi.org/10.1016/j.socnet.2006.04.001.

Stone, Brad L. 2008. "The Evolution of Culture and Sociology." *American Sociologist* 39 (1): 68–85. http://dx.doi.org/10.1007/s12108-008-9031-4.

Svensson, Erik L. 2009. "Understanding the Egalitarian Revolution in Human Social Evolution." *Trends in Ecology & Evolution* 24 (5): 233–35. http://dx.doi.org/10.1016/j.tree.2009.01.004.

Thurston, Tina L. 2001. *Landscapes of Power, Landscapes of Conflict: State Formation in the South Scandinavian Iron Age*. New York: Kluwer Academic/Plenum Press.

Tomasello, Michael. 1999. "The Human Adaptation for Culture." *Annual Review of Anthropology* 28 (1): 509–29. http://dx.doi.org/10.1146/annurev.anthro.28.1.509.

West, Stuart A., Ashleigh S. Griffin, and Andy Gardner. 2007. "Evolutionary Explanations for Cooperation." Review *Current Biology* 17 (16): R661–72. http://dx.doi.org/10.1016/j.cub.2007.06.004.

Wiessner, Polly. 2002. "The Vines of Complexity: Egalitarian Structures and the Institutionalization of Inequality among the Enga." *Current Anthropology* 43 (2): 233–69. http://dx.doi.org/10.1086/338301.

Willer, Rob. 2009. "Groups Reward Individual Sacrifice: The Status Solution to the Collective Action Problem." *American Sociological Review* 74 (1): 23–43. http://dx.doi.org/10.1177/000312240907400102.

Wilson, David Sloan, and Edward O. Wilson. 2007. "Rethinking the Theoretical Foundation of Sociobiology." *Quarterly Review of Biology* 82 (4): 327–48. http://dx.doi.org/10.1086/522809.

Zink, Caroline F., Yunxia Tong, Qiang Chen, Danielle S. Bassett, Jason L. Stein, and Andreas Meyer-Lindenberg. 2008. "Know Your Place: Neural Processing of Social Hierarchy in Humans." *Neuron* 58 (2): 273–83. http://dx.doi.org/10.1016/j.neuron.2008.01.025.

3

War, Collective Action, and the "Evolution" of Human Polities

Paul Roscoe

There are no larger human groups nor any greater challenges to collective action theory than polities, the autonomous political communities that characterize human macrosociality. Their most recently emergent form, the nation-state, represents an especially acute problem because, in these colossal "imagined communities" (Anderson 1991), no member knows more than a tiny fraction of the rest, and yet somehow they manage to cohere and continue through time.

In one guise or another, the nature of polities and the processes that propel their development have occupied anthropology and archaeology from their earliest days. Recent thought, however, dates to the 1950s and the beginnings of political evolution[1] as a research field. Since then, three basic types of theory have emerged, and though each has focused primarily on the evolution of polities, none could avoid either explicitly addressing or implicitly assuming something about their nature—about why they are formed in the first place and how they are socially reproduced.

From the 1950s to around 1980, when the field was still dominated by cultural anthropology, these three approaches became known as the *voluntaristic* (also the *integrative* or *functional*) approach, the *conflict* (or *coercive*) approach, and the *systems* (or *multivariant*) approach. In their essentials, none of these lines of thought was original; all can be traced back to the Enlightenment or even earlier, to thinkers such as Hobbes, Locke, and Rousseau. Their claim to distinction, though, was to modernize, expand, and clarify earlier thought and embed it more securely in the ethnographic and archaeological evidence. Voluntaristic theories (e.g., Service 1975) were based on the proposition that humans form polities to capitalize on some benefit that can best or only be secured through collective action. Members might perceive a common benefit to building or extending irrigation systems, subsistence redistribution mechanisms, surveillance and defensive systems, or some other public work, and they designate or strengthen a political center in order to organize the system for the benefit of the group.

Conflict approaches (e.g., Carneiro 1970) took a far darker view of political evolution and the nature of polities, presenting both as the product of exploitation. In Carneiro's circumscription theory, for instance, political evolution is pictured as stimulated by population growth and enabled by circumscribed conditions (i.e., circumstances that tie people down by making it difficult or undesirable for them to relocate). Polities then expand in size and become politically more centralized as one polity succeeds in conquering, incorporating, and then exploiting the labor of others. By implication, since people do not voluntarily submit to exploitation, the approach assumes that polities are held together not by collective interest but by coercion: a minority military elite oppressing a majority under conditions that prevent the latter from escaping the oppression of the former.

Systems theories sought to combine the voluntaristic and conflict approaches. One version proposed that, at some points in the trajectory of their political evolution, systems develop through voluntaristic processes, while at other points the processes are coercive. Another version proposed that both processes operated simultaneously (for a summary, see Cohen 1978).

By the 1980s cultural anthropology had become more interpretative and had largely surrendered the study of political evolution to archaeology. The voluntaristic approach was transformed into *managerial* (or *adaptational*) models (see review by Diehl 2000). To the extent that human behavioral ecology and "collective action" models address large-group cooperation and political centralization, they too fall into this camp (e.g., Blanton and Fargher 2008; Shennan 2008). The conflict approach became the *political* (or *exploitative*) model (e.g., Arnold 1993; Dye 2009; Hayden 1995). As for the systems approach, archaeologists had already sketched its elementary forms (e.g., Flannery 1972; Rathje and McGuire 1982; Wright 1978), and more recently it is apparent in appeals to combine heterarchy with hierarchy (Crumley 1995) or "bottom-up" with "top-down" approaches (e.g., Carballo, chapter 1). The names may have changed, and there have been many elaborations and modifications in the details, but the core approaches remain largely the same.

Criticisms of these three approaches are numerous. The voluntaristic approach proposes that polity members voluntarily surrender their sovereignty to a political center in return for a set of expected benefits, but it is difficult to understand in what sense individuals can be said to surrender autonomy. The conflict approach views polities as held together by force, but force generates distress in those subjected to it and is therefore an exceedingly costly means of control.

Systems theories claim to synthesize voluntaristic and conflict approaches, but they fail in practice to explain in any detail how such starkly opposed views of groups and processes might be reconciled.

These theoretical differences persist, I think, because we have failed to recognize that terms like *polity*, *chiefdom*, *state*, and the like actually collapse together two political phenomena that are better kept apart. The issue is neatly summarized in Giddens's (1985: 17) observation that the term *state* carries two quite different meanings in ordinary language. It can refer to the "apparatus of government or power," as in "The State." Or, it can refer to "the overall system subject to that government or power"—the totality of individuals who make up a body politic. More generally, the term *polity* gets applied to both a political center (albeit sometimes little more than an elder or two in the case of a band) and a political community. The problem is that voluntaristic and conflict approaches fail to distinguish these two phenomena, and to confuse matters further they implicitly focus on different aspects. Voluntaristic arguments concern themselves primarily with the polity *qua* system or political community. While they do not ignore the polity *qua* governing apparatus or political center, they analytically reduce it to the collective goals of the group, representing it as the means by which the collective benefit is realized. Conflict approaches, by contrast, focus on the polity *qua* center, analytically reducing the polity *qua* community to the consequences of its military or otherwise coercive actions. As a result, the two approaches have always talked past one another. As for the systems approach, it too is unaware of the problem, and as a result necessarily fails to amalgamate them.

In this chapter, I reexamine the whole issue of polities and their evolution in light of this paradigmatic confusion. My central point is that the polity *qua* governing apparatus and the polity *qua* group are different phenomena, and the processes that govern their emergence and development should therefore be analyzed as such. I shall argue that polities as political communities are and were almost everywhere defensive organizations, aimed at securing the collective benefits of mutual protection against enemy attack. In contrast, polities as political apparatuses are hierarchies of power relations created, reproduced, and extended to advance elite agendas. Now, to assert that the vertical and horizontal dimensions of polities are distinct in theory is not to say that they are unrelated in practice. As both a central elite and a defensive group, a polity comprises knowledgeable and capable agents with interests, and to the degree that these interests intersect, the two dimensions will affect one another. Mutual interests will generate cooperative linkages, opposed interests will produce conflict, exploitation, and/or oppression. And in this articulation of interests, I argue, we have a basis on which to reconcile voluntaristic and conflict approaches to polities and political evolution. I then briefly sketch how this perspective can shed light on the constitution of both empires and failed states.

Within the confines of a single chapter it is difficult to develop an argument about the political history of the world without seeming crassly reductive, and it is even more of a challenge to provide adequate empirical support. To demonstrate that the analysis is not without empirical foundation, however, I draw on evidence from both ends of the political complexity spectrum. At one pole, I use data from contact-era New Guinea, an ethnographic theater that was home to "band," "tribal," and even "petty chiefdom" polities, the political forms conventionally viewed as some of the least complex on earth. At the other pole, I shall consider nation-states, in particular the United States of America, a nation-state whose formative rationale is perhaps better documented than any other. Against the charge of reductionism, I argue in conclusion that attempts to identify major imperatives in the formation and evolution of polities are not to be confused with prime-mover or deterministic models of human history.

COLLECTIVE-ACTION THEORY AND HUMAN INTERESTS

Recent years have seen a surge of interest in the analysis of large-scale cooperation (Carballo, chapter 1). Much of the discussion has been devoted to the free-rider problem and how "selection" processes of one sort or another enable large groups to overcome the fissive threat that individual self-interest poses to collective interest (for succinct summaries, see Eerkens, chapter 7; Henrich, 2006; for a valuable critique, see Blanton and Fargher, chapter 5).

This work is important, but more needs to be done to reconcile it with mainstream social theory. The most important issue concerns human interests, which in current collective-action theorizing are essentially reduced to the proposition that humans *have* interests—i.e., they are motivated by "self-interest," except when they are motivated by altruism, which is analyzed either as "self-interest in the long run" or as an evolved property of group-level processes. As established social theory from Marx to recent practice theorists such as Bourdieu (1990) and Giddens (1984) emphasizes, however, humans are not just self-interested. They have multiple, specific interests, a circumstance that has important ramifications for cooperation as a practice.

First, humans cooperate in pursuit of those interests that can only be, or are most efficiently, advanced through collective action. Many of their other interests, however, may bring them into conflict, which threatens their cooperation, and they must find some way of managing the former if they are to procure the benefits of the latter (Roscoe 2009: 71–72). Second, interests have "lifetimes." Some are situational, others more enduring, and the difference is vital to understanding the formation of groups and the reproduction of social structure. Many interests can be advanced through collective action, but if these interests are ephemeral, the result will be a temporary coalition (or "task group") rather than what commonly has been thought of as a group. A pressure group to lobby for a town bypass or a millenarian movement to bring about the end times are cases in point. In the former instance, the coalition usually dissolves once a decision on the bypass has been reached; in the latter, the movement dies once the dead have risen or (more commonly) when they fail to. Other interests, however, are more enduring—interests in reproduction, sustenance, and security are the more obvious examples—and to the extent that an enduring interest can be advanced through collective action, it will give rise to a more stable social entity, one that I shall refer to as a *social group* (or simply *group*). By this I mean a collectivity in which members cooperate in an effort to realize a common, enduring interest, organizing themselves according to tacit or explicit codes and sanctions that regulate aspects of their conduct that they perceive as important to advancing the common interest at stake.

A further advantage of focusing on interests, as opposed just to self-interest, is that it allows us more effectively to address the complexity of human social structure. Current theorizing on cooperation poses the issue as one about how individuals overcome their self-interest in order to pursue collective ends. In so doing, however, it frequently represents cooperation as the *explicandum* and the common or public "good" as flowing from it, as though cooperation were a condition in itself that, once achieved, can be applied to procuring any and all public goods (though see Carballo, chapter 1). The problem is that different interests are best advanced through different scales of cooperation: reproduction (biological and social) would seem optimally to be advanced by the cooperation of just two individuals, generating the nuclear family. Security against (human or nonhuman) predators, by contrast, is best advanced by as large a collective as is feasible (see Spencer, chapter 9). In these differences of scale, I have argued elsewhere (Roscoe 2009: 70–71), we have an explanation for complexity in human social organization.

Finally, a focus on human interests, as opposed just to self-interest, draws attention to their importance, in themselves, as deterrents to free-riding and to the fact that some are more of a deterrent—and hence easier to institute—than others. For example, free-riding may be an attractive option if the goal of collective action is to optimize subsistence production. If I can get away with shirking my obligations to participate in procuring resources and/or sharing the products of my labor, the costs that I run in doing so may be small enough that I am tempted to try; effective collection action may then depend critically on enforcement. It is quite another matter, however, if my survival depends on whether or not I contribute to collective action. If my chances of surviving an enemy attack by fighting alone or trying to flee are minimal, but more promising if I cooperate with comrades in an organized defensive response, then free-riding will seem less attractive. By shirking my defensive responsibilities, I reduce the fighting capacity of my group and hence my own chances of survival. In this case, effective collective action may require little in the way of sanctions in order to work.

WAR AND THE FORMATION OF LARGE-SCALE SOCIAL GROUPS

The idea that war is somehow responsible for human sociality is a long-standing one. It was mooted by Darwin (1871: 162–163), and it has frequently resurfaced, albeit in poorly theorized forms, in both archaeology and social anthropology. In one of the more developed versions of the argument, Richard Alexander (1979: 221–223, 1987: 79) proposed that human macrosocial groups were the product of "balance-of-power" races. Under established social and ecological theory, Alexander observed, it is difficult to account for the scale that social formations have assumed in recent human history. Complex chiefdoms, archaic states, and nation-states are larger by far than anything that might be needed for reproduction or cooperative hunting, and from an evolutionary point of view they may even be deleterious, intensifying competition for mates and resources and elevating the transmission of parasites and infectious diseases. Warfare, he argued, is the only factor capable of explaining the emergence of groups on this scale. By making humans their own predator and prey, "their own principal 'hostile force of nature'" (Alexander 1987: 79), warfare has acted as a formidable selective force, precipitating "balance-of-power races" that, among other consequences, generate ever larger social groupings. The "necessary and sufficient forces to explain the maintenance of every kind and size of human group above the nuclear family, extant today and throughout all but the earliest portions of human history, were (a) war, or intergroup competition and aggression, and (b) the maintenance of balances of power between such groups" (Alexander 1979: 222).[2]

More recently, Choi and Bowles (2007) and Bowles (2008, 2009) have tried to relate war and large-group formation via the emergence and proliferation of "parochial altruists," individuals who are parochial in their genetic disposition to engage in hostile conflict with outsiders and altruistic in their disposition to do so on behalf of their fellow group members.

> Among ancestral humans, parochial altruists may have provoked conflicts between groups over scarce natural and reproductive resources, and at the same time contributed to a group's success in these conflicts. Altruism would have facilitated the coordination of raiding and ambushing on a scale known in few other animals, while parochialism fuelled the antipathy towards outsiders . . . As winning groups gained territory, an increase in reproductive opportunities and political and cultural influence could have overcome the selective disadvantages of parochialism and altruism when occurring separately. (Bowles 2008: 326)

The problem with these approaches is that they consider war as though it were a unitary phenomenon when, in fact, it comprises two, quite distinct phenomena—*offensive* warfare and *defensive* warfare. In advancing his balance-of-power argument, Alexander seems to take the view that large social groups confer both offensive and defensive advantages on their members in competition for resources, although in his earlier work (Alexander 1974) it is clear that he saw defense as the motive behind large-scale grouping behavior. The burden of Choi and Bowles's (2007) argument is that offensive warfare, via the genetic payoffs it brings, is the paramount mechanism generating large-scale groups.

On theoretical and empirical grounds, however, it is important to distinguish between offensive and defensive warfare and to recognize that each has different implications for group formation. For one thing, interests in common defense are not the interests that motivate offensive warfare (the single exception being a preemptive strike to eliminate a looming threat). Defensive warfare is motivated by the desire to survive in the face of attack. Offensive warfare may be motivated by a desire to revenge the killing of a kinsperson or theft of a pig, to annex material resources, to procure a spouse, and so on. In consequence, there is no necessary reason why the circle of individuals motivated to form a defensive organization should comprise all of the people motivated to engage in offensive warfare. The members of a village may all have common interests in defense, for example, but if a member of one of the village's component clans is killed, those motivated to take revenge may be limited to other members of the clan, not the whole village. For another thing, the interests motivating offensive warfare are more situational and ephemeral than those motivating common defense. Once a pig has been retrieved, a kinsperson avenged, land annexed, or a spouse procured, any common interest in offensive warfare dissipates; by contrast, an interest in defensive warfare endures for as long as there are enemies or potential enemies close enough to pose a threat.

Empirical evidence corroborates these theoretical expectations. In New Guinea, we find that the largest autonomous groups—the entities referred to in the ethnographic literature as longhouse communities, hamlets, clans, tribes, villages, and the like—were defensive organizations but by no means always offensive units (Roscoe 1996, 2009: 80–88). Among the Yangoru Boiken, subjects of my own fieldwork, the warriors of a village *never* combined to mount an attack: offensive actions were the precinct of a clan—or, more commonly yet, a subclan—if only because terrain and vegetation restricted the viable size of ambush parties and because the interests motivating attacks were never shared by every village member. Defensive actions, by contrast, *always* brought forth united village action: in the event of an attack, every capable village male rushed to defend those in jeopardy, while every available woman snatched up children and valuables and sought safety.

Likewise, among the neighboring Abelam, the village group was "a defensive but not necessarily an offensive unit" (Forge 1990: 162). The component descent groups of Middle Sepik River villages could, and often did, act quite independently of one another in launching war, but "a village acted as a unit only for defense" (Harrison 1993: 66). Among the Middle-Sepik Manambu, for example, attacks "tended actually to be made by factions of a village, often in combination with external allies, rather than by a village in its entirety," though the village "combined to defend itself as a last resort when under serious threat" (Harrison 1993: 68). The large Arapesh village of Ilahita "was exclusively a *defensive* unit. To be sure, there were offensive operations, but these were always prosecuted by individual wards, or at most two or three wards in temporary partnership" (Tuzin 1976: 59, emphasis in original); "Ilahita's constituent wards unanimously convened only when the village was under direct attack" (Tuzin 1976: 56). Among the Bena Bena, "a clan would respond defensively as a single unit," but it "was a rare man who could actually mobilize an entire clan for a raid, and in the cases I have heard described, the man who wanted revenge usually recruited only a portion of the clan to accompany him on his raid" (Langness 1973: 308).

THE NATION-STATE AND DEFENSIVE ORGANIZATION

It is conceivable that in very marginal environments hunter-gatherer bands might have little need to group for defensive purposes (Fry 2006: 181–183). Living at extremely low densities and united by a common interest in sharing resources to buffer environmental fluctuations, bands might live in a social universe characterized solely by amity rather than amity and enmity. In more productive and populated environments, however, the potential of a social universe to embrace individuals and groups with no common interests rises. The possibility then exists for lethal violence and with it an incentive for defensive grouping. This was the case among New Guinea hunter-gatherers (Roscoe 2002): hamlet-sized groups may have served the purposes of subsistence optimization, but they also operated as defensive groups (Roscoe 2009: 85). Under more intensive subsistence regimes, moreover, the New Guinea polity *qua* group was solely dedicated to common defensive interests (Roscoe 2009: 80–87).

But does the New Guinea case scale? Consider, for example, the opposite end of the social spectrum from New Guinea: the nation-state. Alexander's balance-of-power argument implies that, like the New Guinea village or clan, the nation-state is a defensive organization. But is this actually the case? To many theorists, the answer will seem obvious to the point of triviality: Yes. After all, defense "from the invasion of foreigners" was the principal function that Hobbes granted the Leviathan, along with its role in preventing individuals from "the injuries of one another" (Hobbes 1660: chapter 17). Since then, the idea that "States make war, but war also makes States" (Porter 1994: 1) has surfaced frequently in sociology and political science, though usually the criticality of defense as the formative interest gets obscured by conflating it with offensive military capacity (e.g., Collins 1999; Hintze 1994: 178; Skocpol 1979: 22), or the argument is addressed to the state as governing apparatus rather than as community subject to that apparatus (e.g., Porter 1994; Tilley 1985). For other theorists, though, the idea that the state, as a social group, is a security organization that protects against external threat will seem willfully reductive if not entirely wrong-headed. So complex an entity, it might be argued, cannot be reduced to any single "purpose" or "function." Or, as world-systems theory would have it, nation-states have economic interests and should be interpreted as economic entities in addition to, or instead of, military entities.

I have no wish at all to ignore or overly reduce the complexity of the modern nation-state. It is undeniable, for example, that states are now economic actors. The question, though, is whether they were always so or whether their economic status is an epiphenomenon of their defensive unity. It is impossible properly to evaluate this proposition in a single chapter, but consider two suggestive pieces of evidence concerning the origins and unity of the United States, the nation-state that is not only the most powerful on earth but also one of the exceedingly rare instances for which a documentary record is available of the rationale for its formation.

By 1787, when delegates to the Constitutional Convention met to consider the problem, the central government formed in the wake of the War of Independence had become so enfeebled that everyone agreed it needed to be reformed and strengthened. The Constitution produced by the convention was then put out for ratification by the states, with the *Federalist Papers* providing insight into the founders' philosophy and motivations for proposing the system.

Defense was quite clearly the paramount rationale. Following Hamilton's preamble in *Federalist* 1, the very first argument about the benefit of union was Jay's "Concerning Dangers from Foreign Force and Influence" (*Federalist* 2 through 5). As Jay framed the issue:

> Among the many objects to which a wise and free people find it necessary to direct their attention, that of providing for their SAFETY seems to be the first. The SAFETY of the people doubtless has relation to a great variety of circumstances and considerations . . . At present I mean only to consider it as it respects security for the preservation of peace and tranquillity, as well as against dangers from FOREIGN ARMS AND INFLUENCE, as from dangers of the LIKE KIND arising from domestic causes . . . Let us therefore proceed to examine whether the people are not right in their opinion that a cordial Union, under an efficient national government, affords them the best security that can be devised against HOSTILITIES from abroad. (Jay, *Federalist* 3; emphasis in original)

It will surprise no one that Jay found a cordial union to be a most splendid means of devising security against foreign hostilities.

Langton (1988) has observed, in fact, that 25 of the first 36 articles in *The Federalist* stressed the issue of national security (the remainder, *Federalist* 37 through 85, explained the difficulties the founders encountered in devising the Constitution, its details, and how the proposed federation would work). This compares to just three devoted to economics—or "commerce" as the authors phrased it. (*Federalist* 11 and 12 emphasized that the defensive strength conferred by union would act "as the guardian of our commerce"; *Federalist* 13 dwelled on the savings to be had "in Respect to Economy in Government"). As Christopher Gore, paraphrasing Hamilton (*Federalist* 25), put it at the Massachusetts Ratifying Convention: "Let the gentlemen consider the situation of our country; they will find we are circumscribed with enemies from Maine to Georgia" (quoted in Langton 1988: 491). The point was taken, and the 13 states aggregated under a single supreme power.

The US response to 9/11, one of the few times in history that the nation has suffered an attack on its homeland, provides a more recent indication of the nature of the US state. The weeks that followed saw a florescence in expressions of unity that had not been seen since the Second World War (see also Collins 2004: 56–61). Breaching political divisions across the country, approval ratings for the nation's leader, George Bush, rocketed almost overnight to a record 90 percent. Old Glory—a core symbol that contrives to collapse into a single identity the US nation and "the blood of those who died defending it"—blossomed across the land. The nation was awash in "United We Stand" bumper stickers (subsequently memorialized on a US postal stamp), which later morphed into depictions of the flag alongside the resolute slogan, "These Colors Don't Run!"

Transcending whatever exploitation, inequality, and repression exists within the borders of a nation-state, florescences such as these indicate what motivates the members of a nation-state to "imagine" they are a community (Anderson 1991). In Anderson's (1991: 7) words, it is a conception of "a deep *horizontal comradeship*. Ultimately it is this fraternity that makes it possible . . . for so many millions of people, not so much to kill, as willingly to die for such limited imaginings" (emphasis added).

SCALE AND THE DEFENSIVE COMMUNITY

Let us assume therefore that, whatever their other goals and functions, polities *qua* groups are first and foremost organizations dedicated to the security of their members. Why then have these defensive groups usually (though not always) increased in population size over time, and why at any one moment in time do they vary in size from one place to another? Alexander's "balance-of-power" races provide a plausible engine to explain what drives the process: numerical superiority being an important component of military strength, polities have an ongoing interest in maximizing their numbers and hence their defensive strength relative to neighbors. But Alexander (along with Choi and Bowles 2007) provides no explanation for why, at any one time, groups should assume the size they do.

In theory at least, the answer is straightforward. The maximum size to which a defensive group can—and, in a balance-of-power situation, will—grow is defined by the ability of its members to render one another effective military aid in the event of an advance or attack by an enemy. Individuals located close enough to one another to render military aid have an interest in cooperating in defense, and in a balance-of-power situation we should expect them to do so. Those situated too far apart to provide such aid have no such interest; if they belong to defensive groups, it will be to different groups.

As easy as it is to identify in theory what determines polity size, it is difficult in the extreme to specify the outcome in practice (Feinman, chapter 2). With the data currently available, for instance, it is not a straightforward matter to establish what constitutes "effective" military aid—that is, assistance sufficient to provide a unifying motive for collective action. If warriors live so far apart that they are unable to reach the site of an attack until it is *nearly* over, will the limited aid they can provide be sufficient to ensure defensive reciprocity or not? The size to which a defensive grouping can grow will also be affected by its capacity to "compress" time-space (i.e., its organizational and technological capacity to transcend the physical limits that human physiology sets on the ability to monitor, move, and communicate). *Ceteris paribus*, a political community with highly developed advance-warning systems will have more time to assemble a defensive response—and hence can grow larger—than one with minimal surveillance systems. A polity with electronic means of communication and mechanical means of transport can rally more warriors to—and thereby bring a larger membership under—a common defense than one limited to face-to-face communication and travel by foot.

Small-scale polities with limited technological and organizational capacities, moreover, will be particularly affected by the military implications of the terrain on which they fight. Polities that live on an open landscape such as a prairie will be able to detect a daylight advance by their enemy much earlier than if they inhabited a terrain blanketed in heavy forest. If they live on firm and flat terrain, members will be able to rally more quickly against an attack than if they must traverse soft or heavily dissected land such as a swamp or hills. As a result, we should expect groups living on firm, flat, open landscapes to be larger than those on landscapes that are soft, dissected, or occluded.

Although it is a difficult matter to predict polity size in practice, we can nonetheless draw some broad conclusions. Polities with electronic communications and mechanical forms of transport, for example, will be far larger in numbers as well as areal extent than small-scale societies, where organizational and technological capacity is severely limited. Furthermore, in small-scale societies, where all interaction is face-to-face and transport is by foot, the most important influence on polity size will be population density. Because a defensive group is defined by an area within which members are able to render mutual defensive aid, it follows that the more people that area can bear, the larger their defensive group can be.

This relationship finds empirical support from Table 3.1, which presents data on density and group size in twenty-six contact-era New Guinea communities (Column 1). The sample was selected solely by whether ethnographic information existed that definitively specified the presence or absence of Big Men (see following section). Column 2 estimates the crude, contact-era population density of these communities (see Roscoe 2006: 34, 36–38 for an explanation of methods used to derive these estimates). Column 4 estimates the contact-era size of their largest autonomous polity (LAP), defined as the largest local group that acted in mutual defense in the event of an attack (see Roscoe 2009: 80–88). In the lowlands, the LAP was usually a longhouse community or village group. In the highlands, the common social structure was a tribe made up of clans, and in most cases the LAP was the clan. (In a few cases, it is difficult to determine ethnographically whether the LAP was the clan or the tribe [Roscoe 2009: 87]; to err toward the conservative, however, clans rather than tribes have been taken as the LAPs.) Inspection of the table will confirm that, as crude density rises so too does the size of the LAP, the Pearson correlation coefficient being 0.74 ($p < 0.001$).

THE NATURE AND SCALE OF POLITICAL CENTRALIZATION

The processes that govern the emergence and development of political centralization—the vertical dimension of polity constitution—are quite different from those that control the formation and expansion of the political community, the horizontal dimension. Where the political community comprises relationships of mutual defense, political centralization involves relationships of power. Where the former rests on the ability of individuals to render one another defensive aid, the latter rests on the capacity of a leader or centralized elite to exercise control over the rest of the body politic.

I have considered in some detail elsewhere the nature of power, power relationships, and how political hierarchies are constructed (Roscoe 1993, 2000b: 116–126, 2008: 79–86). Here it is sufficient to note that power relations are built through manipulation in the pursuit of interest; political centralization is the consolidation of power relations in the hands of a few; in the early phases of political centralization, those who consolidate power are those most talented at political manipulation; and as centralization develops, those in political control can draw not just on their political talents but also, and increasingly, on the institutionalization of power relations (i.e., the "sedimentation" into a centralized political structure of control over resources such as land, wealth, armies, and the like that can be deployed to political advantage).

The degree to which a center is able to create and consolidate power—to build hierarchy—depends critically on its ability to interact with the body politic. To state the obvious, would-be leaders have to be able to interact with potential followers if they are to manipulate and extend relationships of power over them. It follows that the costs of bringing about these political interactions are a critical component in the degree to which a centralized power—be it a New Guinea Big Man or the government of an industrialized nation-state—can construct,

TABLE 3.1 Population distribution, community size, and the emergence of Big Men

Group	*Contact density (/km²)*	*LAP[1] (Size)*	*Big-Man Society[2]*
Asabano	0.4	Longhouse (50)	N
Sanio	1.6	Hamlet (25)	n
Onabasulu	1.7	Longhouse (60)	N
Kaluli	2.2	Longhouse (60)	N
Gebusi	3.6	Longhouse (27)	N
Etoro	3.7	Longhouse (36)	N
Wovan	4.6	Homestead (15)	N
Umeda	4.8	Village (205)	N
Amanab	5.2	Village (85)	N
Namie/Lujere	5.9	Village (100)	N
Binumarien	7.2	Village group (110)	y

continued on next page

TABLE 3.1—*continued*

Group	*Contact density (/ km²)*	*LAP*[1] *(Size)*	*Big-Man Society*[2]
Baruya	10.5	?	y
Telefolmin	11.5	Village (133)	N
Gnau	13.4	Village (270)	N
Olo (Wape)	13.6	Village (190)	N
Kopon (Lower)	13.7	?	N
Tauade	15.2	Clan (27)	N
Fore (South)	19.9	Hamlet (19)	y
Au	26.9	Village (202)	N
Enga (Raiapu—Saka)	28.4	Clan (c.350)	Y
Wahgi (North)	29.6	Clan (151)	Y
Wahgi (South)	33.6	Clan (450)	Y
Mendi	41.9	?	y
Enga (Raiapu—Lai)	43.8	Clan (225)	Y
Melpa (North)	55.9	Clan (264)	Y
Melpa (Central)	59.5	Clan (750)	Y
Chimbu (Upper)	61.3	Clan (459)	Y
Enga (Kyaka)	70.1	Clan (330)	Y
Chimbu (Central)	76.8	Clan (645)	Y
Enga (Mae)	111.7	Clan (350)	Y

Notes:

1. **Size of LAP:** LAP = Largest autonomous polity.
2. **Big Man society?:** Y = Explicitly acknowledged or referred to as present; y = Explicitly acknowledged to be present but less developed than in "classic" highland Big Man societies; said to be present, but qualifications made in reference to classic model of (highland) Big Men; n = Said to have no or weakly developed leadership; N = Big Men explicitly stated to be absent.

Sources: **Amanab**—Amanab census registers; Juillerat 1996: xx–xxii**; Asabano**—Lohman 2009: pers. comm.; Oksapmin census registers; OKS 1–70/71:3; **Au**—Lumi census registers; Philsooph 1980: 89; **Baruya**—Godelier 1986: 162–188; **Big Men societies**—Allen 1984: 22; Feil 1987: 6–7, 38, 94, 98, 111–113; Godelier 1986: 162–188; Lederman 1990: 3; Treide 1985: 169; **Binumarien**—Kainantu census registers; Hawkes 1978: 161, 183; **Boiken (Yangoru)**—Roscoe fieldnotes; **Chimbu (Central)**—Brookfield and Brown 1963: 73, Table 3; Brown and Brookfield 1959: 44; Kundiawa census registers; **Chimbu (Upper)**—Brookfield and Brown 1963: 122, Table 10; Criper 1967: Figures 1–3, following pp. 27–29, 63–64; **Enga (Kyaka)**—Bulmer 1960: 469–473; **Enga (Mae)**—Meggitt 1965: 9; Wabag census registers; **Enga (Raiapu—Lai)**—Wapanamanda census registers; Westermann 1968: 69; **Enga (Raiapu—Saka)**—Feachem 1974: 10–11; 1977: 142; Wapanamanda census registers; **Etoro**—Kelly 1977: 28, fn. 20, 138–139; **Fore (South)**—Glasse and Lindenbaum 1973: 376; Kainantu census registers; Sorenson 1976: 30; **Gebusi**—Kelly 1993: 35; Knauft 1985: 2; **Gnau**—Lewis 1975: 28–29, 344, 357; Lumi census registers; **Kaluli**—Kelly 1993: 35; Schieffelin 1976: 38; 1991: 61; **Kopon (Lower)**—Jackson 1975: 186, 193; **Melpa (Central)**—Burton 1988a: 2.8–2.25; Strathern 1971: 230; Vicedom and Tischner n.d.: 8–9, 49, 57; **Melpa (North)**—Strathern 1971: 9; 1972: 58–59; **Mendi**—Lederman 1990: 3, 8; **Namie**—Lumi census registers; Mitchell 1977: 183; 1978: 6; **Olo (Wape)**—Lumi census registers; Mitchell 1978: 6; **Onabasulu**—Kelly 1993: 35; Schieffelin 1981: 2; 1991: 60–61; **Sanio**—Ambunti census registers; Townsend 1969: 8; 2002: pers. comm.; **Tauade**—Hallpike 1977: 54, 61, 86, 139–143, 154; **Telefomin**—Jorgensen 1981: 471; **Umeda**—Gell 1975: 13; Imonda census registers; **Wahgi (North)**—Burton 1988b; Minj census registers; O'Hanlon 1989: 27; **Wahgi (South)**—Burton 1988c; Minj census registers; Reay 1959: 28–33; **Wovan**—Flanagan 1983: 38, 62, 100.

solidify, and extend its power relations (Roscoe 1993: 116–117). Where these costs are high, the extent of its power (i.e., the number of people it controls) and the effectiveness (i.e., the degree) of its control will be limited. Conversely, where the costs of bringing about interactions are low, the center can build a more extensive and entrenched political hierarchy.

No resource is more valuable to political centralization than time, and no cost is more universal to its construction than the time that must be spent in bringing political interactions about. As a result, innovations in transport and communications influence not only the ability of a people to render one another mutual military aid but also the capacity of a political center to augment its power. Developments in transportation—first the canoe and horse, then motorized land and water vehicles, and now airliners—reduce the time costs that the center must spend traveling in order to interact directly with others. Developments in writing, the printing press, and even more so electronic media allow the center to cut these costs further by communicating indirectly with their followers. In addition, developments in modes of address—from a village piazza, through political pamphlets and books, to radio and television—allow it to "scale" these communications, to increase the number of followers it can simultaneously address. Where a New Guinea Big Man, limited to face-to-face interaction, might be able to address 100 or so people at the same time, pamphleteers and political authors can reach many thousands, while contemporary leaders can use electronic media to reach millions.

These developments have had their most striking results on political hierarchy in the nation-state. Using campaign bus tours or barnstorming rallies at airports, political leaders can interact with tens of thousands of people a day, spread over vast distances. Via electronic and print media—through TV political ads, the news media, telephone push-polls, and the like—they interact with millions more remotely. The results are polities in which the political center can exercise profound levels of control over massive numbers of people.

Things are very different, however, in small-scale societies, where communication is exclusively face-to-face and human mobility is by foot alone. Under these circumstances, the crucial influence on the scale and frequency of human interaction is once again density, the number of people on a landscape (Roscoe 1993, 2000a, 2012). Where densities are low, the scale and frequency of interaction is tightly constrained by the time that agents must invest to bring interactions about in the first place. As a result, political entrepreneurs would have to spend so much time traveling between interactions that they would be unable to build any significant political hierarchy. Would-be leaders among the !Kung, for example, are faced with enormous travel costs if they are to interact with and build a significant following from a population scattered across the Kalahari in bands of about twenty-five people at densities of about one person per square kilometer. The effort is hardly worthwhile, and it is no surprise therefore that the !Kung are so egalitarian. Where populations are dense, however, people live on a political entrepreneur's doorstep, so to speak, facilitating interaction and the construction of political relations. In contrast to the !Kung, for instance, political entrepreneurs on contact-era Tahiti were surrounded by potential followers living at a density of some 350–500 people per square kilometer. As we should expect, early European visitors to Tahiti recorded the presence of powerful chiefs, some of whom commanded followings of between 7,000 and 15,000 people (Roscoe 1993: 119).

Again we can look to New Guinea for empirical confirmation. Contrary to Sahlins's (1963) popular stereotype, the Big Man was neither ubiquitous in New Guinea nor just an economic entrepreneur. As Godelier (1986) made clear, most New Guinea leaders were Great Men not Big Men: men who gained status primarily as warriors, hunters, or ritual experts. In a number of other communities, most of them heavily dependent on hunting and gathering, leadership was weak or even nonexistent (e.g., Townsend 1969: 8). Furthermore, in places where a Big Man did emerge, he was not so much an economic entrepreneur as a political one, an organizer or, as Burridge (1975) and Meggitt (1973: 193) put it, the "manager" of his local group's activities. The talents that marked him out from other men—his gift for oratory or "public verbal suasion" (Sahlins 1963: 290); his skills in mediation and conflict resolution; his charisma, diplomacy, ability to plan, industriousness, and intelligence; his abilities in political manipulation (Burridge 1975: 100–102; Lawrence 1973: 17)—brought him renown not only because they allowed him to outdo others in mustering contributions for material distributions and other displays but also, and more importantly, because they qualified him for the central role in bringing about and organizing these displays.[3] For the Big Man, the point of conspicuous material distributions and other collective displays was that they objectified this political ability for all to see (Burridge 1975: 92). They made visible in concrete and comparable terms his organizational and manipulative talents, allowing them to be calibrated against those of other Big Men in the perpetual competition for preeminent status (Roscoe 2012).

A Big Man could only become ethnographically visible, however, if he was able to display his political talents—in other words, if he was able to interact with enough people for a long enough time to be able to create a political hierarchy that would be noticed by ethnographers. But not every community provided these opportunities. The critical variable was population density.

It is not as straightforward as one might think to identify a Big Man society (Roscoe 2012). To circumvent these difficulties, therefore, Table 3.1 tracks the emergence of Big Men using three sets of New Guinea societies (Column 4). The first set (represented as "Y") includes those societies of the western and central highlands whose leaders are universally considered to be classic Big Men. The second set of societies is those for which the ethnographer explicitly denies the presence of Big Men (represented as "N") or, alternatively, explicitly denies the presence of any leader whatsoever ("n"). The final social type (represented as "y") represents societies where the ethnographer refers to leaders as "Big Men" but explicitly distinguishes them as less powerful than the classic Big Men of the western and central highlands (e.g., designating them as "little Big Men").

As Table 3.1 indicates, the emergence of Big Men correlated strongly with population density, the crucial transition occurring between about 20 and 30 people per square kilometer (Columns 2 and 4). Below crude densities of about 20 people per square kilometer, Big Men were not sufficiently visible to be noted in the ethnographic record. If there were leaders in these societies, they were presumably either "little" Big Men or Great Men. Above about 30 people per square kilometer, however, Big Men become clearly visible. The point-biserial correlation coefficient between crude density and the presence or absence of the classic Big Man is 0.81 ($p < 0.001$) (assuming "little Big Men" ["y"] represents the absence of Big Men.)

SYMBIOSIS AND ANTAGONISM: RECONCILING VOLUNTARISTIC AND CONFLICT APPROACHES

To this point, I have treated the horizontal and vertical dimensions of polities and the processes by which they evolve as distinct phenomena. In analytical terms, they are. In practice, however, they also interact. In chapter 5 of this volume, Blanton and Fargher (see also Blanton and Fargher 2008) summarize a number of the ways that this interaction can play out. For the purposes at hand, however, the interactions between the horizontal and vertical dimensions allow us to reconcile the voluntaristic and conflict approaches to political evolution. The key lies in the symbiotic and antagonistic ways in which the interests of the polity-as-defensive-group and the polity-as-political-hierarchy intersect.

On the one hand is a set of mutual interests between center and group that create interdependence and cooperation, the symbiosis on which voluntaristic approaches focus. To the extent that the center provides an effective means of maintaining and improving the military strength of the group—organizing, supplying, and training its defense forces; inventing, developing, and manufacturing weaponry to arm them, communication technologies to improve their interactive capacity, and mechanical means of transport to speed their deployment; policing and adjudicating internal conflict and violence that threaten its defensive unity; and so on—the group has an existential interest in the existence and activities of the center. Conversely, the center has both existential and political interests in providing these defensive services for the group because the alternative, a weak defensive organization, would imperil both its own physical survival and the existence and continuance of the group, on which its political fortunes rest.

Yet the center has other interests besides these, and to the extent that their pursuit impinges on the interests of the rest of the body politic, there is the potential for exploitation and conflict. This is the antagonistic dynamic between center and group, on which conflict approaches focus, and with its superior political capacity, the center usually has the advantage in these struggles. Ironically, some of the most powerful vehicles that the center has at its disposal for advancing its interests at the expense of others are precisely the military and policing services that it contributes to the community's defensive strength. Weaponry developed to counter external threat can also, if monopolized by the center, be turned to exploiting the body politic. Police and judicial systems instituted to mitigate internal dissension and enhance defensive unity can be directed in addition to suppressing dissent about the center's activities, and so on.

In chapter 2 of this volume, Feinman reviews a large body of scholarship on the "size-complexity rule," a cross-cultural relationship that has been widely documented between polity size and organizational complexity. As complexity increases, so does polity size, and vice versa. This relationship may have played a significant part in obscuring the qualitatively different status of the vertical and horizontal dimensions of polities. The fact that the size of political communities and the levels of their political centralization increase together easily fosters an impression that they are one and the same thing or, at the very least, that the one is reducible to the other.

If the preceding arguments are correct, however, we can explain why this relationship between polity size and political complexity exists. In a "balance-of-power race," as we have just seen, the political community and the political center have shared interests in augmenting one another, creating a positive feedback mechanism between the two. More important, the two are positively correlated because the exogenous factors that enable expansions in polity size simultaneously facilitate political centralization. In societies that lack electronic communications and mechanical forms of transport, the exogenous variable is population density. Rising density increases the number of people who can render one another mutual military aid; at the same time, it reduces the travel costs of political entrepreneurs, producing an expansion in the extent and effectiveness of their power. Where developments in communication and transportation allow humans to transcend the limitations of face-to-face communication and movement on foot, the exogenous variable is the "time-space compression" they enable, which stimulates yet further expansions in both community scale and the centralization of power.

CONQUEST WARFARE, POLITICAL INTEGRATION, AND EMPIRE

To the extent that the political center controls a community's defense forces, it can also deploy them for offensive purposes beyond homeland borders. The offensive use of military force was crucial to Carneiro's circumscription theory of political evolution and Choi and Bowles's (2007; Bowles 2008, 2009) parochial altruism, and my disagreement with their positions has been stated above. Nevertheless, in the form of conquest and subjugation, offensive warfare has functioned as an important political tool throughout history, and it has had significant consequences for polity structure and political evolution.

On the one hand, offensive warfare can function as a tool for expanding the polity *qua* community. For example, innovations in the capacity to "compress" time-space may allow several polities that once were separate defensive organizations to unite into a single, more effective defensive unit. In a broader, balance-of-power context it will be to their advantage to do so, but there may be impediments to translating collective interest into collective action. If they are to unify, for instance, all but one of their political centers will have to surrender their supremacy, and there is no guarantee that these centers will be enchanted by the prospect. Ethnic differences can also impede integration in the common interest. In these cases, offensive warfare instigated by elites may be instrumental in the formation of a larger political community.

So long as conquest warfare does not overstep the viable boundaries of mutual defensive aid, the result of such wars will be a stable and expanded polity *qua* community. But offensive warfare can overstep these boundaries, and when it does it introduces a disconnect between the horizontal and vertical axes of polity organization. It expands the reach of the polity-*qua*-political hierarchy, but it does not expand the areal extent of the polity-*qua*-political community, which responds only to developments in the ability of individuals to render one another effective military aid. In these cases, the result of conquest warfare is not a single, expanded polity but rather a dominant-subordinate relationship between two polities, one holding the other in its military thrall.

Consider, for example, the most familiar manifestation of this dominant-subordinate relationship: the empire. In the standard definition, an empire results when one state manages to extend its control over other polities, either directly through force of arms or indirectly through their threat. In so doing, however, the dominant state remains a state; the political reach of its center expands but its viable defensive borders do not. In his bloody rampage across Eurasia, Alexander the Great created an empire that extended from Egypt in the south to modern day Afghanistan in the east, but the boundaries of the Macedonian Kingdom did not expand or disappear in the process. Rather, a new political structure was created, in which Macedon remained an entity in itself that, via satraps, enjoyed some degree of control over the polities Alexander had conquered. Using horse-borne forces against sedentary agricultural polities vulnerable to a mobile strike force, Genghis Khan likewise amassed an empire stretching from the Caspian Sea to the Sea of Japan. He reigned not over a single, colossal polity, however, but a vast array of vassal polities that professed fealty to the Mongol Confederation.

The structure that defines empire, however, is also the problem of empire. In contrast to a state, the component polities of an empire are in no easy position to render one another mutual military aid in the event of an attack, and therefore they are not united by a common interest in defense. By default, the empire is kept together by the imperial polity's threat or use of force, which is a costly means of control. As a result, empires are typically less stable than their constituent polities. Alexander was still alive when the eastern reaches of his empire began to crumble, and though Hellenic cultural influences lasted longer, the rest of his empire, at least as a political entity, was in collapse just a few years after his death. The Mongol Empire fractured into four sectors within just a few decades of its foundation, and within a century it too had started to disintegrate.

EMPIRE AND THE FAILURES OF MODERNITY

Although I hesitate to push this analysis of state and empire too far, I do think it offers an interesting perspective on the present world as well as that of the past. At the end of the first decade of the twenty-first century, a large number of the world's nation-states have been classed as dysfunctional or failing. It may be no coincidence that most are postcolonial states. Take, for example, the nations of the British Commonwealth, remnants of the British Empire, several of which (e.g., Nigeria, Pakistan) periodically verge on dysfunction, with some (e.g., Sierra Leone, Zimbabwe) arguably dysfunctional.

Like the Macedonian and Mongol empires, the British Empire was not a single polity, rooted in the British Isles with borders flung across the globe, but rather a large set of considerably less complex polities controlled by a dominant state centered in Britain. Buttressed by Britain's formidable military force, it lasted rather longer than its Hellenic and Asian counterparts, some three centuries, and its decline was the result more of economic considerations and international politics than of the insurrections and civil wars that destroyed the Macedonian and Mongol empires. Here, though, the structural problem of empire is more apparent in its legacy, in the dysfunctional nature of the independent nations that imperial withdrawal left in its wake. Prior to colonization, each of these states had comprised a set of autonomous indigenous polities, most of them periodically or permanently at war with their neighbors. With colonization, these polities were then grouped together as a "colony" within borders that were arbitrarily defined by accidents of geography and history and later solidified under imperial geopolitical processes. Finally, at independence, the autonomous polities within these accidental borders were pronounced a single entity by fiat and constituted as a "nation-state."

As history has shown, these postcolonial states have proven chronically unstable. Their problem is structural. Rather than a single defensive organization that has evolved gradually under a balance-of-power logic, these "states" comprise a set of autonomous political communities that, with a wave of the colonial wand, were supposed almost overnight to conceptualize themselves and behave as a single, united entity (Roscoe 2004). To be sure, colonialism brought some developments in communications and transportation that likely expanded the area of territory within which people could render one another mutual military aid. There is little reason to expect, though, that these territories would happen to coincide with boundaries set by colonial fiat. The modern postcolonial state is an entity like an empire, held together by the (relatively weak) political and military capacity of a central governing apparatus rather than by a conception among its members that they constitute a unity grounded in a common interest in mutual defense. It should hardly surprise us, then, that they have proved so fragile.

CONCLUSION

Anthropologists and archaeologists have rightly grown wary of prime-mover accounts of political evolution, theories that purport to identify a single force or process, be it irrigation, redistribution, population growth on circumscribed resources, warfare, or whatever, as *the* factor that moves human systems from small-scale polities like bands and villages to more complex ones such as states. The principal argument against prime-mover explanations is that they are reductive; the secondary complaint is that they deny humans their agency, presenting them as though they were automatons, hostage to social and historical processes over which they have no control.

I should hope that my emphasis on humans as knowledgeable and capable agents who pursue interests is sufficient to deflect the latter criticism. The analytical challenge for a theory of society and its evolution is to encompass both structure and agency (Saitta, chapter 6), and if I have largely bracketed the latter it is only because I am for the moment interested in the former. Even so, as an analysis of structural and organizational development, I am acutely aware that my argument still may appear reductive. In proposing that polities can be analytically reduced to horizontal and vertical axes, that they constitute defensive organizations along the former axis and hierarchies of power relations along the latter, and that their development through time is governed by demographic conditions and technological and organizational innovations in communications and transportation, it may seem that I am appealing, if not to a single prime-mover, then at least to just a handful. In conclusion, therefore, I want to emphasize that this is not the case. If my argument seems reductive, it is primarily the result of attempting explanatory clarity in a limited number of words.

Reduction, in itself, is no offense; all worthwhile explanation is reductive. The error lies in being overly reductive, in supposing that complex phenomena such as human polities or the processes governing their political evolution are simple when they are not. This is not my position. Humans act to advance numerous interests, they respond to any number of conditions, and as a result they may create many kinds of polities that follow multiple evolutionary trajectories. Religious interests, for instance, appear to be extremely influential in refracting both the theological character of polities and their political dynamics. The provision of disaster relief may serve a solidifying function in polity evolution, albeit influenced by culturally specific views of individual responsibility. As I have noted, political rivalries and ethnic divisions can interfere with the expansion of a political community even though technological advances might favor its defensive advantages.

What I have tried to do, however, is to identify what I consider the most salient interests and conditions, along with their implications. Humans have numerous interests, but some are analytically more important than others, notably existential interests in defending against attack and, among some humans at least, in augmenting their access to valued resources. In addition, some conditions—in particular the asymmetric distribution of political talents across a population, the density of a population, and communications and transportation capacities—are more influential than others in enabling or constraining human efforts to advance these interests. It is these interests and conditions, I contend, that shape polities and their dynamics more profoundly than others, producing at least some regularities in the political structures and trajectories that have characterized the otherwise kaleidoscopic variability of human political society.

ACKNOWLEDGMENTS

For comments on previous versions of this chapter, I thank David Carballo, Ulrike Claas, Tim Waring, and two anonymous reviewers.

NOTES

1. I am uncomfortable with the term *political evolution*. Quite apart from its Darwinian overtones, which are not necessarily useful in thinking about political as opposed to genetic processes, it has too often in the past been associated with ideas of linear political transformation, of "progress," or of some kind of arrow of inevitability in sociopolitical transformations. Despite these reservations, I retain the term for want of a sufficiently specific alternative.

2. It is analytically unfortunate that Alexander chose to refer to "balance-of-power races" because what is really at issue is not "power" in the general sense but a particular aspect of it: *military* power, the capacity to secure outcomes through *military* action. From an anthropological perspective, it would be more precise, albeit far less elegant, to refer to *balance-of-fighting-capacity* or *balance-of-fighting-strength* races.

3. Spencer's assessment (chapter 9) of similar leader-follower strategies in El Gaván, a modestly sized South American chiefdom, suggests a political continuity between Big Men and chiefdom societies.

REFERENCES

Alexander, Richard D. 1974. "The Evolution of Social Behavior." *Annual Review of Ecology and Systematics* 5 (1): 325–83. http://dx.doi.org/10.1146/annurev.es.05.110174.001545.

Alexander, Richard D. 1979. *Darwinism and Human Affairs*. Seattle: University of Washington Press.

Alexander, Richard D. 1987. *The Biology of Moral Systems*. Hawthorn NY: de Gruyter.

Allen, M. 1984. "Elders, Chiefs, and Big Men: Authority Legitimation and Political Evolution in Melanesia." *American Ethnologist* 11 (1): 20–41. http://dx.doi.org/10.1525/ae.1984.11.1.02a00020.

Anderson, Benedict R. O'G. 1991. *Imagined Communities: Reflections on the Origin and Spread of Nationalism*. London: Verso.

Arnold, J. E. 1993. "Labor and the Rise of Complex Hunter-gatherers." *Journal of Anthropological Archaeology* 12 (1): 75–119. http://dx.doi.org/10.1006/jaar.1993.1003.

Blanton, Richard, and Lane Fargher. 2008. *Collective Action in the Formation of Pre-Modern States*. New York: Springer.

Bourdieu, Pierre. 1990. *The Logic of Practice*. Stanford: Stanford University Press.

Bowles, Samuel. 2008. "Conflict: Altruism's Midwife." *Nature* 456 (7220): 326–27. http://dx.doi.org/10.1038/456326a.

Bowles, Samuel. 2009. "Did Warfare Among Ancestral Hunter-Gatherers Affect the Evolution of Human Social Behaviors?" *Science* 324 (5932): 1293–98. http://dx.doi.org/10.1126/science.1168112.

Brookfield, H. C. and Paula Brown. 1963. *Struggle for Land: Agriculture and Group Territories among the Chimbu of the New Guinea Highlands.* Melbourne: Oxford University Press.

Brown, Paula, and H. C. Brookfield. 1959. "Chimbu Land and Society." *Oceania* 30: 1–75.

Bulmer, Ralph N. H. 1960. "Leadership and Social Structure among the Kyaka People of the Western Highlands District of New Guinea." Ph.D. dissertation, The Australian National University, Canberra.

Burridge, Kenelm O. L. 1975. "The Melanesian Manager." In *Studies in Social Anthropology: Essays in Memory of E. E. Evans-Pritchard,* ed. J. M. Beattie and R. G. Lienhardt, 86–104. Oxford: Oxford University Press.

Burton, John. 1988a. *Local Group Structures and Territories: Hagen Census Division.* Western Highlands Research Monograph (Gazetteer Series) No.3. Port Moresby: Department of Anthropology and Sociology, University of Papua New Guinea.

Burton, John. 1988b. *Local Group Structures and Territories: North Wahgi Census Division.* Western Highlands Research Monograph (Gazetteer Series) No.1. Port Moresby: Department of Anthropology and Sociology, University of Papua New Guinea.

Burton, John. 1988c. *Local Group Structures and Territories: South Wahgi Census Division.* Western Highlands Research Monograph (Gazetteer Series) No.2. Port Moresby: Department of Anthropology and Sociology, University of Papua New Guinea.

Carneiro, Robert L. 1970. "A Theory of the Origin of the State: Traditional Theories of State Origins Are Considered and Rejected in Favor of a New Ecological Hypothesis." *Science* 169 (3947): 733–38. http://dx.doi.org/10.1126/science.169.3947.733.

Choi, Jung-Kyoo, and Samuel Bowles. 2007. "The Coevolution of Parochial Altruism and War." *Science* 318 (5850): 636–40. http://dx.doi.org/10.1126/science.1144237.

Cohen, Ronald. 1978. "Introduction." In *Origins of the State: The Anthropology of Political Evolution,* ed. Ronald Cohen and Elman Service, 1–20. Philadelphia: Institute for the Study of Human Issues.

Collins, Randall. 1999. *Macro-history: Essays in Sociology of the Long Run.* Stanford: Stanford University Press.

Collins, Randall. 2004. "Rituals of Solidarity and Security in the Wake of Terrorist Attack." *Sociological Theory* 22 (1): 53–87. http://dx.doi.org/10.1111/j.1467-9558.2004.00204.x.

Criper, Clive. 1967. "The Politics of Exchange: A Study of Ceremonial Exchange amongst the Chimbu." Ph.D. dissertation. Australian National University, Canberra.

Crumley, Carole L. 1995. "Heterarchy and the Analysis of Complex Societies." *Archeological Papers of the American Anthropological Association* 7: 1–5.

Darwin, Charles. 1871. *The Descent of Man, and Selection in Relation to Sex.* vol. 1. London: John Murray.

Diehl, Michael W. 2000. "Some Thoughts on the Study of Hierarchies." In *Hierarchies in Action: Cui Bono?* ed. Michael W. Diehl, 11–30. Carbondale, IL: Center for Archaeological Investigations.

Dye, David H. 2009. *War Paths, Peace Paths: An Archaeology of Cooperation and Conflict in Native Eastern North America.* Lanham, MD: AltaMira Press.

Feachem, Richard. 1974. "The Raiapu Enga of the New Guinea Highlands: Some Ethnographic Notes." *Asian & Pacific Quarterly of Cultural and Social Affairs* 6 (3): 9–24.

Feachem, Richard. 1977. "Environmental Health Engineering as Human Ecology: An Example from New Guinea." In *Subsistence and Survival: Rural Ecology in the Pacific,* ed. Timothy P. Bayliss-Smith and Richard G. Feachem, 129–82. London: Academic Press.

Feil, D. K. 1987. *The Evolution of Highland Papua New Guinea Societies.* Cambridge: Cambridge University Press. http://dx.doi.org/10.1017/CBO9781139084994

Flanagan, James G. 1983. "Wovan Social Organization." Ph.D. dissertation. University of Pennsylvania, Philadelphia.
Flannery, Kent. 1972. "The Cultural Evolution of Civilizations." *Annual Review of Ecology and Systematics* 3 (1): 399–426. http://dx.doi.org/10.1146/annurev.es.03.110172.002151.
Forge, Anthony. 1990. "The Power of Culture and the Culture of Power." In *Sepik Heritage: Tradition and Change in Papua New Guinea*, ed. Nancy Lutkehaus, Christian Kaufmann, Wlliam E. Mitchell, Douglas Newton, Lita Osmundsen, and Meinhard Schuster, 16–170. Durham, NC: Carolina Academic Press.
Fry, Douglas P. 2006. *The Human Potential for Peace: An Anthropological Challenge to Assumptions about War and Violence.* New York, Oxford: Oxford University Press.
Gell, Alfred. 1975. *Metamorphosis of the Cassowaries: Umeda Society, Language and Ritual.* London: Athlone Press.
Giddens, Anthony. 1984. *The Constitution of Society.* Berkeley: University of California Press.
Giddens, Anthony. 1985. *The Nation State and Violence.* Vol. 2. A Contemporary Critique of Historical Materialism. Cambridge: Polity Press.
Glasse, Robert, and Shirley Lindenbaum. 1973. "South Fore Politics." In *Politics in New Guinea: Traditional and in the Context of Change: Some Anthropological Perspectives*, ed. Ronald M. Berndt and Peter Lawrence, 362–80. Nedlands and Seattle: University of Western Australia Press and University of Washington Press.
Godelier, Maurice. 1986. *The Making of Great Men: Male Domination and Power among the New Guinea Baruya.* Cambridge: Cambridge University Press.
Hallpike, C. R. 1977. *Bloodshed and Vengeance in the Papuan Mountains: The Generation of Conflict in Tauade Society.* Oxford: Oxford University Press.
Harrison, Simon. 1993. *The Mask of War: Violence, Ritual and the Self in Melanesia.* Manchester: Manchester University Press.
Hawkes, Kristen. 1978. "Big-men in Binumarien." *Oceania* 48:161–87.
Hayden, Brian. 1995. "Pathways to Power: Principles for Creating Socioeconomic Inequalities." In *Foundations of Social Inequality*, ed. T. Douglas Price and Gary M. Feinman, 105–27. New York: Plenum Press.
Henrich, Joseph. 2006. "Cooperation, Punishment, and the Evolution of Human Institutions." *Science* 312 (5770): 60–61. http://dx.doi.org/10.1126/science.1126398.
Hintze, Otto. 1994. "Military Organization and the Organization of the State." In *The State: Critical Concepts*, ed. John A. Hall, 181–202. New York: Routledge.
Hobbes, Thomas. 1660. *The Leviathan.* http://oregonstate.edu/instruct/phl302/texts/hobbes/leviathan-contents.html.
Jackson, Graham. 1975. "The Kopon: Life and Death on the Fringes of the New Guinea Highlands." Ph.D. dissertation. University of Auckland, Auckland.
Jorgensen, D. 1981. "Order or disorder in Melanesian religions?" *Man (N.S.)* 16: 470–72.
Juillerat, Bernard. 1996. *Children of the Blood: Society, Reproduction and Cosmology in New Guinea.* Trans. Nora Scott. Oxford: Berg.
Kelly, Raymond C. 1977. *Etoro Social Structure: A Study in Structural Contradiction.* Ann Arbor: University of Michigan Press.
Kelly, Raymond C. 1993. *Constructing Inequality: The Fabrication of a Hierarchy of Virtue among the Etoro.* Ann Arbor: University of Michigan Press.
Knauft, Bruce M. 1985. *Good Company and Violence: Sorcery and Social Action in a Lowland New Guinea Society.* Berkeley: University of California Press.
Langness, L. L. 1973. "Bena Bena Political Organization." In *Politics in New Guinea: Traditional and in the Context of Change: Some Anthropological Perspectives,* ed. R. M. Berndt and

P. Lawrence, 298–316. Nedlands and Seattle: University of Western Australia Press and University of Washington Press.
Langton, John. 1988. "Publius and Political Anthropology: Explaining the Transformation of the United States, 1781–89." *American Behavioral Scientist* 31 (4): 484–96. http://dx.doi.org/10.1177/000276488031004009.
Lawrence, Peter. 1973. "Introduction." *Politics in New Guinea: Traditional and in the Context of Change: Some Anthropological Perspectives*, ed. R.M. Berndt and P. Lawrence, 1–34. Nedlands and Seattle: University of Western Australia Press and University of Washington Press.
Lederman, Rena. 1990. "Big Men, Large and Small? Towards a Comparative Perspective." *Ethnology* 29 (1): 3–15. http://dx.doi.org/10.2307/3773478.
Lewis, Gilbert. 1975. *Knowledge of Illness in a Sepik Society: A Study of the Gnau, New Guinea.* London, New Jersey: The Athlone Press and Humanities Press.
Meggitt, Mervyn J. 1965. *The Lineage System of the Mae-Enga of New Guinea.* Edinburgh: Oliver and Boyd.
Meggitt, Mervyn J. 1973. "The Pattern of Leadership among the Mae-Enga of New Guinea." In *Politics in New Guinea: Traditional and in the Context of Change: Some Anthropological Perspectives*, ed. R.M. Berndt and P. Lawrence, 191–206. Nedlands and Seattle: University of Western Australia Press and University of Washington Press.
Mitchell, William E. 1977. "Sorcellerie chamanique: 'Sanguma' chez les Lujere du Cours supérieur de Sépik." *Journal de la Société des Oceanistes* 33 (56–57): 179–89. http://dx.doi.org/10.3406/jso.1977.2955.
Mitchell, William E. 1978. "On Keeping Equal: Polity and Reciprocity among the New Guinea Wape." *Anthropological Quarterly* 51 (1): 5–15. http://dx.doi.org/10.2307/3317120.
O'Hanlon, Michael. 1989. *Reading the Skin: Adornment, Display and Society among the Wahgi.* London: British Museum.
OKS 1–70/71. 1970. *Oksapmin Patrol Report No.1, 1970/71.* National Archives, Port Moresby.
Philsooph, H. A. 1980. "A Study of a West Sepik People, New Guinea, with Special Reference to Their System of Beliefs, Kinship and Marriage, and Principles of Thought." Ph.D. dissertation. University of Edinburgh, Edinburgh.
Porter, Bruce D. 1994. *War and the Rise of the State: The Military Foundations of Modern Politics.* New York: Simon & Schuster.
Rathje, W. L., and R. H. McGuire. 1982. "Rich Men . . . Poor Men." *American Behavioral Scientist* 25 (6): 705–15. http://dx.doi.org/10.1177/000276482025006008.
Reay, Marie. 1959. *The Kuma: Freedom and Conformity in the New Guinea Highlands.* Victoria, Carlton: Melbourne University Press on behalf of The Australian National University.
Roscoe, Paul B. 1993. "Practice and Political Centralisation: A New Approach to Political Evolution." *Current Anthropology* 34 (2): 111–24. http://dx.doi.org/10.1086/204149.
Roscoe, Paul B. 1996. "War and Society in Sepik New Guinea." *Journal of the Royal Anthropological Institute* 2 (4): 645–66. http://dx.doi.org/10.2307/3034301.
Roscoe, Paul B. 2000a. "New Guinea Leadership as Ethnographic Analogy: A Critical Review." *Journal of Archaeological Method and Theory* 7 (2): 79–126. http://dx.doi.org/10.1023/A:1009512726844.
Roscoe, Paul B. 2000b. "Typologies and Hierarchies in Action." In *Hierarchies in Action: Cui Bono?* ed. Michael W. Diehl, 113–33. Carbondale: Center for Archaeological Research, University of Southern Illinois.

Roscoe, Paul B. 2002. "The Hunters and Gatherers of New Guinea." *Current Anthropology* 43 (1): 153–62. http://dx.doi.org/10.1086/338289.

Roscoe, Paul B. 2004. "Crime and 'Tribal' Warfare in Contemporary Papua New Guinea." In *Globalization and Culture Change in the Pacific Islands*, ed. Victoria S. Lockwood, 59–71. Upper Saddle River, New Jersey: Pearson Education.

Roscoe, Paul B. 2006. "Fish, Game, and the Foundations of Complexity in Forager Society: The Evidence from New Guinea." *Cross-Cultural Research* 40 (1): 29–46. http://dx.doi.org/10.1177/1069397105282432.

Roscoe, Paul B. 2008. "Catastrophe and the Emergence of Political Complexity: Some Models from Social Anthropology." In *Climate and Catastrophe*, ed. Daniel Sandweiss and Jeffrey Quilter, 77–100. Washington, DC: Dumbarton Oaks.

Roscoe, Paul B. 2009. "Social Signaling and the Organization of Small-Scale Society: The Case of Contact-Era New Guinea." *Journal of Archaeological Method and Theory* 16 (2): 69–116. http://dx.doi.org/10.1007/s10816-009-9062-3.

Roscoe, Paul B. 2012. "Before Elites: The Political Capacities of Big Men." In *Before Elites: Alternatives to Hierarchical Systems in Modelling Social Formations*, ed. Tobias L. Kienlin and Andreas Zimmerman, 41–54. Universitätsforschungen zur prähistorischen Archäologie. Bonn: Rudolp Habelt.

Sahlins, Marshall D. 1963. "Poor Man, Rich Man, Big-Man, Chief: Political Types in Melanesia and Polynesia." *Comparative Studies in Society and History* 5 (03): 285–303. http://dx.doi.org/10.1017/S0010417500001729.

Schieffelin, Edward L. 1976. *The Sorrow of the Lonely and the Burning of the Dancers*. St. Martin's Press: New York.

Schieffelin, Edward L. 1981. "The End of Traditional Music, Dance and Body Decoration in Bosavi, Papua New Guinea." In *The Plight of Peripheral People in Papua New Guinea*, Vol. 1: *The Inland Situation*, ed. Robert J. Gordon, 1–22. Cambridge: Cultural Survival.

Schieffelin, Edward L. 1991. "The Great Papuan Plateau." In *Like People You See in a Dream: First Contact in Six Papuan Societies*, ed. Edward L. Schieffelin and Robert Crittenden, 58–87. Stanford: Stanford University Press.

Service, Elman R. 1975. *Origins of the State and Civilization: The Process of Cultural Evolution*. New York: Norton.

Shennan, Stephen. 2008. "Evolution in Archaeology." *Annual Review of Anthropology* 37 (1): 75–91. http://dx.doi.org/10.1146/annurev.anthro.37.081407.085153.

Skocpol, Theda. 1979. *States and Social Revolutions: A Comparative Analysis of France, Russia, & China*. Cambridge: Cambridge University Press.

Sorenson, E. Richard. 1976. *The Edge of the Forest: Land, Childhood and Change in a New Guinea Protoagricultural Society*. Washington, DC: Smithsonian Institution Press.

Strathern, Andrew. 1971. *The Rope of Moka: Big-Men and Ceremonial Exchange in Mount Hagen, New Guinea*. Cambridge: Cambridge University Press. http://dx.doi.org/10.1017/CBO9780511558160

Strathern, Andrew. 1972. *One Father, One Blood*. London: Tavistock.

Tilley, Charles. 1985. "War Making and State Making as Organized Crime." In *Bringing the State Back*, ed. Peter Evans, Dietrich Rueschemeyer, and Theda Skocpol, 169–191. Cambridge: Cambridge University Press. http://dx.doi.org/10.1017/CBO9780511628283.008

Townsend, Patricia K.W. 1969. "Subsistence and Social Organization in a New Guinea Society." Ph.D. dissertation. University of Michigan, Ann Arbor.

Treide, Barbara. 1985. "Big-men im Prozeß sozialökonomischer Differenzierung." *Jahrbuch des Museum für Völkerkunde zu Leipzig* 39: 169–204.

Tuzin, Donald F. 1976. *The Ilahita Arapesh: Dimensions of Unity*. Berkeley: University of California Press.

Vicedom, Georg F. and Herbert Tischner. N.d. *The Mbowamb: The Culture of the Mount Hagen Tribes in East Central New Guinea, Vol. 2. I. Social Organisation*. Trans. F. E. Rheinstein and E. Klestadt. Canberra: Menzies Library, Australian National University.

Westermann, Ted. 1968. *The Mountain People: Social Institutions of the Laiapu Enga*. Wapenamanda: New Guinea Lutheran Mission.

Wright, Henry T. 1978. "Towards an Explanation of the Origin of the State." In *Origins of the State: The Anthropology of Political Evolution*, ed. Ronald Cohen and Elman Service., 49–68. Philadelphia: Institute for the Study of Human Issues.

4

The Ritualized Economy and Cooperative Labor in Intermediate Societies

Charles Stanish

What the collection of essays in this book represents, in my view, is a profoundly new and powerful way to understand *what* precisely evolves in human society from an anthropological and archaeological perspective. Based upon recent work in game theory, we can now reconceptualize the somewhat vague terms of "cultural evolution" and "adaptation" into the more specific one of the "evolution of cooperation." By focusing on cooperation we immediately align ourselves with the latest and most exciting theoretical work in game theory in anthropology, economics, and economic history.

Game theory and its allied disciplines of evolutionary game theory and evolutionary psychology seek to study conflict and cooperation between rational decision makers (see Myerson 1991: 1). In classic economic game theory, there is the assumption that people will maximize utility—this is considered rational economic behavior. *Utility* is the amount of satisfaction an agent derives from a good or service. In Western research traditions outside of anthropology, utility is usually described in monetary terms and "rational" is understood as the maximizing of advantage in each economic interaction. This concept of "*Homo economicus*" is described as the "canonical" model of human behavior and represents the basic assumption of Western economic theory for over a century (Henrich et al. 2001).

Anthropology and modern game theory has directly challenged this canonical model. In anthropological applications we conceive of this optimizing behavior in much broader terms such as "maximizing material and social well-being." As modified in this sense, utilities are culturally contingent and may not even be "priced" in a Western economic sense. Needless to say, this anthropological perspective is particularly relevant in intermediate or "chiefly" societies that function well without money or standardized units of account or exchange and where it is difficult to quantitatively measure utility. In this sense, the definition of a rational agent as simply a decision maker who acts in pursuit of his or her objectives, both short and long term, can be universally applied. This conception recognizes that "objectives" are culturally and historically conditioned and these objectives are at times not framed as simple, immediate self-interest.[1]

Evolutionary game theory derives from classic game theory but stresses the process of natural selection of behaviors in populations of interacting individuals and groups over time. Experimental research has demonstrated that humans can be irrationally prosocial, demand fairness, and can cooperate under the appropriate conditions in what is counter to the assumptions of the canonical model (Bowles et al. 1997; Henrich et al. 2010). There is a substantial debate in the literature as to whether this prosocial human behavior derives from an innate psychology that evolved in the Paleolithic that was then applied to non-kin, or on the contrary, whether people developed over time and through trial and error "social norms and informal institutions that are capable of domesticating our innate psychology for life in ever-expanding populations" (Henrich et al. 2010: 1480). These institutions are then subject to selection, but not in a classic biological sense. Rather, cultural transmission models based on group selection mechanisms are required to understand the evolution of human cooperation. As Boyd et al. (2003: 3531) point out, "Unlike any other species, humans cooperate with non-kin in large groups. This behavior is puzzling from an evolutionary perspective because cooperating individuals incur individual costs to confer benefits on unrelated group members."

This new experimental and theoretical work therefore directly challenges the core principle of the canonical model: rational choice theory or the notion that people constantly maximize economic advantage in each and every human interaction. While we can still maintain that people maximize utility as described above, that culturally conditioned and contingent "utility" in many small-group interactions is economically "irrational" even though these are commonsense responses to the exigencies of daily life. We can therefore redefine this maximization principle in anthropological terms to assume that people are motivated "to seek a better way of living." As the evolutionary economist Ulrich Witt (2003: 10) observes: "'Nature's parsimony,' as Ricardo once put it, has always imposed constraints on the human existence. And at all times, humans have been eager to use their creativity and problem-solving activity to relax some of those constraints or, at least, to alleviate their consequences."

From this perspective, the origin, evolution, and change in nonmarket economies over time stands as a core issue in anthropology. Of course, archaeology is well suited to contribute to this fascinating debate. In this essay, however, I will avoid this deeper theoretical issue and instead focus on a particular aspect of cooperative behavior relevant to anthropological and archaeology theory building—the role of ritual in keeping groups cooperating in ways that contradict the canonical model.

COOPERATION

From a long time frame, or archaeological perspective, we can model the transitions of small agrarian societies of a few score population to larger political entities as a process in which increasingly more people cooperate, either voluntarily or otherwise, to produce those features that traditionally define complex society in anthropological archaeology: monuments, religious institutions, temples, heightened commodity production, intensified agricultural production, warrior classes, palaces, and the like. Likewise, we can understand shifts to simpler forms of sociopolitical organization as people choose or are forced into a less cooperative form of organization in favor of village or local ones.

The key to this evolutionary process is *cooperation* among individuals and households to increase economic production and provide a competitive advantage against other groups. Economic production and political cooperation are central components for the success of groups. I follow Richerson and Boyd and other cultural transmission theorists and argue that, in a competitive environment, less successful groups will either imitate successful ones or they will be absorbed by politically and economically stronger groups. As Boyd et al. (2003: 3531) put it in starkly selective terms, "more cooperative groups are less likely to become extinct" while less cooperative groups are subject to disappearing in a competitive environment. Or, as Henrich (2004: 11) aptly says, "Between-group selection . . . favors groups with more prosocial individuals because such groups can outcompete groups dominated by free riders." This is a process that I have elsewhere called "economic selection" (Stanish 2004). The concept is structurally similar to natural selection in biology but, as applied to cultural evolution, relies on models of multilevel selection. The concept of group selection in biology is theoretically controversial. However, along with cultural transmission, it stands as a central concept in *cultural* evolution, and I align with the large body of theoretical literature that has demonstrated its theoretical feasibility and utility (Boyd et al. 2003; Boyd and Richerson 1992). We simply cannot account for the rapid evolution of complex society (from hunter/foragers to states in just a few millennia) with biological models. Cultural evolution is analogous to, but more complex than, biological evolution, requiring its own set of concepts. In short, the creation of economic surplus for the group becomes the key factor in cultural evolution as it is reformulated as a question of the evolution of cooperation.

Game theorists have developed a number of concepts that help us understand how cooperation is maintained and how it evolves, particularly in competitive environments among small societies. From this perspective, the most successful group is that which cooperates to create the most resources, which are used, in turn, to compete with other groups.

Critical concepts for maintaining cooperation center on maintaining and signaling one's reputation as a cooperator, fairness in the interactions between individuals, and the punishment of free-riders or defectors in social interactions. Reputational effects are those that increase cooperation through the establishment of trust between individuals, allowing higher payoffs for both parties over what could be achieved by each acting in their immediate self-interest (Ostrom 2003). Trust is essential in achieving levels of repeated reciprocity (Ostrom and Walker 2003). In the experimental game known as the Prisoner's Dilemma (PD), people have to choose between defecting or cooperating with each other. In the one-time game, the optimal strategy is always to defect even though cooperation would result in a better outcome for both. This is because the cost of being a cooperator while the other defects is far higher than the "suboptimal" strategy of both defecting.

This is the stable strategy if played as a one-shot game. However, if players continue this game over time (known as the Iterated Prisoner's Dilemma [IPD]), people develop reputations of being defectors or cooperators. Reputation alters the cost-benefit calculation from the simple PD where defection is the norm to a situation more reflective of human social interaction. In running simulations of the IPD, Axelrod (1984) found that the highest payoff for players was a simple strategy called "tit-for-tat." The strategy was to simply cooperate on the first move and do whatever the other player did on each subsequent move. As he describes it, tit-for-tat sustains cooperation based on a very simple strategy of reciprocity. Cooperation emerges between competing individuals because it was in each player's self-interest. A simple "social history" of the players develops, and this is the basis for reciprocal behavior over time. The IPD demonstrates that cooperation can develop between selfish agents and that coercion need not be present for hierarchy to emerge among autonomous agents.

Other work has demonstrated that people act irrationally if there are perceptions of unfairness in human interactions. The Ultimatum Game (UG), in which one person can propose a certain amount of money to another, illustrates this principle. If the responder declines the offer, then each person loses all of the money. The rational behavior that conforms to the canonical model is to offer very little and to accept very little. However, in practice, responders are willing to give up "free" money if the offer is not deemed "fair." Nowak, Page, and Sigmund (2000) ran a series of simulations using an iterated UG. What they discovered is that there is an "irrational" human emphasis on fairness in games that most approximate social reality. As Clark and Sefton (2001: 52) summarize it, "Subjects are willing to forego material gain in order to reward those who have treated them favourably." There is a substantial and convincing body of research that demonstrates how prosocial punishment can maintain cooperation (Boyd et al. 2003; Boyd and Richerson 1992).

Punishment, in game theoretic terms, is a type of prosocial, irrational (costly) behavior that appears to be a surprisingly strong attribute of human sociality and which contradicts classic economic assumptions of the canonical model

(Henrich et al. 2001). While the experimental work also suggests that behavior tends to approach the canonical model as the stakes go up, there still is a significant amount of irrational behavior even at the highest payoffs.[2]

In short, research in game theory and evolutionary game theory tells us that when information on players' reputations is included with the potential for prosocial punishment, cooperation emerges as a very stable evolutionary strategy. As Bowles and colleagues (1997: 5–6) put it, "If costly retaliation opportunities are combined with communication opportunities almost no defection occurs and, therefore, no resources are wasted for retaliation." Or in the words of Henrich and colleagues (2010: 1480), "norms can facilitate trust, fairness, and cooperation in a diverse array of interactions, thereby allowing the most productive use of unevenly distributed skills, knowledge, and resources, as well as increasing cooperation in exchange, public goods, and warfare." From an economic selection perspective, the group that is able to maintain cooperation through these principles will, on average, be more successful than groups that do not.

MAINTAINING ECONOMIC COOPERATION THROUGH RITUAL AND TABOO

I have argued elsewhere that in intermediate societies (otherwise known as chiefdoms), the positions of elite power are transitory and are not coercive (Stanish 2009). Managerial leadership emerges in chiefly societies as the benefits of cooperation outweigh the costs of cooperation to the individuals in that group. I have also argued elsewhere that the key to heightened economic production in intermediate societies is to create rudimentary economies of scale that increase efficiencies in production (Stanish 2004). Such labor organization increases production by reducing transaction costs and not by increasing labor inputs. As a result, a highly cooperative group can increase their resources without working more on a per capita basis.

There is, however, a cost to economic cooperation, particularly in specialized production: the loss of autonomy by the primary producers over the goods or other resources that they create. In effect, people who work in specialized labor organizations are able to produce substantially more goods than those working strictly as individuals or as households due to the well-known principles of economies of scale (the latter issue explored in Stanish 2004). But the individuals in the group have to give up autonomy over the disposition of those goods since individuals in a specialized labor organization cannot lay claim to resources that they shared in creating. The question facing any society in such a circumstance is how to maintain norms of fairness and prevent defection. In both chiefly and state contexts, failure to keep people cooperating results in political collapse. Ethnography and history teach us that chiefly societies are notoriously politically unstable, a result of managerial elites' inability to maintain their factions.

From a game theoretic perspective, in order to maintain cooperation in chiefly societies, norms that reinforce cooperation must be developed and be enforced. These norms are both costly to the individuals who enforce them and are costly for those who comply. In a recent publication (Stanish 2009), I argue that one very effective way to create and maintain cooperative labor organizations in intermediate societies (i.e., those without the institution of coercive force as seen in state societies) is to embed the production process in set schedules, defined by political ritual, conducted in periodic feasts and sanctioned by taboo or customary law. It is necessary to establish norms of punishment with clear costs to free-riders. Ritual, feasting, and taboo helps avoid the dangers of free-riding that will destroy cooperative behavior in groups.

The ritualized economy of chiefly societies therefore can be understood from a game theoretic perspective as effectively keeping groups cooperating by establishing norms of work embedded in a ritual schedule and enforced through taboo. Ritual schedules labor; it provides a series of benchmarks that people can count on to guarantee a return on their labor investment in the form of known periodic feasts. Ritual likewise establishes taboos that punish defectors, establishes norms by which reputations are measured, and effectively allows for much greater cooperation by creating a known social universe where individual behavior can be channeled into more effective group cooperation.

In the next section, I will focus on one aspect of the norms and strategies to keep cooperative groups together. This is the use of ritual to schedule and regulate cooperative work above the level of the household in agricultural labor. I use ethnographic data from Malinowski and Firth to illustrate how these strategies actually work on the ground among small-group societies and how the principles of game theory can be clearly seen operating in these historical data from intermediate societies.

SCHEDULING OF DOMESTIC LABOR AND THE PRINCIPLE OF PROPORTIONAL RITUALIZATION

> Success in most economic pursuits depends in the eyes of the natives upon the performance of effective magic. Gardening, the most important, is bound through a series of rites. Gardening is ritually regulated to a great degree. (Malinowski 1918: 52)

From the perspective of group cooperation, one of the key functions of ritual and taboo is to organize and schedule complex labor tasks in intermediate societies. For most of the human history of settled village life, the bulk of economic production was organized at the household level. With the advent of larger villages and towns, craft specialization was adopted by groups as a way to increase production, both in the aggregate and per capita. For instance, specialized production increases per capita resources by creating rudimentary economies of scale that eliminate redundant transaction costs (Stanish 2004).

Once production moves beyond the household level of organization, however, a series of potential political and social problems threaten group cooperation. The most immediate problem centers on the distribution of the product of a specialized labor group. That is, when production is limited to the household, then it is the household itself that controls the entire process of production and exchange of goods. However, if there are many households or independent individuals involved, then the distribution of the end product has to be equitable (see Eerkens, chapter 7; Pluckhahn, chapter 8). As we have seen, fairness is one of the principal concerns of a cooperating group, and any perception of unfairness will result in irrational, prosocial behavior by group members. Such behavior would threaten the viability of the productive group. Fairness principles can be maintained if members of the cooperative group are fully aware of their duties and rewards. In the case of specialized production over time, it is necessary to have benchmarks for when and where the goods will be received. Such benchmarks also serve to schedule the labor tasks in predictable and fair ways.

Ritual behavior in which all group members implicitly and explicitly know a strict sequence of rites is ideally suited for the scheduling of labor and the equitable distribution of goods. The classic ethnographic work of Malinowski (e.g., 1921, 1922) in the Trobriand Islands of Melanesia, for instance, is full of references to the role of ritual and magic in regulating labor and scheduling the timing of redistribution. In *Argonauts*, he clearly states that the chief and garden magician, through "rites and spells . . . initiate each stage of work and each new stage of development of plant life" (Malinowski 1922: 59). There are, in fact, so many references to the role of ritual leaders in scheduling labor in the Trobriand Islands that is beyond doubt that the control of agricultural production was seen as central to their duties. For instance, "The magician inaugurates successfully all the various stages which follow one another [in planting and harvesting]" (Malinowski 1922: 59).

Similar agricultural cycle ritual is found in Tikopia, another Melanesian group studied by Raymond Firth. Here again the most important crop and the one with the most significant ritual is the yam. "The yam is the premier food product in ritual, and its rites alone belong to the seasonal cycle of the Work of the Gods" (Firth 1967). Similar and separate rites are conducted for taro, coconut, and breadfruit. The rituals involve elaborate preparations that are timed to coincide with the agricultural labor cycle. The feasts involved with the rituals bring the entire village together. As a result, the labor force is guaranteed to be together during the critical periods when cooperation is necessary. The feasts also provide the opportunity to determine who is working well and who is avoiding their responsibilities. These public and community events allow for chiefs and religious specialists to criticize those who are not pulling their weight and to assess the needs of agricultural labor.

Malinowski also points out that ritual is critical in managing other nonagricultural labor tasks as well, such as shark and mullet fishing. He writes: "Thus fishing, an activity of great economic importance and favorite sport of the Trobriands, ranges from a purely economic pursuit to almost a magico-religious ceremony. In fact, the mullet fishing in Labai is surrounded with more stringent taboos and is bound up with tradition and ceremony than any other social activity in the Trobriands" (Malinowski 1918: 92). The same is true for shark fishing, which involves numerous rituals, taboos, and ritual-specialist control.

Firth's work among the Maori of New Zealand reveals a similar level of ritual control over fishing, farming, and crafts. "Magic, in fact, in one shape or another, permeates all the economic life of the native. Every craft has its spells and incantations, its rites and omens, its regulations of supernatural import" (Firth 1929: 234). Shark fishing was a very important economic and social activity that was strongly regulated by ritual and enforced with taboo and outright physical punishment. Firth notes that anyone who violated the rules (fishing out of turn, for instance) had "their goods plundered or their canoes split up" (Firth 1929: 227).

We find, however, along with these highly ritualized fishing and farming behaviors, similar activities that are not associated with much ritual. Fishing in ponds and lagoons, for instance, is not subject to ritual or magic at all. Likewise, working the family plots did not involve chiefly or ritual specialist intervention, except when a family's fences were untended and let out pigs into other peoples' gardens. There is a clear distinction between labor conducted under ritual schedules and taboos and labor that is not.

A review of both the Trobriand and Maori data indicate a clear pattern: *The degree to which an activity requires complex coordination and involves risk is the degree to which ritual regulation and taboo will be important to that activity.* Pond fishing, for instance, involved lone individuals using poisons to bring the fish to the surface. The same is true for net fishing by individuals. Malinowski (1918: 88) says: "Now there is absolutely no magic in connection with fish poisoning, and very little in connection with the ordinary fishing by means of nets." There was very little risk and investment and no coordination with others. The same is true for agricultural work on individual plots. In effect, ritual regulation corresponds to the degree to which cooperation was essential to the success of the task.

A critical observation that supports this pattern is that the labor involved in all of these activities is not technically difficult. In fact, people do the kinds of tasks alone all the time. A specialized knowledge is rarely required. Rather, the main challenge is keeping people cooperating for their perceived self-interest in these specialized labor organizations. The group must ensure that potential defectors do not defect. It must ensure that fairness pervades the activities to avoid the rise of antigroup behavior. These conditions are met by ritual that exists outside of any individual but which expresses the needs and ideology of the group.

SUMMARY

If we seek to understand the evolution of cooperation in intermediate societies, we must understand how groups function and maintain their integrity. Ethnographic data show us that ritualizing certain aspects of specialized labor avoids defection and free-riding by ensuring an individual's return on his or her labor and instilling group-enforced (via taboo) norms against noncooperative behavior. Ritual provides the framework to keep people working together consistent with game theoretic understandings of group cooperation. Ritual provides a series of benchmarks by which a cooperative group fully understands when they will receive their payoff for laboring in a specialized organization, and ritual and taboo reinforce norms of fairness and reciprocity. Ritual provides guarantees to all members of the labor organization that they will receive a fair share of their production. It furthermore provides sanctions against noncooperators, and prescribes the social rewards individuals receive for cooperating over a long period of time. Ideology and its concomitant ritual is the social means by which a group is guaranteed the exchange of surplus wealth. This helps keep the organization viable and is an essential component of the evolution of complexity.

NOTES

1. In the last decade or so, economists have been incorporating this assumption into their models of human behavior. One of many examples would be: "All this evidence suggests that people are not motivated solely by material self-interest." (Dufwenberg and Kirchsteiger 2004: 269).

2. For example: "While the amount of positive reciprocation falls as subjects play further repetitions of the one-shot game, non-negligible rates of co-operation are observed even in the final repetition" (Clark and Sefton 2001: 52).

REFERENCES

Axelrod, Robert. 1984. *The Evolution of Cooperation*. New York: Basic Books.

Bowles, Samuel, Robert Boyd, Ernst Fehr, and Herbert Gintis. 1997. "Homo reciprocans: A Research Initiative on the Origins, Dimensions, and Policy Implications of Reciprocal Fairness." Manuscript at http://www.umass.edu/preferen/gintis/homo.pdf

Boyd, Robert, Herbert Gintis, Samuel Bowles, and Peter J. Richerson. 2003. "The Evolution of Altruistic Punishment." *Proceedings of the National Academy of Sciences of the United States of America* 100 (6): 3531–5. http://dx.doi.org/10.1073/pnas.0630443100.

Boyd, Robert, and Peter J. Richerson. 1992. "Punishment Allows the Evolution of Cooperation (or Anything Else) in Sizable Groups." *Ethology and Sociobiology* 13 (3): 171–95. http://dx.doi.org/10.1016/0162-3095(92)90032-Y.

Clark, Kenneth, and Martin Sefton. 2001. "The Sequential Prisoner's Dilemma: Evidence on Reciprocation." *Economic Journal* 111 (468): 51–68. http://dx.doi.org/10.1111/1468-0297.00588.

Dufwenberg, Martin, and Georg Kirchsteiger. 2004. "A Theory of Sequential Reciprocity." *Games and Economic Behavior* 47 (2): 268–98. http://dx.doi.org/10.1016/j.geb.2003.06.003.

Firth, Raymond. 1929. *Primitive Economics of the New Zealand Maori*. With a preface by R. H. Tawney. New York: Dutton and Co.

Firth, Raymond. 1967. *The Work of the Gods in Tikopia*. London: Athlone Press.

Henrich, Joseph. 2004. "Cultural Group Selection, Coevolutionary Processes and Large-Scale Cooperation." *Journal of Economic Behavior & Organization* 53 (1): 3–35. http://dx.doi.org/10.1016/S0167-2681(03)00094-5.

Henrich, Joseph, Robert Boyd, Samuel Bowles, Colin Camerer, Ernst Fehr, Herbert Gintis, and Richard McElreath. 2001. "In Search of Homo Economicus: Behavioral

Experiments in 15 Small-Scale Societies." *American Economic Review* 91 (2): 73–78. http://dx.doi.org/10.1257/aer.91.2.73.

Henrich, Joseph, Jean Ensminger, Richard McElreath, Abigail Barr, Clark Barrett, Alexander Bolyanatz, Juan Camilo Cardenas, Michael Gurven, Edwins Gwako, Natalie Henrich, et al. 2010. "Markets, Religion, Community Size, and the Evolution of Fairness and Punishment." *Science* 327 (5972): 1480–84. http://dx.doi.org/10.1126/science.1182238.

Malinowski, Bronislaw. 1918. "Fishing in the Trobriand Islands." *Man* 18: 87–92. http://dx.doi.org/10.2307/2788612.

Malinowski, Bronislaw. 1921. "The Primitive Economics of the Trobriand Islanders." *Economic Journal* 31 (121): 1–16. http://dx.doi.org/10.2307/2223283.

Malinowski, Bronislaw. 1922. *Argonauts of the Western Pacific*. E.P. Dutton: New York.

Myerson, Roger B. 1991. *Game Theory: Analysis of Conflict*. Cambridge: Harvard University Press.

Nowak, M. A., K. Page, and K. Sigmund. 2000. "Fairness versus Reason in the Ultimatum Game." *Science* 289 (5485): 1773–75. http://dx.doi.org/10.1126/science.289.5485.1773.

Ostrom, Elinor. 2003. "Toward a Behavioral Theory linking Trust, Reciprocity, and Reputation." In *Trust and Reciprocity: Interdisciplinary Lessons from Experimental Research*, ed. Elinor Ostrom and James M. Walker, 19–79. New York: Sage Publications.

Ostrom, Elinor, and James M. Walker. 2003. "Introduction." In *Trust and Reciprocity: Interdisciplinary Lessons from Experimental Research*, ed. Elinor Ostrom and James M. Walker, 3–18. New York: Sage Publications.

Stanish, Charles. 2004. "The Evolution of Chiefdoms: An Economic Anthropological Model." In *Archaeological Perspectives on Political Economies*, ed. Gary M. Feinman and Linda Nicholas, 7–24. Salt Lake City: The University of Utah Press.

Stanish, Charles. 2009. "The Evolution of Managerial Elites in Intermediate Societies." In *The Evolution of Leadership. Transitions in decision making from small-scale to middle-range societies*, ed. Kevin Vaughn, Jelmer Eerkens, and John Kanter, 97–119. Santa Fe: School of American Research.

Witt, Ulrich. 2003. "Evolutionary Economics and the Extension of Evolution to the Economy." Paper presented at the Workshop on Evolutionary Economics, Schloss Wartensee, Switzerland, June 2003.

5

Reconsidering Darwinian Anthropology

With Suggestions for a Revised Agenda for Cooperation Research

RICHARD E. BLANTON AND LANE F. FARGHER

We agree with Richerson, Boyd, and Henrich (2003: 361) that a well-formed evolutionary theory for cooperation will benefit humans as they adapt to rapid technological and economic change in the contemporary world, but what kind of evolutionary theory should this be? One candidate that has gained much recent attention in anthropology and sister disciplines (Aunger 2000) is a Darwinian approach, originally termed *sociobiology*, but now referred to variously as *evolutionary psychology*, *evolutionary anthropology*, *Darwinian anthropology*, and *Darwinian social science*, that aims to understand the bioevolutionary origins of what is considered to be, in part, an instinctive basis for human prosociality (e.g., human "ultra prosocial tendencies" in Simpson and Beckes [2010: 36]).[1] The perceived prosociality is thought to provide an explanation for why human societies are among the largest and most complex on earth. Further, a pervasive prosociality is thought to explain why we see tendencies to cooperate and to punish noncooperators in experimental games (e.g., Gintis et al. 2005: 8).

Should we understand the evolution of large and complex human societies as evidence for an inherent "ultra" human prosociality? Our response to this question is strongly negative. For one thing, while human societies are sometimes large and complex, not all of them are based on very much cooperation at all; instead, in many cases they are based on social and cultural domination by a governing elite. Even in those instances in which human societies are formed around the shared intentions of cooperating individuals, we argue that the current thrust of Darwinian anthropology fails to provide a useful pathway to explanation or theory building. In this chapter we first critique the current arguments of Darwinian anthropology and point to an alternative approach grounded in primate and social neurobiological research on Theory of Mind and related topics that provide a basis for understanding a form of cooperation that is far more contingent than allowed for by most current Darwinian anthropologists. This approach identifies the human as a social actor capable of prosocial action, as well as deception and defection, so that when we find cooperation it is the result of the instrumentality of institutional, organizational, technological, and cultural mechanisms that solve the inevitable cooperation problems humans face. To illustrate the latter point, we provide an example of a pattern of cultural change that represents a response to cooperation problem solving in the development of states. Our example also provides a template for an empirical research epistemology, based on hypothesis testing, of the sort largely lacking in the Darwinian anthropology paradigm.

PROSOCIAL ACTION IS ARGUED TO RESULT (IN PART) FROM BIOLOGICALLY EVOLVED MENTAL MODULES AND SOCIAL INSTINCTS

A commonly expressed assumption in much of the Darwinian anthropology literature is that human behavior reflects the operation of functionally specialized, neurally encoded mental modules ("mental rules" and "neural machinery" in Cosmides and Tooby [1992] and Tooby and Cosmides [1997: 292]; "special-purpose 'minicomputers'" in Buller [2005: 127]) or social instincts (e.g., Richerson and Boyd 1999) that evolved biologically among Pleistocene foragers (e.g., Boehm 2004; Kaplan and Gurven 2005).[2] In response to critics who argued against the excessive biological reductionism of earlier sociobiological writing (e.g., as described in Carruthers 2006: 37), contemporary Darwinian anthropologists, while still maintaining that mental modules (Carruthers 2006) and social instincts (e.g., Gintis et al. 2005) are the ultimate causes of human social behavior, have admitted that the more proximate factors of agency and creative problem solving should be considered alongside innate behavioral tendencies (Carruthers 2006: 37, 155–156, 191, 277–278). A cultural component has been incorporated into some Darwinian evolutionary scenarios ("dual inheritance theory" or "culture-gene coevolution"), most notably by Robert Boyd, Peter J. Richerson, and Joseph Henrich (e.g., 2005a; cf. Boyd and Richerson 1985; Henrich and Henrich 2007: 9), although not all Darwinian anthropologists agree that culture matters (e.g., Betzig 1997: 17, fn. 49). However, when culture is incorporated into Darwinian schemes it is argued to consist of "inherited strategies" and is analyzed in a manner consistent with Darwinian theory. This approach follows Richard Dawkins's (1982: 109) idea of the "meme," a discrete unit of culture similar to a gene, that is socially transmitted by imitation and subject to natural selection. Based on the assumption that more successful inherited strategies will be copied ("biased transmission"), they will become widespread in a population;

hence, culture change over time in the relative frequencies of different inherited strategies can be modeled in exactly the same way that change in gene frequencies is modeled in natural selection scenarios (Boyd and Richerson 2000: 161).

Culture, even in this "Darwinized" variant (e.g., Aunger 2000), is relegated to a limited role because, ultimately, cooperation is argued to reflect the force of highly specialized, domain-specific mental modules or social instincts that have evolved through natural selection (Cosmides and Tooby 1992; Gintis et al. 2005: 22–23; Richerson, Boyd, and Henrich 2003; Tooby and Cosmides 1992), including "tribal instincts" (Boyd, Richerson, and Henrich 2005a: 265), "innate psychology," "innate content biases," "cerebral dispositions," "proximate psychologies," "social norm psychologies," and "reciprocator genes" (e.g., Henrich and Henrich 2007: 43). Social instincts and mental modules promote cooperation (e.g., modules providing proximate psychologies for "cooperators," or "contributors,"), provide for a "genetic predisposition to behave altruistically" (Bowles 2006; cf. Boehm 2004: 277–278), allow for detection of cheating (Cosmides and Tooby 1992), and provide an indirect incentive to cooperative action through punishment of noncooperators (e.g., Herrmann, Thöni, and Gächter 2008) (e.g., as inherently "altruistic punishers," "timid punishers," etc.). However, a human population will also feature carriers of mental modules and social instincts that prompt amoral actions (e.g., "defectors," "cheaters," "norm violators," etc.) or that bring avoidance of punishment (Boyd and Richerson 2005: chapter 9; Gintis et al. 2005). The key question for cooperation researchers (the "core dilemma of cooperation" in Henrich and Henrich [2007: 43]) is that, while (as estimated from computer simulations) genes inhibiting cooperation, such as a tendency to defect or avoid the costs of punishment, can become well established in populations, innate cooperators (and those who punish for noncooperation) will have relatively lower fitness and hence will tend to occur in lower frequencies.

With their suite of methods, contemporary Darwinian anthropologists claim to have resolved their core dilemma as follows, in three parts, although there is not complete agreement among them about which of the causal factors should be considered most important. First, given the primarily small-group framework for Pleistocene cooperation, altruistic sharing is argued to have evolved in part because altruistic action toward kin enhances the biological fitness of the altruist, following the logic of kin selection (from Hamilton [1964]). According to the theory, kin selection resulted in a pattern of "kin psychology" (Henrich and Henrich 2007: 43) or "familial sociality" (Richerson and Boyd 1978), or even an "axiom of amity" toward kin, no matter how distantly related (Jones 2000), that are thought to serve as important building blocks for more-inclusive forms of prosociality (e.g., Kaplan and Gurven 2005: 105; van den Bergh and Barash 1977). Second, also in small groups, cooperation may be fostered among non-kin when altruist sharing is based on balanced reciprocal gifting ("reciprocal altruism" from Trivers [1971]) or when a potential cooperation partner has a favorable reputation for cooperation ("indirect reciprocity" in Alexander [1987], e.g., in Bowles and Gintis [2005: 29]). Cooperation is also argued to result from the evolution of mental modules for punishment of nonreciprocators so long as punishment is coordinated (Boyd, Gintis, and Bowles 2010; Boyd, Richerson, and Soltis 2005: 216) or when there is a culturally established pattern of reciprocity, for example, when religious beliefs and ritual serve to enhance in-group solidarity even with non-kin. In this case, according to the theory, a mental module for punishment can thrive and become fixed since little punishment will be required (Boyd and Mathew 2007; Boyd and Richerson 2005: chapter 9).[3]

Lastly, we refer to group selection as one more proposed mechanism favoring the evolution and spread of prosocial behavior (e.g., Boyd, Richerson, and Soltis 2005; Wilson 2002: chapter 1). A group selection approach purports go beyond kin selection and reciprocity to "model a complex regulatory system that binds members of a group into a functional unit" (Wilson 2002: 25); for example, when a group adheres to a religious belief that functions to coordinate activities and solves the free-rider problem by preventing cheating. In this case, shared cultural values are thought to provide a framework for cooperation in which individuals tending to be less cooperative (or less willing to punish noncooperators) would be shunned or excluded; in this way, gene-culture coevolution builds "cultural imperatives into our genes" (Richerson, Boyd, and Henrich 2003: 371). Further, if degrees of intragroup cooperation vary between populations, those groups featuring heightened levels of cooperation (and tendency to punish) are argued to have biological evolutionary advantages (higher average fitness) by comparison with populations featuring lower levels of cooperation and punishment (e.g., Boyd and Mathew 2007; Nowak 2006: 1561). Prosocial mental modules then proliferate because cooperative groups are more successful and replace groups that have lower frequencies of culturally based or cooperation-driving social instincts, and because cultural systems of cooperation will be emulated by other, less cooperative groups (e.g., Gintis et al. 2005: 22; Wilson 2002).

METHODOLOGICAL PROBLEMS AND OTHER CHALLENGES TO DARWINIAN ANTHROPOLOGY

> "In much sociobiology . . . almost no effort is made to describe how cultural norms are formed and represented in the minds that supposedly produce them, or to causally spell out how they actually work in producing behaviors. It is simply assumed that, in some as yet wholly mysterious ways, specific combinations of unidentified genes are responsible" (Atran 2002: 271–272).

Darwinist anthropology and evolutionary psychology have met with strongly critical responses as is evident from Atran's comments (cf. Buller 2005; Gould 1980; Rose and Rose 2000; Sahlins 1976). While contemporary Darwinian anthropologists and evolutionary psychologists attempt to distance themselves from the earlier sociobiology, we find their attempt to be flawed in many respects. To begin, although Darwinian natural selection ideas are central to theory building, the kind of rich immersion in empirical data that characterized Darwin's project is not a model for much of the recent Darwinian cooperation research.[4] Instead, the predominant methods consist of, in addition to experimental games,[5] the use of mathematical simulations of variable interactions as they might have played out in natural selection scenarios ("evolutionary storytelling" in Gould [1980]; e.g., Boyd and Richerson 1985; Gintis et al. 2005). Because the variables (such as the relative frequencies of biologically evolved social instincts within and between populations) cannot be directly observed or measured, hypotheses cannot be empirically tested (Richerson and Boyd 1978). And since variable values used for computer simulations typically are assumed or conjectured (for example, as is often the case in Boyd and Richerson 2005), the results can be made to work out however desired, pointing to the possibility that we are witnessing a research program built around what Peter Lipton (2005) calls "purpose-built" scenarios that cannot fail, and, we would add, by researchers who show a strong commitment to a particular paradigm in the Kuhnian (1970) sense. We conclude that much of current Darwinian anthropology embraces a strongly deductivist epistemology (Salmon 1988: 2) that is situated well outside the accepted standards of contemporary scientific practice because, as Hull (1988: 357) expresses it, "the domain of science is limited to those areas of human inquiry that lend themselves to testing."

In addition to the nonmeasurability of their key variables and lack of valid scientific epistemology, we find additional problems with the proposed Darwinian anthropological theory. We regard the Darwinizing memetic approach to culture as sparse and limiting when culture is argued to consist of what Boyd, Richerson, and Henrich refer to as inherited strategies, acquired through social learning, that predispose persons to highly specific social behaviors or decision rules, for example, "reluctant cooperators" who will always defect from obligations until punished (Boyd and Richerson 2005: 139). This "shallow psychology" (Sperber 2006) is lacking in any sense that rational choice or other factors could also influence a behavioral choice, reducing human brains to what Rose and Rose (2000: 12) describe as "lumbering robots required for memic transmission just as our bodies are, for Dawkins, for our genes." We also oppose a method of cultural analysis devised with highly simplifying assumptions in order to facilitate simulation analyses of culture traits compatible with biological natural selection methods, and we find the idea of culture consisting of only discrete bits of information not at all useful since cultural traits often constitute complex patterns of related traits (e.g., Kuper 2000).

Another problematic aspect of Darwinian cooperation theory is its tendency to consider cooperation primarily in small-group contexts. An important assumption of cooperation research is that the ultimate, biologically based propensity for human cooperation evolved during the Pleistocene among foraging ancestors, and, in particular, resulted from the advantages of food sharing in risky environments (e.g., Kaplan and Gurven 2005: 105).[6] This assumption provides us with few insights for understanding how humans are able to build very large and complex social groups, and, in fact, even fails to provide an adequate theory of small-group behavior. This is true even in families, where conflicts often are noted over the management of resources and from other problems such as free-riding (e.g., Blanton 1995; Hart 1992; Sung-hsing 1985), which require monitoring of family members' conduct (Ben Porath 1982). To the degree that kin selection might have played a role in the behavioral evolution of humans,[7] rather than constituting an important foundation for cooperative group building, any moral sentimentalism favoring kin would have to be understood as a source of cooperation problems (Sahlins 1976: 21). As Banfield (1958) discovered in the course of his ethnographic research in a southern Italian village, a cultural preference for kin-based cooperation ("amoral familism") limited the possibilities for community-level cooperation and economic development. "Kin psychology" would be an especially acute problem in human groups that form around either common-property management or collective polity building, where high levels of cooperation and moral accountability must be maintained (Ostrom 2007)—we discuss such groups in our analyses to follow. In this kind of situation, expressions of kin favoritism may threaten participants' trust that persons in positions of authority and control of group resources will treat kin and non-kin equally. In fact, in these cases, nepotism among governing officials often is carefully monitored and is considered a punishable offense against the collectivity (e.g., in the highly cooperative Venetian Republic mentioned later [McClellan 1904: 164–165]). We argue that, in spite of what Mary Douglas (1986: 24) describes as the "powerful emotive idea" in Western social science that small groups face few cooperation problems, it is simplistic to think that small groups are less subject to cooperation problems than larger groups (cf. Douglas and Wildavsky 1982: 111–112).

The claim that humans have a psychology fostering cooperation cannot be confirmed in light of human social behavior as known from ethnographic accounts, including food sharing among contemporary foragers—presumably a useful analogy following the logic that human cooperativeness evolved originally during the Pleistocene. Although Richerson, Boyd, and Henrich (2003: 367–368) conclude that the "social instincts" that evolved during the Pleistocene include a tendency to cooperate with kin and even with unrelated individuals, from a survey of ethnographic sources we see, instead, a predominance of strongly rule-driven obligate food sharing not necessarily entailing any "cerebral disposition" to share (Marlowe 2004). In addition, much of the observed food sharing is not premised on expectations of balanced reciprocity, or any reciprocity at all—for example, in situations such as "demand sharing" and "tolerated scrounging" (e.g., Boehm 2004: 276; cf. Harris 1975: 284; Kaplan and Hill 1985: 234; Marlowe 2004: 72). In addition, food transfers can be understood in part as a way to maintain social and wealth equality between hunters with varying hunting abilities (Kishigami 2004: 344, passim), making "sharing" in these cases a form of enforced egalitarianism rather than a biologically mediated adaption to life-threatening environmental conditions.

Kin selection, reciprocity, and group selection scenarios also appear problematic when investigated in light of other kinds of ethnographic data. For example, in the well-described "Big Man" systems of the western New Guinea Highlands (e.g., Feil 1987), pig exchanges that aim to enhance social standing are at odds with kin benefit, since Big Men may drive their households to produce surpluses for the purpose of competitive exchanges while, as a result, family members suffer from food shortages or even protein malnutrition (McArthur 1977; Sinnett 1977). Not all highland New Guinea societies feature the highly competitive Big Man pattern, and a comparison of the more and less competitive groups is of interest in light of group selection theories.[8] In the less competitive pattern, found frequently in the eastern New Guinea Highlands, by comparison with the western highlands, there are higher levels of intragroup balanced reciprocity and shared religious beliefs coupled with elaborate ritual cycles that enforce comparatively high levels of intragroup cooperation and social solidarity (e.g., Feil 1987: 175). In these cases cooperation and competition evidently do not result from specific proximate psychologies, since in some cases it appears that groups cycle between the two patterns (Blanton and Taylor 1995: 138–140). An additional problem stems from the fact that, although the sociobiological theory of group selection predicts adaptive advantages to the more cooperative variant, there is no evident group selection disadvantage to the western, or less cooperative, pattern. Instead, it is in the east where there are more pronounced adaptive problems stemming from the intense in-group versus out-group animosities in a situation in which, to paraphrase Wilson (2002: 136), in-group cooperation is the only game in town (cf. Norenzayan and Shariff 2008: 62). This places the smaller populations of the eastern region at more risk of social, and perhaps biological, extinction owing to the pervasiveness of a destructive pattern of warfare that Feil (1987: 70–73) refers to as "total" (i.e., with few rules of engagement). The total warfare pattern results in generally higher per capita male (and female) mortality rates, by comparison with the western region, up to as high as 32 percent males per generation killed in battle (Feil 1987: Table 4). In addition, in the eastern region average group size is smaller than in the west and overall population density lower (Feil 1987: 46). In the western region, wars are viewed as disruptive to the exchange cycles that serve as showcases for prestige competition, and, correspondingly, Big Men promote institutions such as wergild-like war compensation payments that "deflate" the intensity of war (Feil 1987: 81, 121).

Cooperation Is Not Necessarily a Product of Mental Modules or Social Instincts

Even in the evolutionary psychology and Darwinian evolutionary literatures (e.g., Carruthers 2006: chapter 5; Richerson, Boyd, and Henrich 2003), humans are noted to have the capacity for creativity and the construction of social formations and cultural norms that enhance cooperation, an admission counter to arguments about behaviorally deterministic mental modules and social instincts. Does the potential for social and cultural production imply a necessary separation of social from biological sciences, or that cooperation researchers can ignore the results of primate research and evolutionary psychology? We would reply no on both counts because, ultimately, the cognitive capacities required to behave cooperatively and to build social and cultural systems that can foster cooperation are products of primate biological evolution (Nesse 2001; Rose 2000: 303). There will always be a level of social neurobiological explanation required to understand human cooperation (and conflict) in the construction of societies and cultures, and the results of recent research on hominoid cognition point to how this can be understood without having to resort to assumptions about "tribal instincts" or the like.

Primate research suggests that a main result of social neurobiological evolutionary change has been the emergence of a generalized and creative problem-solving brain with new kinds of cognitive abilities described in some living great apes and humans as "social intelligence" (Dunbar 2003) or "Machiavellian intelligence" (Byrne and Whiten 1997).[9] According to social intelligence theory, rather than highly determinative mental modules and social instincts, humans and great apes such as chimpanzees share several key cognitive abilities, including what Byrne (1997: 298) refers to as a "representational understanding of the world," that is, the ability to envision the probable consequences of courses of action, including cooperation as well as deception (the latter is described by Enfield and Levinson [2006: 8] as the "defining Rubicon in human evolution"). Representational understanding is coupled with the cognitive capacity known as Theory of Mind that provides normal adult individuals with the ability to distinguish between the mental perspectives of self and others (Byrne and Whiten 1997: 9, passim; Carruthers and Smith 1996; Povinelli 1996).[10] In apes and humans these aspects of social intelligence are combined with elaborate neural pathways that allow for the detection of the intentions of others through the analysis of gaze and facial expressions (Schmidt and Cohn 2001: 20; Stone 2006), and the ability to recall past actions of others (social memory) to anticipate the likely behavioral predispositions of others in social contexts (Brothers 1997: 41).

Theory of Mind and other hominoid cognitive abilities imply the capacity for making rational choices about social actions either as contingent cooperators (or punishers) or as contingent defectors (or nonpunishers), by allowing for a sensitivity to the potential advantages (or disadvantages) of cooperation and punishment, but also an awareness of the possible gains (or costs) to be derived from free-riding or other selfish or deceptive behaviors. Hominoid primate research is complemented by the recent research of neurobiologists on neural plasticity and epigenetic processes (gene-environment interactions) occurring as a result of experience during brain development (or even in adulthood) and that reflect the evolution of a highly plastic human (and great ape) neural structure (e.g., Buller and Hardcastle 2006; Gibson 2005: 29–33). This feature is consistent with the observed behavioral plasticity in humans and, and, to a lesser extent, in great apes—for example, the ability to enhance Theory of Mind skills through learning and experience (Smith 1996)—and brings into question the assumption often stated by Darwinian anthropologists that behavior reflects the operation of mental modules or social instincts that evolved biologically during the Pleistocene.

Theory of Mind and other cognitive capacities allow the individual to be sensitive to the possible advantages and disadvantages of their social actions, but we extend this to include sensitivity to the potential outcomes of social and cultural production that aims to foster cooperation. Social and cultural production are made possible by combining a prevailing "folk psychological theory of the structure and functioning of the mind" (Carruthers 1996: 22) with knowledge of prevailing norms, forms of organization, and cultural beliefs. In this case Theory of Mind is expanded from anticipating the actions of particular others to a broader sense of anticipating the actions of other socially intelligent minds in a community. Of course, because folk theories of mind are likely to be incomplete or misleading, and because there often will be social opposition to social and cultural change, the building of cooperative groups is likely to initiate a dynamic iterative process when unintended consequences necessitate new or modified forms of social and cultural production and perhaps a modified folk psychological theory of mind to accomplish cooperation goals. Blanton (2011) noted an intensification of philosophical and other discourses concerning the nature of the moral person in polities undergoing change to more cooperative forms.

An approach emphasizing the importance of Theory of Mind and social and cultural production is unlike many of the reductionist arguments of Darwinian anthropology for two main reasons. First, a contingent cooperator engages in cooperative social action when there is a sense of trust that benefits will likely ensue for self and group (e.g., Hart 1988: 187; Kahan 2005), based on past experience as well as a sense that a particular institutional and organizational structure will serve its intended purpose—we address an example of the latter in our analyses. Hence, for our purposes, we define cooperation as social action predicated on an actor's trust that prosocial action will likely have beneficial material, social, emotional, or other desired outcomes for self and/or for group. From this perspective there are no "work-arounds" that humans can make use of to build large-scale societies by taking advantage of a human nature featuring "a psychology evolved to cooperate" (Richerson, Boyd, and Henrich 2003: 367–368). Instead, work-arounds involve the production of institutional structure (rules), cultural meaning (symbols and rituals), and organizational capacity that will limit free-riding and agency, thus instilling trust among group members that cooperation will bring personal and group benefits. We return to this issue in relation to how humans have constructed highly cooperative, large-scale, collective political regimes.

Steps to a Multiscalar Perspective on Contingent Cooperation

To set the stage for our discussion of cooperation in state formation, we first make a crucial distinction between patterns of cooperative action at three main social interactive scales. We start by addressing the bases for cooperation in small "primary groups" in which potential cooperators are able to gauge the likely consequences of social actions based on their history of interactions ("mutualistic cooperation" in Bowles and Gintis [2005: 22–25]), direct social encounters with others that allow for the analysis of intention from gaze and facial expression (Frank 1988: 134–136), and from patterned interactional practices such as greeting rituals (Schegloff 2006) that certify the vitality of an ongoing relationship. Also, from prior experience, potential cooperators are aware that asocial actions may limit the possibility for beneficial future in-group interactions. Owing to memory limitations, the maximum primary group size is not well documented, but is anticipated to be less than 200 persons based on primate data (Dunbar 2003; cf. Johnson 1983, who bases his conclusions on ethnographic data; Feinman, chapter 2). However, we suggest that primary group size will be dependent on the degree to which it is necessary to remember large amounts of information about each person's past actions and to carefully monitor behavior, for example, in a common-property situation such as a water-sharing irrigation system. In the latter case, memory limits of the human brain might imply more severe scalar stress (e.g., in the sense of Johnson 1983) and perhaps smaller primary group size.

Larger primary groups are possible when language enhances communication of shared intentionality (Tomasello 2008) and allows for the communication of information about past social actions through second parties (reputational knowledge) rather than direct social experience (Smith 2003). Language communication also appears to be important in supporting the development of Theory of Mind skills (Smith 1996). Group ritual may also serve to expand group size when participation in costly ritual cycles serves as a signal of commitment to group goals (Iannaccone 1992; cf. Sosis and Alcorta 2003), and when social conventions promoting prosocial behaviors are sentimentalized through ritual (Durkheim 1965 [1912]: 474–475; Rappaport 1979: 211–216) and through other sources of numinous experience including built environments (Blanton 1989; Moore 1996: chapter 3).[11]

Even though factors such as reputational knowledge, costly signaling, and numinous experience allow for an extension of the size of primary groups, we are still left with another of the dilemmas facing cooperation research: in very large groups, cooperation may take place between strangers who may not share in common language, ritual cycles, and numinous experience, and are not known to each other through either direct social knowledge or reputational knowledge, all greatly enhancing opportunities for free-riding and agency in more impersonal social interactions. State formation (Blanton and Fargher 2008) and the development of large-scale market systems (Blanton 2013) are the two most important contexts where we see that humans have devised solutions to the large-group cooperation problem, and in what follows we address the role of religion in the development of the state to illustrate one example of a contingent cooperator approach.

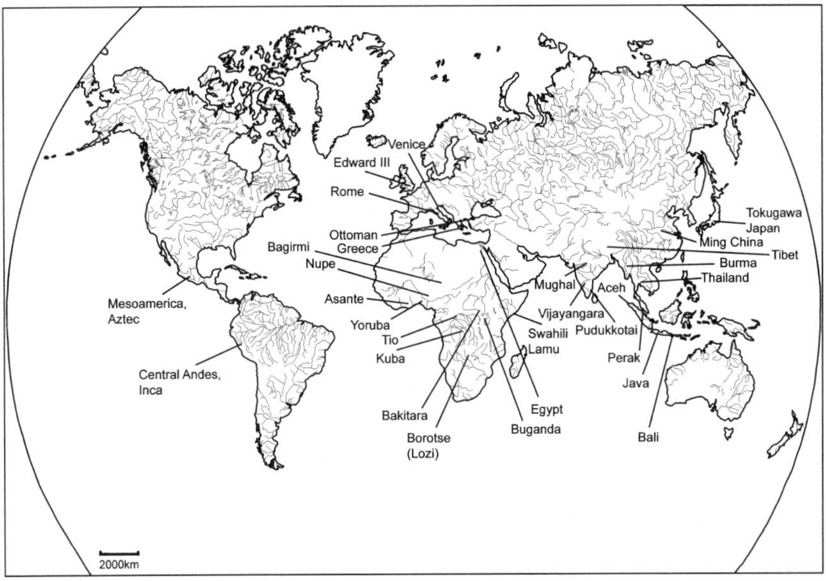

FIGURE 5.1 *Comparative cases discussed in text*

AN EMPIRICAL AND SCIENTIFIC APPROACH FOR COOPERATION RESEARCH APPLIED TO PREMODERN STATES

We disagree with the claim that mathematical modeling and game-based research provide a "firm empirical ground" (Norenzayan and Shariff 2008: 61) for cooperation research. In the following application of a contingent cooperation approach, we do not make use of the results of experimental games and mathematical simulations and make no assumptions that state builders have available to them "a psychology evolved to cooperate." Instead, we place our work firmly within the framework of a scientific epistemology based on the development of testable hypotheses derived from sources such as collective action theory (e.g., Levi 1988; Ostrom 2007). We then engage in an objective and methodologically astute attempt at hypothesis (and theory) falsification based on the statistical analysis of a large body of data collected from a culturally and geographically varied comparative sample of premodern states (Blanton 2010, 2011, 2013; Blanton and Fargher 2008, 2009, 2010; Fargher and Blanton 2007) (Figure 5.1), using the well-established methods of cross-cultural comparative research (Ember and Ember 2001). Our approach allows us to identify those social and cultural processes that underlie the evolution of cooperation in large groups across the great diversity in culture, social form, and technology found in our comparative sample. Our approach opposes the argument that scientific research is valid only when it aims to find ultimate (e.g., biological) causes because, while we address more proximate causes of social and cultural change resulting from conscious human choice and social action, our processually oriented research program is grounded in systematic comparative method and the scientific epistemological mandate of hypothesis testing.

Since the Enlightenment revival of the ideas of Plato, Aristotle, Cicero, and Livy, Western philosophers and social scientists have compared the relative merits of different forms of the state. Our comparative analysis of premodern state formation (Blanton and Fargher 2008) allows us to elaborate on this discourse from a contemporary cooperation perspective by investigating and measuring the varying degrees to which state formation is based on cooperation. Our method follows up on the suggestions of rational choice and collective action theorists such as Margaret Levi (1988) and Mark Lichbach (1996), but we extend our comparative reach beyond their primary interest in Western history to include a geographically and temporally broad world-wide sample of thirty premodern states.

The Darwinian anthropologist Boehm (1997) argues that the evolution of "political intelligence" in humans has provided us only with the ability to decide whether to dominate others or submit to others. Collective action theory provides an alternative theory, as it proposes that the basis of relatively high levels of cooperation in large polities is found where there are mutual obligations of rulers and taxpayers. In this setting for polity building, institutions and organizations allow for a trust basis of cooperation beyond the social knowledge and/or reputational knowledge found in groups that are nested within the larger structure.[12] That polity participants will comply with moral obligations is of utmost importance for the success of this kind of collectivity (e.g., Kahan 2005; Ostrom 2007: 200–201). State builders must develop policies aimed at increasing the likelihood that its officials will act in the best interest of the collectivity, and we found that to do this, rather than being simply expressed as vague ideals, moral obligations are enacted in social intercourse in relation to specific institutions (rules such as fiscal constitutions) that specify what is expected of the moral person and what constitutes moral abrogation. Institutions are then tied to organizational structures that have the capacity to enact rules in social practice, for example, by monitoring for rule compliance and having the authority to punish for noncompliance.

From our comparative research, we concluded that premodern states exhibited quite variable degrees of overall cooperation, even though some concept of a social contract between rulers and ruled typically is expressed. However, the degree of institutional and organizational development that can uphold social contracts was found to vary depending on the degree to which principals (rulers or their equivalent) are dependent on revenues derived from a broad population of taxpayers, although other factors also prompted collective regime building (Blanton and Fargher 2008: chapter 6). In these cases, taxpayers stand in a position to demand public goods and moral governance, and principals who fail to provide these services experience loss of taxpayer trust, leading to high rates of noncompliance, free-riding, and defection. When ruling groups depend on resources largely apart from those produced by taxpayers—for example, when a polity's revenues derive primarily from the control of foreign trade or when a ruler personally controls vast estates that can fund a regime—we found less evidence that state formation was based on high levels of expected cooperation. In these cases, taxpayers stand in a weaker position to demand high levels of cooperation from rulers and their administrative agents,[13] so the social organization of the polity is based on some combination of political and military domination and religious sanctification of rulers (Collins 1975: 367). Accordingly, principals are predicted to behave more autocratically, to provide comparatively few public goods, and to govern society without an elaborate administrative apparatus. Our method for assessing comparative degrees of cooperation depended on the systematic measurement of the following scale variables (based on Blanton 1998 and Levi 1988):

(1) "Modes of Control of Principals" (Blanton and Fargher 2008: chapter 9)

This scale measures the degree to which principals (rulers or other key policy makers) are obliged to acknowledge moral obligations to the collectivity and accept restrictions on their agency. We also assessed the degree to which there exists an organizational structure that allows for effective monitoring of principal behavior and has the authority to punish principals for amoral actions.

(2) "Bureaucratization" (Blanton and Fargher 2008: chapter 8)

This scale measures the degree to which state builders have implemented an administrative structure able to equitably disseminate public goods and able to confirm that the state's governing cadres adhere to moral mandates and fiscal constitutions while at the same time detecting and punishing taxpayer free-riding. In this respect, we also looked for evidence that compliance failure on the part of administrative officials can be detected and punished and that commoners have suitable communication channels through which they are dependably able to express complaints about administrative agency and free-riding and to appeal unfair judicial decisions.

(3) "Public Goods" (Blanton and Fargher 2008: chapter 7)

This scale assesses the degree to which the state allocates some portion of tax revenues to public goods such as water control, transportation infrastructure, and the maintenance of civil order.

We found high levels of statistical correlation among these three scale measures, as well as statistically significant positive correlations between the scale measures and a measure of sources of tax revenues (Blanton and Fargher 2008: chapter 10), both of which are consistent with collective action theoretical predictions.[14] By summing the three scale measures, we arrived at an overall measure of degree of cooperation ("collective action total," Blanton and Fargher 2008: Table 10.1), and in Table 5.1 we provide this variable as well as additional information such as the focal (coding) period for each polity.

TABLE 5.1 The coded societies, indicating values for Collective Action Total (Blanton and Fargher 2008: Table 10.1), Public Goods, Bureaucratization, and Principal Control, the focal period (the specific period for which the coded data were coded), evidence for problematization of religion and power (no evidence = 1, evidence for = 2), and key references for problematization.

AFRICA

West Africa to East-Central Sudan

1. Nupe (Fulani-Nupe), 25.5, 10, 7.5, 8; CE 1837–1897, 1; Nadel (1942: 141)
2. Yoruba (Oyo Empire), 36.5, 16, 9.5, 11; CE 1750–1800, 2; Law (1977: 66)
3. Asante (Akan), 44.5, 18.5, 10.5, 15.5; CE 1800–1873, 2; McCaskie (1995: 47–48, 124, 127); Rattray (1923: 151, 170, 289; 1929: 83)
4. Bagirmi, 27.5, 13, 8.5, 6; CE 1800–1900, 1; Reyna (1990: 59)

Central Equatorial

5. Kuba (Bushoong), 32, 13.5, 10, 8.5; CE 1880–1892, 1; Vansina (1978: 130, 203–204, 208)
6. Tio, 27, 12.5, 6, 8.5; CE 1800–1899, 1; Vansina (1973: 378)

Interlacustrine

7. Buganda, 37, 15.5, 11, 10.5; CE 1800–1880, 2; Ray (1991: 15)
8. Bakitara (Bunyoro-Kitara, Nyoro), 23.5, 10, 6.5, 7; CE 1860–1890, 1; Roscoe (1923: 90–113, 120)

Southern and East Coastal

9. Lozi (Barotseland), 49, 22, 12, 15; CE 1864–1900, 2; Gluckman (1961: 26, 27, 30–31); Prins (1980: 120–121, 123–129)
10. Swahili Lamu, 36, 10, 11.5, 14.5; CE 1800–ca. 1870, 2; Horton and Middleton (2000: 160), Prins (1971)

SOUTHEAST ASIA

Mainland

11. Thailand (Early Bangkok Period, Chakkri Dynasty, esp. Rama III), 36, 18.5, 8, 9.5; CE 1782–1873, 1; Tambiah (1976: 185–189)
12. Burma (Early Kon-baung Period), 41, 20, 12, 9; CE 1752–ca. 1800, 1; Koenig (1990: 83–84)

Insular

13. Bali (the Later Mengwi Polity), 28, 14, 6, 8; CE 1823–1871, 1; Geertz (1980: 102, 104–106)
14. Aceh (Aceh Sultanate), 25, 10, 6, 9; CE 1850–1900 (little data on problematization); Hurgronje (1906)
15. Perak, 25.5, 12.5, 5.5, 7.5; CE 1800–1870, 1; Gullick (1958: 8, 20, 45)
16. Java (Late Mataram Period), 38, 18.5, 10, 9.5; CE 1700–1900, 1; Moertono (1981: 26, 34)

continued on next page

TABLE 5.1—*continued*

SOUTH ASIA	
17.	Vijayanagara (esp. the reign of Deva Raya II), 37, 18, 9.5, 9.5; CE 1350–1564, 1; Fritz (1986: 48, 49–53); Fritz, Michell, and Nagaraja Rao (1984); Sinopoli and Morrison (1995: 87)
18.	Pudukkottai, 31.5, 17, 7, 7.5; CE 1700–1800, 1; Dirks (1987: 130, 156, 166–167)
19.	Mughal (reigns of Akbar, Jahangir, and Shah Jahan), 45, 23.5, 12, 9.5; CE 1556–1658 (not coded because the sources are inconclusive); Richards (1998: 298–303, 306); Sarkar (1963: 4, 134–135, 137)
EAST ASIA	
20.	Ming Dynasty; Early and Middle Ming, 51, 22, 14.5, 14.5; emphasis on CE 15th century, 2; Taylor (1998)
21.	Japan (Tokugawa Period, Edo Shogunate), 31.5, 16.5, 7, 8; CE 18th century, 1; Hall (1991: 160)
22.	Tibet, 34, 19.5, 8.5, 6; CE 1792–1951, 1; Bell (1992); Carrasco (1959: 122)
NORTH AFRICA/MEDITERRANEAN/EUROPE	
23.	Ancient Egypt (New Kingdom, esp. 18th and 19th dynasties), 38, 20, 10, 8; BCE 1479–1213, 1; Bryan (2000: 232), O'Connor (1983: 186, 189)
24.	Athens (the 'New Democracy' or 'Age of Demosthenes'), 52, 20, 14, 18; BCE 403–322, 2; Camp (1986), Hölscher (1991: 371)
25.	Roman Empire ('High Empire'), 48, 24, 12, 12; CE 69–192, 2; Many sources were cited, but much of the information is summarized in Griffin (2000a, 2000b)
26.	Venice, 51.5, 21, 14, 16.5; CE 1290–1600, 2; Norwich (1982: 282)
27.	England, 28, 11, 8.5, 8.5; CE 1327–1336, 1; Morris (1940: 10)
28.	Ottoman Empire ("Classical Period," but emphasizing the reign of Suleiman I), 34.5, 16, 9.5, 9; CE 1300–1600, 2; Lybyer (1966: 151), Wright (1935: 22)
NEW WORLD	
29.	Aztec Empire (Triple Alliance), 45, 21, 11.5, 12.5; CE 1428–1521, 2; Davies (1987: 101), Durán (1994: 217, 484, 486, 488), van Zantwijk (1985: 97)
30.	Inca Empire, 40, 22, 10, 8; CE 1438–1532, 1; D'Altroy (2002: 91)

THE PROBLEMATIZATION OF RELIGION AND POWER IN MORE COOPERATIVE POLITIES

In what follows we test a hypothesis concerning the role of religion in state formation. We do this to enlarge on a theory relating religion and the evolution of human cooperation proposed by Wilson (2002; cf. Norenzayan and Shariff 2008). Wilson's main argument is that religion has played a pivotal role in the evolution of cooperation in humans because, as he puts it, as a unifying system, "Religions exist primarily for people to achieve together what they cannot achieve alone" (Wilson 2002: 159). And, he argues, groups whose religious beliefs sanctify propositions about cooperation would have fitness advantages by comparison with other groups (Wilson 2002: chapter 4). Even before doing the research reported here, we found this claim to be questionable in light of the sociology of religion literature as well as comparable anthropological treatment of the topic that document abundant cooperation problems in religious groups (e.g., Douglas and Wildavsky 1982: 102–125). Earlier we critiqued the group selection aspect of the argument, based on data from highland New Guinea that showed that groups featuring high levels of shared religious belief and ritual cycles appeared more prone to social and biological extinction by comparison with less religiously organized groups.

Does religion facilitate cooperation in state formation? Judging from what has been written about the role of ruler sanctification in the more autocratic states, this seems unlikely. As Roy Rappaport (1978) argues, when interjected into religious ritual, social conventions that legitimize power may become sanctified and thus are more likely to be unquestioningly accepted. Because religion and religious ritual may serve to sanctify social conventions enhancing the power of governing authorities, we predict that religion would play a key role in state formation principally in those polities scoring lower on our cooperation measures in which ruler accountability is not an important issue. In these polities, ruler power is exercised principally through the use of coercive force coupled with extensive control over symbolic resources, including religion, that situate rulers at the center of the cosmic order (e.g., Wolf 1999). By contrast, we hypothesize that in more cooperative states rulers are obligated to uphold moral codes and fiscal constitutions and to provide public goods, and a failure to do so results in a loss of taxpayer trust in the collective system. Religion in these cases—at least in the sense that it serves to sanctify rulers—might inhibit the potential for endurable cooperation at the level of the polity by reducing accountability and trust.

In those states exhibiting higher levels of cooperation, we predict that state builders would strategize ways to increase the likely level of taxpayer trust and compliance by diminishing the degree of religious sanctification of power holders. Although not based in collective action theory, a similar idea was proposed by sociologists, and we acknowledge a predecessor to these kinds of ideas in the writings of Karl Jasper and others (e.g., Eisenstadt 1986). These authors point out that, beginning in the latter first millennium BCE, what they call Axial Age "breakthroughs" in several civilizations brought new cultural codes specifying that rulers and the state be placed within mundane categories and thus were subject to the higher precepts of a transmundane religious dimension. This redirected political evolution away from the traditional and sacred legitimation of rulers of earlier "pagan" civilizations ("king-gods"), and paved the way toward the open recruitment of political elites and toward modes of government similar to Max Weber's rational-legal bureaucracy. Our approach is different from Axial Age theory in not necessarily seeing problematization as an evolutionary step from "pagan" civilizations to modern democracies. Rather, while our hypothesis overlaps with certain Axial Age ideas, we expand their suggestions by incorporating them into a collective action theory of state formation to account for the fact that religious problematization occurred in a greater variety of polities than they originally envisioned (e.g., Blanton and Fargher 2008: 291–294) because it is a product of collective action social process rather than constituting a social evolutionary stage.

How Is the Relationship of Power and Religion Problematized?

To evaluate the role of religion in state formation, we developed a coding scheme that dichotomized our sample of thirty societies by both collective action total score and by a variable assessing the degree to which religion, in some way, was problematized or could be a symbolic source amplifying state and/or ruler power (Table 5.1).[15] The presence of one or more of the following features was coded as "1" (religion is a symbolic source supporting autocratic rule) for the "Evidence of Problematization" variable in Table 5.1:

(a) Important religious cults or orders were controlled by or were subservient to the ruler, for example, England (Morris 1940: 10), Tokugawa Japan (Hall 1991: 160), Nupe (Nadel 1942: 141), Thai (Tambiah 1976: 185–189), Burma (Koenig 1990: 83–84), and Tibet (Carrasco 1959: 122).

(b) Rulers were strongly sanctified, were thought to have supernatural powers, and/or served as a conduit of supernatural power to the populace, for example, Bagirmi (Reyna 1990: 59), Tio (Vansina 1973: 378), New Kingdom Egypt (Bryan 2000: 232), and Inca (D'Altroy 2002: 91).

(c) Rulers gained power through their control of supernaturally potent objects, for example, Kuba (Vansina 1978: 130, 203–204, 208).

(d) The political system was thought to have been established by a creator god, for example, Ancient Egypt (O'Connor 1983: 186, 189).

Evidence for problematization of rulership and religion was evident in the sample even while demonstrating considerable cross-cultural variation in practice. A "2" (in Table 5.1) was coded when one or more of the following elements were present (this section provides a more detailed discussion of several of the more collective societies to clarify our coding scheme):

(a) Ruler is not considered divine, for example, Yoruba (Law 1977: 66) and Buganda (Ray 1991: 15) (and see Ming China, Aztec, High Roman Empire, Ottoman Empire, and Asante).

(b) Rulership itself is sanctified, but not the office holders, for example, Asante.

(c) Major ritual centers or shrines are located separately from the main political centers, for example, Asante, Lozi, and Athens.

(d) Religious cults and orders were largely separate from state control, for example, Asante.

(e) The polity was highly secularized, for example, Venice and Swahili Lamu.

Two polities were not coded. For Aceh, we found the information incomplete. Mughal presented a coding problem for this kind of exercise because there are elements of both the use of religion as a symbolic source and religious problematization (from Richards 1998: 298–303, 306; Sarkar 1963: 4, 134–135, 137). For example, the ruler claimed some aspects of divinity, influenced by Persian mystic Sufism and Hindu practice, and was considered head of both state and religion, in theory. At the same time the mullas and mosques operated, in part, separately from ruler's control, even criticizing Akbar's attempts to promote his sanctification as contrary to principals of Islamic theology.

Asante

As part of the state-building efforts of Osei Tutu and the renowned priest, Komfo Anotche (Rattray 1929: chapter 14; Wilks 1993: 41), the Golden Stool (*sika dwa kofi*) came to signify legitimate political authority and power of the Asante state (McCaskie 1995: 47–48, 127). The Golden Stool was a potent marker of rulership given its perceived supernatural origin and its role as a repository of the collective essence of the Asante people (Rattray 1923: 289); interestingly, however, it was not associated with any particular ruler or matriclan. For example, if destooled, a ruler was regarded as an ordinary person who could be criticized and punished for his actions while holding the stool (Rattray 1929: 83). In addition to a concept that allowed for the sanctification of office but not the holder of office, we see problematization also in the fact that in the Asante domain, major ritual sites and temples were not located in Kumasi, the political capital, and instead were located in zones to the north of the capital (such as Tekiman) (Rattray 1923: 151, 170). Priests (*abosom*) had little political influence in the polity, and, in fact, state officials regarded them as being potentially politically subversive (McCaskie 1995: 124). Interestingly, a new ritual cycle was developed associated with Asante state formation, the Odwira or annual yam harvest festival (McCaskie 1995: 145; Rattray 1929: 279), that was attended by large numbers. While the ritual had some religious themes, it served primarily political functions such as affirming the status of conquered groups and providing a venue for the assembly of office holders (McCaskie 1995: 146–147).

Lozi

The Lozi polity did feature some aspects of ruler sanctification, for example, the ruler was transformed, through an installation ritual, into a powerful spiritual being (Prins 1980: 120–121). Royal burial sites and cenotaphs were sacred sites, deceased rulers retained spiritual potency, and offerings were made at the burial sites (the "cult of the royal graves" [e.g., Gluckman 1961: 26, 30–31; Prins 1980 123–129]). However, we see an element of religious problematization in the fact that rulership was expressed in dualistic terms, as is evident in a dual-capital system with northern and southern capitals. While secular power, the basis of Lozi governance, was vested in the north ruler (and capital), the southern capital and ruler were vested with greater spiritual power but little actual authority to govern (Gluckman 1961: 27).

Swahili Lamu

In this republican form of government, an elaborate symbolic system structured the didemic (dual) organization of the polity (Prins 1971) but did not sanctify governing officials (and the main symbol of high office was an antique brass trumpet that appears to have no connection to religion [Horton and Middleton 2000: 160]).

Early and Middle Ming Dynasty

The Ming emperors aimed for a "vigorous promotion of an ideological orthodoxy based upon neo-Confucianism" (Farmer 1976: 6). This Confucian theory of government is held up as one of the foremost examples of Axial Age transformation, based on the concept of a Mandate of Heaven that placed authority in a supreme being who could demand that rulers be pious and righteous (Hsu 1986: 308). Rulers received their mandate to rule from heaven, but the mandate was confirmed only through sacrifice and "by living as a man of piety" (Taylor 1998: 861). While in pre-Confucian times sacrificial rituals to significant clan ancestors were a source of power for rulers who were heads of powerful clan aristocracies, in the Confucian system religion was democratized by making household ritual a source of access to ancestor spirit forces; as a result, "ancestor worship developed into a thoroughly democratic institution, and the effect was that ancestors shrank in value" (Laufer 1965: 449). The Ming Dynasty founder established new ritual cycles at the local community level. Rather than to promote religion, however, or to sanctify the ruler, these were intended to promote local cooperation (Heijdra 1998: 469–470).

Athens

The main concentration of temples and shrines in Athens was located on the Acropolis and in other locations such as the sanctuary of Demeter at Eleusis. While the Acropolis to some degree embodied the spirit of the Athenian community, as Whitley (2001: 340) points out, the polity's government was situated at some remove from this concentration of symbols and religious practice, in the Athenian agora (market), which, beginning with the 5th and 6th century BCE democratic reforms, emerged as the civic and commercial downtown of Athens (Camp 1986). Extensive building programs in the agora and adjacent zones made them the focal point for the development of a new civic identity for the people of Athens. According to Hölscher (1991: 371), the agora "suddenly became a center which attracted from all over Attica hundreds and soon thousands of citizens for political and judicial functions," aimed at the development of a "*présence civique.*" New public rituals were introduced as part of the civic life of the newly designed Athenian polity (Snodgrass 1981: 118). However, these had no relationship to religious sanctification of the state or its principals; as Carlton (1977: 235) put it, religion was not "harnessed for the needs of the state."

Roman High Empire Period

This polity presented some coding problems but does, we think, illustrate a strategy of problematization. Although the emperor was not considered a god (at least not during his lifetime), certainly the principals flirted with the idea of sanctification, as we see when Titus arranged for the deification of his father soon after assuming rulership and in his dedication of a temple to Vespasian (Griffin 2000a: 47). Yet it was still the case that the legitimacy of the Roman principals of the focal period appears to have been based primarily on an ethic of *civilitas* (submitting himself to the law) and *moderatio*, for example, Trajan's desire to be understood as "the soldier-emperor serving Rome in an unostentatious way, labouring alongside his soldiers, accessible to his officers, striving for peace rather than glory in war" (Griffin 2000b: 103).

Venice

The Venetian polity of the focal period was highly secularized. As Norwich (1982: 282) put it:

> The Church was kept rigidly in its place, its duties exclusively pastoral, barred from the slightest interference in affairs of state . . . The families of Venetians holding ecclesiastical positions were also suspect. Their members . . . who belonged to the various governing bodies of the state, by a series of laws beginning in 1411 were regularly excluded from all deliberations concerning ecclesiastical matters. Some appointments, such as the coveted embassy to Rome, were closed to these families altogether. No member of the clergy was allowed to serve the Venetian state in any capacity, even as a clerk or notary, or to have access to the public archives.

In the case of Venice, we see a situation where ritual cycles promoted the sense of civic devotion, including an inauguration called the Ducal Procession (e.g., Norwich 1982: 167). As in the Asante Odwira, these rituals were not always highly religiously charged, and instead also had civic functions, as is evident in the detailed representations of the processions painted by Carpaccio (Brown 1988).

Ottoman Empire

The moral code governing rulers in the Ottoman Empire was the Sacred Law of Islam (shari'a), which forbade the sanctification of ruler or rulership, and, instead, gave the sultan only the functions of administration and justice (Lybyer 1966: 151; Wright 1935: 22).

Aztec

In the Aztec arrangement, as van Zantwijk (1985) calls it, rulers participated in ritual events and ritual cycles carried out in a symbolically charged precinct (van Zantwijk 1985: 213–16, 261), yet they were not considered divine (Davies 1987: 101). Although rulers could be priests and occupied an office associated with considerable symbolic force, still, "the Tenochca *tlatoani* (the central imperial office-holder) was not himself considered a god, but rather, was the god's representative or substitute" (Davies 1987: 101), and the gods are described at times as critical of the ruler (Durán 1994: 217, 484, 486, 488; van Zantwijk 1985: 97).

Conclusions from the Analysis of Dichotomized Data on Polity and Religion

When we cross-tabulated "evidence for problematization of religion and power" by the dichotomized collective action total score, the result showed a strong association between problematization and higher levels of collective action (Fisher's Exact Test, two-tail, $p = .0183$, $n = 28$). This strong statistical result points to an important processual regularity in the development of highly cooperative states, namely, that state builders in a variety of cultural and geographical settings theorized that to build a more collective state some kind of separation of religion and power holders would be necessary. This kind of conscious strategic regime building aimed at solving cooperation problems is nothing like the supposed biologically evolved "political intelligence" that argues that people are capable only of deciding whether to dominate others or submit to others (Boehm 1997), because the evolution of polities may also involve high levels of agreed-upon rules and organizational structures that foster trust and cooperation but involve neither domination nor submission. Our results also bring into question Darwinist claims that religion can be thought of as an evolutionary force fostering human cooperation; instead, for cooperation to flourish in states, religion is problematized so as to disconnect religion from rule. Further, as we noted for Asante, Athens, Venice, and Ming China, attaining higher levels of cooperation also entailed the institution of new ritual cycles that in some respects achieved cooperation goals through civic rather than religious ritual.

DISCUSSION AND CONCLUSIONS

Earlier we expressed our agreement with Richerson and others that a well-formed evolutionary theory will benefit humans as they adapt to rapid change in the contemporary world, but we questioned what kind of evolutionary theory would produce the appropriate kinds of knowledge. In this chapter we described the memetic analysis of culture and the biological reductionism embraced by contemporary Darwinian anthropology, and contrasted those with a contingent cooperator approach. These views of the foundations for human cooperation have quite different implications for problem solving. Darwinian anthropology implies that cooperative groups should be relatively easy to form and maintain, given that humans are capable imitators with strongly prosocial tendencies, while a contingent cooperator approach allows for the possibility that cooperative groups are possible, but is more sensitive to the challenges humans face in solving cooperation problems. We conclude that a Darwinian anthropology perspective is not the more suitable of the two approaches owing to its numerous inadequacies. That human social action reflects mental modules, social instincts, and inherited strategies is limited by failing to capture the highly contingent character of human social action. For example, that populations will be made up of persons exhibiting clear behavioral propensities on each variable dimension (e.g., contributors versus defectors, punishers versus nonpunishers, etc.) is a drastically simplified conception of the social human necessitated by the analytical method of choice for Darwinist anthropology, namely, mathematical simulation of evolutionary scenarios. In addition to its unrealistically simple depiction of human thought and social action, this method directs research toward a sterile deductivism and pseudo-quantitative methodology and away from systematic empirical work, while completely failing to provide a way to theorize about how humans consciously solve cooperator problems. We also argue that Darwinian anthropological theories are subject to the kinds of critiques that have been leveled at structuralist or similar agentless theories in sociology and anthropology. In these theories, human social action is seen to result from already constituted forms of reason, so the thinking subject is unacceptably absent from the analysis (e.g., Doja 2005). The more widely accepted theoretical position in much of contemporary social science states that, as expressed by Sewell (1992: 20), "a capacity for agency . . . is inherent in all humans."

The Contingent Character of Human Social Action Brings Cooperator Dilemmas

The dilemma for the cooperator is that social formations that could provide mutual benefits to participants are threatened by selfish but individually rational action (Lichbach 1996). Cooperation dilemmas are resolved through the development of problem-oriented institutional and organizational innovations such as the problematization of power and religion we discussed. The ultimate psychological capacity to resolve cooperator dilemmas is found in Theory of Mind and other hominoid cognitive capacities seen in some great apes and humans. Thus we would suggest that, in addition to the "three Rs" (reciprocity, reputation, retribution) (Carballo, chapter 1), we find the evidence compelling that the capacity for cooperation ultimately results from "STAR" (social memory, Theory of Mind, analysis of intentions, and representational understanding). We point to the fact that STAR is well known from field observations and laboratory experiments with apes and humans, as well as the research in social neurobiology alluded to above, while the claim that human behavior is importantly driven by social instincts cannot be empirically demonstrated.

The closed and rigid nature of the Darwinian anthropology paradigm is evident in their failure to cite the apposite primate research on Theory of Mind and social neurobiology—for example, Henrich and Henrich (2007: 41) ignore these elements of primate cognition research, and, in fact, claim that cooperation is not due to the "superior intelligence" of humans. It is odd that while largely ignoring ape/human cognition, biologically reductionist cooperation researchers often refer to other possible analogs for human cooperation. Yet given the unique cognitive ability found in some hominoid species, we would advise caution when making use of nonhominoid animal models for human cooperation, which has included species ranging from yeast and slime molds (as described in Pennisi 2009) to guppies (Wilson 2002: 193). Monkeys, although constituting a large and diverse primate category, generally lack high levels of representational understanding, learning ability, and Theory of Mind (Byrne and Whitten 1997; Russon 1997), and hence are not always useful cognitive analogs of the higher primates.

Final Comments

Based on their claim that human cooperation is based in part on social instincts, Boyd, Richerson, and Henrich (2005a: 271) doubt that humans could solve cooperation problems only through what they derogatorily refer to as "clever institutions," a phrase that we think fails to properly capture the difficulties typically encountered in building successful and endurable cooperative groups. What we have found from our comparative research on state formation is that, rather than through "clever institutions," in the polities we studied, cooperation often was established only with difficulty and against great odds, facing opposition from a traditional elite threatened with a loss of influence, privilege, and power, and often requiring cultural and social restructuring of society from top to bottom (Blanton and Fargher 2008: 280–289). Solving cooperator dilemmas is difficult and humans have had, and no doubt will continue to have, difficulties in building and maintaining beneficial cooperative social formations. Scientifically grounded empirical and comparative research, framed by a suitable theory of human cognition, can contribute to an understanding of the social and cultural processes that underlie human cooperation as we strive to understand it in the broadest sense.

ACKNOWLEDGMENTS

Our work on collective action has been funded by the National Science Foundation (BCS–0204536) and the Center for Social and Behavioral Sciences, Purdue

University. We thank Gary M. Feinman for assistance with this chapter and David Carballo for his useful comments on an earlier draft. Two anonymous reviewers provided useful comments. The authors are responsible for all errors and omissions.

NOTES

1. We should note that sociobiologists and Darwinian anthropologists have only recently turned to the question of the cooperative human in the second of two phases in their twentieth-century intellectual history. An earlier sociobiology variant reflected interpretations of Darwinian theory that emphasized competition and conflict as the driving forces of bioevolutionary change (e.g., Wilson 1978). Faced with criticisms that such a theory places more emphasis on conflict than on cooperation (Sahlins 1976: 290; Sussman and Chapman 2004; Sussman and Garber 2004), more recently evolutionary anthropology has completely switched gears, and, although still drawing insights from Darwinian theory (Simpson and Beckes 2010: 35), now emphasizes how natural selection could have shaped human propensities toward prosocial behaviors to explain, they believe, why human societies are so "extraordinarily cooperative compared with those of most other animals" (Boyd and Richerson 2006: 453).

2. The idea that the mental modules and social instincts evolved in the context of Pleistocene hunting-gathering lifeways is argued extensively in Barkow, Cosmides, and Tooby (1992) but this and similar sources provide few direct links to actual data from Pleistocene environments, archaeological sites, technologies, etc.

3. Of course this begs the question: What selective pressures would favor an increase in the frequency of a module for a behavior that provides no benefit in a pervasively cooperative cultural milieu? We also find these kinds of culturally based arguments limiting because culture (and social) change is not properly theorized. In Boyd, Richerson, and Henrich (2005b: 198), for example, "cultural evolutionary processes" explain why groups would develop differing degrees of cooperation, but such processes are invoked in *deus ex machina* fashion without adequate theorization.

4. Some research draws on data analysis from natural settings (e.g., from brief ethnographic accounts [Henrich and Henrich 2007]). Henrich et al. (2004) provide the only large-scale empirical study related to cooperation issues that we find to be sufficiently cross-cultural and methodologically sound to be convincing.

5. Our pessimism about Darwinian cooperation research as it is currently practiced extends to the extensive dependence on experimental games that are based on unrealistic assumptions (such as precluding communication between players of the Prisoner's Dilemma game [Barry and Hardin 1982]), and, hence, provide little insight into the difficulties humans face in solving cooperation problems in society (Hechter 1990). And in spite of all that has been written about the "power" of the experimental approach (e.g., Gintis et al. 2005: 5), it is not legitimate to draw broad conclusions about our species as a whole from experiments carried out in a limited range of cultural settings (e.g., as is seen in Herrmann, Thöni, and Gächter 2008: 1363), typically involving college students in commercialized cultures such as the United States where cultural norms favor cooperation and punishment (Fukuyama 1995; Henrich et al. 2010). Recent research extending experimental games to diverse cultural and social settings has shown more variation than expected in degrees of cooperation and punishment (Henrich et al., editors, 2004), demonstrating that the consistent tendency to cooperate and punish seen in most of the

experimental game literature is culture specific, not pan-human. From a survey of ethnographic sources for foragers, Baumard (2010) found only weak support for the idea that punishment plays an important role in fostering cooperation.

6. There is abundant evidence of food sharing in ethnographically known foragers, but Kaplan and Gurven cite Winterhalder (1986) as evidence for the importance of food sharing derived only from computer simulations.

7. Simpson and Beckes (2010: 44–45) provide a summary of psychological sources on nepotistic bias, which, as they point out, has not always been accepted by evolutionary psychologists. Anthropologists have not found much evidence for biologically based nepotism (e.g., Jones 2000; Kaplan and Hill 1985: 227).

8. The idea that societies with more cooperative social and cultural systems will outcompete less cooperative societies and will tend to be emulated (e.g., in Gintis et al. 2005: 22) seems to be based on an unsubstantiated claim made by Charles Darwin (1874: 150), that is often cited by contemporary cooperation researchers (e.g., Wilson 2002: 5).

9. The early evolutionary expression of such a brain can be traced to as early as the Miocene 10 to 12 million years ago with the emergence of species of apes whose cranial morphology indicates direct ancestry to humans and contemporary African apes such as chimpanzees (e.g., Begun 2010: 75–76; Gibson 2005). These data suggest that the ultimate bioevolutionary framework for human social intelligence dates to a period long prior to the Pleistocene and was more likely associated with the evolution of hominoid "protoculture" (based on the superior capacity of great ape learning [Russon 1997]) rather than the characteristic human cultural capacity (based on symbols and language), which is thought to have evolved fully only within the last 150,000 years (Henshilwood et al. 2002; Vanhaeren et al. 2006).

10. While humans and some great apes share Theory of Mind that allows them to gauge the intentions of others, humans comprehend intentionality at a higher level of abstraction than apes (Gibson 2005: 35). In addition, human language ability is argued to provide a more effective form of mental representation than apes are capable of in the form of "inner speech" (Carruthers 2006: 307–312, passim; Smith 1996).

11. Religion, numinous experience, and built environments do not always support cooperation; as we note, they may also serve to legitimate political dominance rather than high levels of cooperation.

12. For example, informal devices such as paragovernmental organizations may foster cooperation beyond the abilities of the official governing system (Cook, Hardin, and Levi 2005: 83–103).

13. This need not imply an entirely passive subaltern class. In the less cooperative states there were comparatively high levels of social disruption such as taxpayer revolts (Blanton 2010: 48) and subaltern cultural production of antistate ideologies (Blanton 2011).

14. All the collective action scale measures are positively correlated and significant at the .05 level or below (public goods by bureaucratization, $r = 0.68$; by principal control, $r = 0.4$; bureaucratization by principal control, $r = 0.76$), and are thus consistent with the predictions of collective action theory. The degree to which state revenues derive from taxpayers is also highly positively correlated with all three collectivity measures and is significant at or below the 0.05 level. These results are presented in Blanton and Fargher (2008: Table 10.3).

15. We dichotomize the data for purposes of statistical analysis, but this does not imply two "types" of states in terms of either dichotomized variable.

REFERENCES

Alexander, Richard D. 1987. *The Biology of Moral Systems*. New York: Aldine de Gruyter.

Atran, Scott. 2002. *In Gods We Trust: The Evolutionary Landscape of Religion*. Oxford: Oxford University Press.

Aunger, Robert. 2000. "Introduction." In *Darwinizing Culture: The Status of Memetics as a Science*, ed. Robert Aunger, 1–24. Oxford: Oxford University Press.

Banfield, Edward C. 1958. *The Moral Basis of a Backward Society*. Glencoe, IL: The Free Press.

Barkow, Jerome H., Leda Cosmides, and John Tooby, eds. 1992. *The Adapted Mind: Evolutionary Psychology and the Generation of Culture*. New York: Oxford University Press.

Barry, Brian, and Russell Hardin. 1982. "Epilogue." In *Rational Man and Irrational Society? An Introduction and Sourcebook*, ed. Brian Barry and Russell Hardin, 367–90. Beverly Hills, CA: Sage Publications.

Baumard, Nicolas. 2010. "Has Punishment Played a Role in the Evolution of Cooperation? A Critical Review." *Mind and Society* 9 (2): 171–92. http://dx.doi.org/10.1007/s11299-010-0079-9.

Begun, David R. 2010. "Miocene Hominids and the Origins of the African Apes." *Annual Review of Anthropology* 39 (1): 67–84. http://dx.doi.org/10.1146/annurev.anthro.012809.105047.

Bell, Charles. 1992. *The People of Tibet*. Delhi: Motilal Banarsidass Publishers.

Ben Porath, Yoram. 1982. "Economics of the Family: Match or Mismatch? A Review of Becker's A Treatise on the Family." *Journal of Economic Literature* 20: 52–64.

Betzig, Laura. 1997. "Introduction: People Are Animals." In *Human Nature: A Critical Reader*, ed. Laura Betzig, 1–17. New York: Oxford University Press.

Blanton, Richard E. 1989. "Continuity and Change in Public Architecture, Periods I Through V of the Valley of Oaxaca, Mexico." In *Monte Albán's Hinterland, Part II: Settlement Patterns in Tlacolula, Etla, and Ocotlán, Valley of Mexico, Volume 1*. By Stephen Kowalewski, Gary M. Feinman, Richard E. Blanton, Laura Finsten, and Linda Nicholas, 409–447. Museum of Anthropology, The University of Michigan, Memoir 23, Ann Arbor.

Blanton, Richard E. 1995. The Cultural Foundations of Inequality in Households. In *Foundations of Social Inequality*, ed. T. Douglas Price and Gary M. Feinman, 105–127. Plenum Press, New York.

Blanton, Richard E. 1998. "Beyond Centralization: Steps Toward a Theory of Egalitarian Behavior in Archaic States." In *Archaic States*, ed. Gary M. Feinman and Joyce Marcus, 135–172. Santa Fe: School of American Research Press.

Blanton, Richard E. 2010. "Collective Action and Adaptive Socioecological Cycles in Premodern States." *Cross-Cultural Research* 44 (1): 41–59. http://dx.doi.org/10.1177/1069397109351684.

Blanton, Richard E. 2011. "Cultural Transformation, Art, and Collective Action in Polity-Building." *Cross-Cultural Research* 45 (2): 106–27. http://dx.doi.org/10.1177/1069397110393145.

Blanton, Richard E. 2013. "Cooperation and the Moral Economy of the Marketplace." In *Merchants, Trade, and Exchange in the Pre-Columbian World*, ed. Kenneth Hirth and Joanne Pillsbury, 23–48. Dumbarton Oaks, Washington, D. C.

Blanton, Richard, and Lane Fargher. 2008. *Collective Action in the Formation of Pre-Modern States*. New York: Springer.

Blanton, Richard, and Lane Fargher. 2009. "Collective Action in the Evolution of Pre-Modern States." *Social Evolution & History* 8:133–66.

Blanton, Richard, and Lane Fargher. 2010. "Evaluating Causal Factors in Market Development in Premodern States: A Comparative Study, with Critical Comments on the History of Ideas About Markets." In *Archaeological Approaches to Market Exchange in Ancient Societies*, ed. Christopher P. Garraty and Barbara L. Stark, 207–226. Boulder: University of Colorado Press.

Blanton, Richard E., and Jody Taylor. 1995. "Patterns of Exchange and the Social Production of Pigs in Highland New Guinea: Their Relevance to Questions about the Origins and Evolution of Agriculture." *Journal of Archaeological Research* 3 (2): 113–45. http://dx.doi.org/10.1007/BF02231435.

Boehm, Christopher. 1997. "Egalitarian Behavior and the Evolution of Political Intelligence." In *Machiavellian Intelligence II: Extensions and Evaluations*, ed. Andrew Whiten and Richard W. Byrne, 341–364. Cambridge: Cambridge University Press. http://dx.doi.org/10.1017/CBO9780511525636.014

Boehm, Christopher. 2004. "Large-Scale Game Hunting and the Evolution of Human Sociality." In *The Origins and Nature of Sociality*, ed. Robert W. Sussman and Audrey R. Chapman, 270–287. New York: Aldine de Gruyter.

Bowles, Samuel. 2006. "Group Competition, Reproductive Leveling, and the Evolution of Human Altruism." *Science* 314 (5805): 1569–72. http://dx.doi.org/10.1126/science.1134829.

Bowles, Samuel, and Herbert Gintis. 2005. "Can Self-Interest Explain Cooperation?" *Evolutionary and Institutional Economics Review* 2:21–41.

Boyd, Robert, Herbert Gintis, and Samuel Bowles. 2010. "Coordinated Punishment of Defectors Sustains Cooperation and Can Proliferate When Rare." *Science* 328 (5978): 617–20. http://dx.doi.org/10.1126/science.1183665.

Boyd, Robert, Herbert Gintis, Samuel Bowles, and Peter J. Richerson. 2005. "The Evolution of Altruistic Punishment." In *Moral Sentiments and Material Interests: The Foundations of Cooperation in Economic Life*, ed. Herbert Gintis, Samuel Bowles, Robert Boyd, and Ernst Fehr, 215–227. Cambridge: MIT Press.

Boyd, Robert, and Sarah Mathew. 2007. "A Narrow Road to Cooperation." *Science* 316 (5833): 1858–9. http://dx.doi.org/10.1126/science.1144339.

Boyd, Robert, and Peter J. Richerson. 1985. *Culture and the Evolutionary Process.* Chicago: University of Chicago Press.

Boyd, Robert, and Peter J. Richerson. 2000. "Memes: Universal Acid or a Better Mousetrap?" In *Darwinizing Culture: The Status of Memetics as a Science*, ed. Robert Aunger, 143–162. Oxford: Oxford University Press.

Boyd, Robert, and Peter J. Richerson. 2005. *The Origin and Evolution of Cultures.* Oxford: Oxford University Press.

Boyd, Robert, and Peter J. Richerson. 2006. "Culture and the Evolution of Human Social Instincts." In *Roots of Human Sociality: Culture, Cognition, and Interaction*, ed. N. J. Enfield and Steven Levinson, 453–477. Berg, Oxford.

Boyd, Robert, Peter J. Richerson, and Joseph Henrich. 2005a. "Cultural Evolution of Human Cooperation." In *The Origin and Evolution of Cultures*, by Robert Boyd and Peter J. Richerson., 251–282. Oxford: Oxford University Press.

Boyd, Robert, Peter J. Richerson, and Joseph Henrich. 2005b. "Why People Punish Defectors: Weak Conformist Transmission Can Stabilize Costly Enforcement Norms in Cooperative Dilemmas." In *The Origin and Evolution of Cultures*, by Robert Boyd and Peter J. Richerson., 189–203. Oxford: Oxford University Press.

Boyd, Robert, Peter J. Richerson, and Joseph Soltis. 2005. "Can Group-Functional Behaviors Evolve by Cultural Group Selection?" In *The Origin and Evolution of Cultures*,

ed. Robert Boyd and Peter J. Richerson, 204–226. Oxford: Oxford University Press.
Brothers, Leslie. 1997. *Friday's Footprint: How Human Society Shapes the Human Mind.* New York: Oxford University Press.
Brown, Patricia F. 1988. *Venetian Painting in the Age of Carpaccio.* New Haven: Yale University Press.
Bryan, Betsy. 2000. "The 18th Dynasty Before the Amarna Period. (c. 1550–1352 BC)." In *The Oxford History of Ancient Egypt*, ed. Ian Shaw, 218–71. Oxford: Oxford University Press.
Buller, David J. 2005. *Adapting Minds: Evolutionary Psychology and the Persistent Quest for Human Nature.* Cambridge: MIT Press.
Buller, David J., and Valerie Gray Hardcastle. 2006. "Modularity." In *Adapting Minds: Evolutionary Psychology and the Persistent Quest for Human Nature.*, 127–200. Cambridge: MIT Press.
Byrne, Richard W. 1997. "The Technical Intelligence Hypothesis: An Additional Evolutionary Stimulus to Intelligence?" In *Machiavellian Intelligence II: Extensions and Evaluations*, ed. Andrew Whiten and Richard W. Byrne, 289–311. Cambridge: Cambridge University Press. http://dx.doi.org/10.1017/CBO9780511525636.012
Byrne, Richard W., and Andrew Whitten. 1997. "Machiavellian Intelligence." In *Machiavellian Intelligence II: Extensions and Evaluations*, ed. Andrew Whiten and Richard W. Byrne, 1–23. Cambridge: Cambridge University Press. http://dx.doi.org/10.1017/CBO9780511525636.002
Camp, John M. 1986. *The Athenian Agora: Excavations in the Heart of Classic Athens.* London: Thames and Hudson.
Carlton, Eric. 1977. *Ideology and the Social Order.* London: Routledge and Kegan Paul.
Carrasco Pizana, P. 1959. *Land and Polity in Tibet.* Seattle: The American Ethnological Society and the University of Washington Press.
Carruthers, Peter. 1996. "Simulation and Self-Knowledge: A Defense of Theory-Theory." In *Theories of Theories of Mind*, ed. Peter Carruthers and Peter K. Smith, 22–38. Cambridge: Cambridge University Press. http://dx.doi.org/10.1017/CBO9780511597985.004
Carruthers, Peter. 2006. *The Architecture of the Mind: Massive Modularity and the Flexibility of Thought.* Oxford: Clarendon Press.
Carruthers, Peter, and Peter K. Smith. 1996. "Introduction." In *Theories of Theories of Mind*, ed. Peter Carruthers and Peter K. Smith, 1–8. Cambridge: Cambridge University Press. http://dx.doi.org/10.1017/CBO9780511597985.002
Collins, Randall. 1975. *Conflict Sociology: Toward an Explanatory Science.* New York: Academic Press.
Cook, Karen S., Russell Hardin, and Margaret Levi. 2005. *Cooperation Without Trust?* New York: Russell Sage Foundation.
Cosmides, Leda, and John Tooby. 1992. "Cognitive Adaptations for Social Exchange." In *The Adapted Mind: Evolutionary Psychology and the Generation of Culture*, ed. Jerome H. Barkow, Leda Cosmides, and John Tooby, 163–228. New York: Oxford University Press.
D'Altroy, Terence N. 2002. *The Incas.* Malden, MA: Blackwell Publishers.
Darwin, Charles. 1874. *The Descent of Man and Selection in Relation to Sex.* 2nd ed. New York: A. L. Burt. http://dx.doi.org/10.5962/bhl.title.54341
Davies, Nigel. 1987. *The Aztec Empire: The Toltec Resurgence.* Norman: University of Oklahoma.

Dawkins, Richard. 1982. *The Extended Phenotype*. Oxford: Oxford University Press.
Dirks, Nichiolas B. 1987. *The Hollow Crown: Ethnohistory of an Indian Kingdom*. Cambridge: Cambridge University Press.
Doja, Albert. 2005. "The Advent of Heroic Anthropology in the History of Ideas." *Journal of the History of Ideas* 66 (4): 633–50. http://dx.doi.org/10.1353/jhi.2005.0054.
Douglas, Mary. 1986. *How Institutions Think*. Syracuse: Syracuse University Press.
Douglas, Mary, and Aaron Wildavsky. 1982. *Risk and Culture: An Essay on the Selection of Technical and Environmental Dangers*. Berkeley: University of California Press.
Dunbar, R. I. M. 2003. "The Social Brain: Mind, Language, and Society in Evolutionary Perspective." *Annual Review of Anthropology* 32 (1): 163–81. http://dx.doi.org/10.1146/annurev.anthro.32.061002.093158.
Durán, Diego. 1994. *The History of the Indies of New Spain*. Trans. Doris Heyden. Norman: University of Oklahoma Press.
Durkheim, Emile. 1965 [1912]. *The Elementary Forms of the Religious Life*. Trans. J. W. Swain. New York: Free Press.
Eisenstadt, S. N., ed. 1986. *The Origins and Diversity of the Axial Age Civilizations*. Albany: State University of New York Press.
Ember, Carol R., and Melvin Ember. 2001. *Cross-Cultural Research Methods*. Lanham, MD: AltaMira Press.
Enfield, N. J., and C. Stephen Levinson. 2006. "Introduction: Human Sociality as a New Interdisciplinary Field." In *Roots of Human Sociality: Culture, Cognition, and Interaction*, ed. N. J. Enfield and Stephen C. Levinson, 1–35. Berg, Oxford.
Fargher, Lane F., and Richard E. Blanton. 2007. "Revenue, Voice, and Public Goods in Three Pre-Modern States." *Comparative Studies in Society and History* 49 (04): 848–82. http://dx.doi.org/10.1017/S0010417507000783.
Farmer, Edward L. 1976. *Early Ming Government: The Evolution of Dual Capitals*. Cambridge: Harvard University Press.
Feil, D. K. 1987. *The Evolution of Highland Papua New Guinea Societies*. Cambridge: Cambridge University Press. http://dx.doi.org/10.1017/CBO9781139084994
Frank, Robert H. 1988. *Passions Within Reason: The Strategic Role of the Emotions*. New York: W. W. Norton.
Fritz, John M. 1986. "Vijayanagara: Authority and Meaning of a South Indian Imperial Capital." *American Anthropologist* 88 (1): 44–55. http://dx.doi.org/10.1525/aa.1986.88.1.02a00030.
Fritz, John M., George Michell, and M. S. Nagaraja Rao. 1984. *Where Kings and Gods Meet: The Royal Centre at Vijayanagara, India*. Tucson: University of Arizona Press.
Fukuyama, Francis. 1995. *Trust: The Social Virtues and the Creation of Prosperity*. New York: Simon and Schuster.
Geertz, Clifford. 1980. *Negara: The Theatre State in Nineteenth-Century Bali*. Princeton: Princeton University Press.
Gibson, Kathleen. 2005. "Epigenesis, Brain Plasticity, and Behavioral Versatility: Alternatives to Standard Evolutionary Psychology Models." In *Complexities: Beyond Nature and Nurture*, ed. Susan McKinnon and Sydel Silverman, 23–42. Chicago: University of Chicago Press.
Gintis, Herbert, Samuel Bowles, Robert Boyd, and Ernst Fehr. 2005. "Moral Sentiments and Material Interests: Origins, Evidence, and Consequences." In *Moral Sentiments and Material Interests: The Foundations of Cooperation in Economic Life*, ed. Herbert Gintis, Samuel Bowles, Robert Boyd, and Ernst Fehr, 3–39. Cambridge: MIT Press.

Gluckman, Max. 1961. "The Lozi of Barotseland in North-Western Rhodesia." In *Seven Tribes of British Central Africa*, ed. Elizabeth Colson and Max Gluckman, 1–93. Manchester: Manchester University Press.

Gould, Stephen Jay. 1980. "Sociobiology and the Theory of Natural Selection." In *Sociobiology: Beyond Nature/Nurture?* ed. George W. Barlow and James Silverberg, 257–269. Washington, D. C.: American Association for the Advancement of Science.

Griffin, Miriam. 2000a. "The Flavians." In *The Cambridge Ancient History, Second Edition, Volume XI: The High Empire, A.D. 70–192*, ed. A. K. Bowman, P. Garnsey, and D. Rathbone, D., 1–83. Cambridge University Press, Cambridge.

Griffin, Miriam. 2000b. "Nerva to Hadrian." In *The Cambridge Ancient History, Second Edition, Volume XI: The High Empire, A.D. 70–192*, ed. A. K. Bowman, P. Garnsey, and D. Rathbone, 84–131. Cambridge University Press, Cambridge.

Gullick, J. M. 1958. *Indigenous Political Systems of Western Malaya*. London: University of London and Athlone Press.

Hall, John W. 1991. "The *Bakuhan* System." In *The Cambridge History of Japan, Volume 4, Early Modern Japan*, ed. John W. Hall, 128–182. Cambridge University Press, Cambridge.

Hamilton, William D. 1964. "The Genetical Evolution of Social Behavior, I and II." *Journal of Theoretical Biology* 7 (1): 1–16, 17–52. http://dx.doi.org/10.1016/0022-5193(64)90038-4.

Harris, Marvin. 1975. *Culture, People, Nature: An Introduction to General Anthropology*. 2nd ed. New York: Thomas Y. Crowell.

Hart, Gillian. 1992. "Imagined Unities: Constructions of 'The Household' in Economic Theory." In *Understanding Economic Process*, ed. Sutti Ortiz and Susan Lees, 111–130. Society for Economic Anthropology, Monographs in Economic Anthropology 10. Lanham, MD: University Press of America.

Hart, Keith. 1988. "Kinship, Contract, and Trust: The Economic Organization of Migrants in an African City Slum." In *Trust: Making and Breaking Cooperative Relationships*, ed. Diego Gambetta, 176–193. Oxford: Basil Blackwell.

Hechter, Michael. 1990. "On the Inadequacy of Game Theory for the Solution of Real-World Collective Action Problems." In *The Limits of Rationality*, ed. Karen Schweers Cook and Margaret Levi, 240–249. Chicago: University of Chicago Press.

Heijdra, Martin. 1998. "The Socio-Economic Development of Rural China during the Ming." In *The Cambridge History of China*, Volume 8, *The Ming Dynasty, 1368–1644, Part 2*, ed. Dennis Twitchett and F. W. Mote, 417–578. Cambridge: Cambridge University Press.

Henrich, Joseph, Robert Boyd, Samuel Bowles, Colin Camerer, Ernst Fehr, and Herbert Gintis, eds. 2004. *Foundations of Human Sociality: Economic Experiments and Ethnographic Evidence from Fifteen Small-Scale Societies*. Oxford: Oxford University Press.

Henrich, Joseph, Jean Ensminger, Richard McElreath, Abigail Barr, Clark Barrett, Alexander Bolyanatz, Juman Camilo Cardenas, Michael Gurven, Edwins Gwako, Natalie Henrich, et al. 2010. "Markets, Religion, Community Size, and the Evolution of Fairness and Punishment." *Science* 327 (5972): 1480–4. http://dx.doi.org/10.1126/science.1182238.

Henrich, Joseph, and Natalie Henrich. 2007. *Why Humans Cooperate: A Cultural and Evolutionary Explanation*. New York: Oxford University Press.

Henshilwood, Christopher S., Francesco d'Errico, Royden Yates, Zenobia Jacobs, Chantal Tribolo, Geoff A. T. Duller, Norbert Mercier, Judith C. Sealy, Helene Valladas, Ian Watts, et al. 2002. "Emergence of Modern Human Behavior: Middle Stone Age Engravings from South Africa." *Science* 295 (5558): 1278–80. http://dx.doi.org/10.1126/science.1067575.

Herrmann, Benedikt, Christian Thöni, and Simon Gächter. 2008. "Antosocial Punishment Across Societies." *Science* 319 (5868): 1362–7. http://dx.doi.org/10.1126/science.1153808.

Hölscher, Tonio. 1991. "The City of Athens: Space, Symbol, and Structure." In *City States in Classical Antiquity and Medieval Italy*, ed. A. Molho, K. Raaflaub, and J. Emlen, 355–380. Ann Arbor: University of Michigan Press.

Horton, Mark, and John Middleton. 2000. *The Swahili: The Social Landscape of a Mercantile Society*. Oxford: Blackwell.

Hsu, Cho-Yun. 1986. "Historical Conditions of the Emergence and Crystallization of the Confucian System." In *The Origins and Diversity of the Axial Age Civilizations*, ed. S. N. Eisenstadt, 306–324. Albany: State University of New York Press.

Hull, David L. 1988. *Science as a Process: An Evolutionary Account of the Social and Conceptual Development of Science*. Chicago: University of Chicago Press.

Hurgronje, C. Snouk. 1906. *The Achehnese*. Leyden: E. J. Brill.

Iannaccone, Laurence. 1992. "Sacrifice and Stigma: Reducing Free-Riding in Cults, Communes, and Other Collectivities." *Journal of Political Economy* 100 (2): 271–91. http://dx.doi.org/10.1086/261818.

Johnson, Gregory A. 1983. "Decision-Making Organization and Pastoral Nomad Camp Size." *Human Ecology* 11 (2): 175–99. http://dx.doi.org/10.1007/BF00891742.

Jones, Doug. 2000. "'Group Nepotism' and Human Kinship." *Current Anthropology* 41 (5): 779–809. http://dx.doi.org/10.1086/317406.

Kahan, Dan M. 2005. "The Logic of Reciprocity: Trust, Collective Action, and Law." In *Moral Sentiments and Material Interests: The Foundations of Cooperation in Economic Life*, ed. Herbert Gintis, Samuel Bowles, Robert Boyd, and Ernst Fehr, 339–378. Cambridge: MIT Press.

Kaplan, Hillard, and Michael Gurven. 2005. "The Natural History of Human Food-Sharing and Cooperation: A Review and a New Multi-Individual Approach to the Negotiation of Norms." In *Moral Sentiments and Material Interests: The Foundations of Cooperation in Economic Life*, ed. Herbert Gintis, Samuel Bowles, Robert Boyd, and Ernst Fehr, 75–113. Cambridge: MIT Press.

Kaplan, Hillard, and Kim Hill. 1985. "Food Sharing Among Foragers: Tests of Explanatory Hypotheses." *Current Anthropology* 26 (2): 223–46. http://dx.doi.org/10.1086/203251.

Kishigami, Nobuhiro. 2004. "A New Typology of Food-Sharing Practices Among Hunter-Gatherers, With a Special Focus on Inuit Examples." *Journal of Anthropological Research* 60:341–58.

Koenig, William J. 1990. *The Burmese Polity, 1752–1819: Politics, Administration, and Social Organization in the Early Kon-baung Period*. Center for South and Southeast Asian Studies, Michigan Papers on South and Southeast Asia 34. Ann Arbor.

Kuhn, Thomas S. 1970. *The Structure of Scientific Revolutions*. Chicago: The University of Chicago Press.

Kuper, Adam. 2000. "If Memes Are the Answer, What is the Question?" In *Darwinizing Culture: The Status of Memetics as a Science*, ed. Robert Aunger, 175–188. Oxford: Oxford University Press.

Laufer, Berthold. 1965. "The Development of Ancestral Images in China." In *Reader in Comparative Religion*, ed. William Lessa and Evon Z. Vogt, 445–450. New York: Harper and Row.

Law, Robin. 1977. *The Oyo Empire c.1600-c.1836: A West African Imperialism in the Era of the Atlantic Slave Trade*. Oxford: Clarendon Press.

Levi, Margaret. 1988. *Of Rule and Revenue*. Berkeley: University of California Press.
Lichbach, Mark I. 1996. *The Cooperator's Dilemma*. Ann Arbor: University of Michigan Press.
Lipton, Peter. 2005. "Testing Hypotheses: Prediction and Prejudice." *Science* 307 (5707): 219–21. http://dx.doi.org/10.1126/science.1103024.
Lybyer, Albert H. 1966. *The Government of the Ottoman Empire in the Time of Suleiman the Magnificent*. New York: Russell and Russell.
Marlowe, Frank W. 2004. "What Explains Hadza Food Sharing?" *Research in Economic Anthropology* 23:69–88. http://dx.doi.org/10.1016/S0190-1281(04)23003-7.
McArthur, Margaret. 1977. "Nutritional Research in Melanesia: A Second Look at the Tsembaga." In *Subsistence and Survival: Rural Ecology in the Pacific*, ed. Timothy Bayliss-Smith and Richard G. Feachem, 91–128. London: Academic Press.
McCaskie, T. C. 1995. *State and Society in Pre-Colonial Asante*. Cambridge: Cambridge University Press.
McClellan, George B. 1904. *The Oligarchy of Venice: An Essay*. Boston: Houghton-Mifflin.
Moertono, Soemarsaid. 1981. *State and Statecraft in Old Java: A Study of the Later Mataram Period, 16th to 19th Centuries*. Cornell University, Modern Indonesia Project, Monograph Series 43, Ithaca, NY.
Moore, Jerry D. 1996. *Architecture and Power in the Ancient Andes: The Archaeology of Public Buildings*. Cambridge: Cambridge University Press. http://dx.doi.org/10.1017/CBO9780511521201
Morris, William A. 1940. "Introduction." In *The English Government at Work, 1327–1336, Volume 1: Central and Prerogative Administration*, ed. William A. Morris, 3–81. The Mediaeval Academy of America, Cambridge.
Nadel, S. F. 1942. *A Black Byzantium: The Kingdom of Nupe in Nigeria*. London: Oxford University Press.
Nesse, Randolph M. 2001. "Natural Selection and the Capacity for Commitment." In *Evolution and the Capacity for Commitment*, ed. Randolph M. Nesse, 1–44. New York: Russell Sage Foundation.
Norenzayan, Ara, and Azim F. Shariff. 2008. "The Origin and Evolution of Religious Prosociality." *Science* 322 (5898): 58–62. http://dx.doi.org/10.1126/science.1158757.
Norwich, John J. 1982. *A History of Venice*. New York: Alfred A. Knopf.
Nowak, Martin A. 2006. "Five Rules for the Evolution of Cooperation." *Science* 314 (5805): 1560–3. http://dx.doi.org/10.1126/science.1133755.
O'Connor, David B. 1983. "New Kingdom and Third Intermediate Period, 1552–664 B. C." In *Ancient Egypt: A Social History*, by Bruce G. Trigger, Barry Kemp, David B. O'Connor, and A. B. Lloyd, 183–278. Cambridge: Cambridge University Press.
Ostrom, Elinor. 2007. "Collective Action Theory." In *The Oxford Handbook of Comparative Politics*, ed. Carles Boix and Susan C. Stokes, 186–208. Oxford: Oxford University Press.
Pennisi, Elizabeth. 2009. "On the Origin of Cooperation." *Science* 325 (5945): 1196–9. http://dx.doi.org/10.1126/science.325_1196.
Povinelli, Daniel. 1996. "Chimpanzee Theory of Mind? The Long Road to Strong Inference." In *Theories of Theories of Mind.*, 293–329. Cambridge: Cambridge University Press. http://dx.doi.org/10.1017/CBO9780511597985.019
Prins, A. H. J. 1971. *Didemic Lamu: Social Stratification and Spatial Structure in a Muslim Maritime Town*. Gronigen: Institut voor Culturele Anthropologie der Rijksvuniversiteit,.
Prins, Gwyn. 1980. *The Hidden Hippopotamus: Reappraisal in African History: The Early Colonial Experience in Western Zambia*. Cambridge: Cambridge University Press.

Rappaport, Roy A. 1978. "Maladaptation in Social Systems." In *The Evolution of Social Systems*, ed. J. Friedman and M. J. Rowlands, 49–72. Pittsburgh: University of Pittsburgh Press.

Rappaport, Roy A. 1979. *Ecology, Meaning, and Ritual.* Richmond, CA: North Atlantic Books.

Rattray, R. S. 1923. *Ashanti.* Oxford: Clarendon Press.

Rattray, R. S. 1929. *Ashanti Law and Constitution.* Oxford: Clarendon Press.

Ray, Benjamin C. 1991. *Myth, Ritual, and Kingship in Buganda.* New York: Oxford University Press.

Reyna, S. P. 1990. *Wars Without End: The Political Economy of a Precolonial African State.* University Press of New England, Hanover.

Richards, J. F. 1998. "The Formulation of Imperial Authority Under Akbar and Jahangir." In *Kingship and Authority in South Asia*, ed. J. F. Richards, 285–326. Delhi: Oxford University Press.

Richerson, Peter J., and Robert T. Boyd. 1978. "A Dual Inheritance Model of Human Evolutionary Process: Basic Postulates and a Simple Model." *Journal of Social and Biological Structures* 1 (2): 127–54. http://dx.doi.org/10.1016/S0140-1750(78)80002-5.

Richerson, Peter J., and Robert T. Boyd. 1999. "Complex Societies: The Evolutionary Origins of a Crude Superorganism." *Human Nature (Hawthorne, N.Y.)* 10 (3): 253–89. http://dx.doi.org/10.1007/s12110-999-1004-y.

Richerson, Peter J., Robert T. Boyd, and Joseph Henrich. 2003. "Cultural Evolution of Human Cooperation." In *Genetic and Cultural Evolution of Cooperation*, ed. Peter Hammerstein, 357–388. Cambridge, Berlin: MIT Press and Preie Universität.

Roscoe, John. 1923. *The Bakitara or Banyoro.* Cambridge: Cambridge University Press.

Rose, Hilary, and Steven Rose. 2000. "Introduction." In *Alas Poor Darwin: Arguments against Evolutionary Psychology*, ed. Hilary Rose and Steven Rose, 1–16. New York: Harmony Books.

Rose, Steven. 2000. "Escaping Evolutionary Psychology." In *Alas Poor Darwin: Arguments Against Evolutionary Psychology*, ed. Hilary Rose and Steven Rose, 299–320. New York: Harmony Books.

Russon, Anne E. 1997. "Exploiting the Expertise of Others." In *Machiavellian Intelligence II: Extensions and Evaluations*, ed. Andrew Whiten and Richard W. Byrne, 174–206. Cambridge: Cambridge University. http://dx.doi.org/10.1017/CBO9780511525636.008

Sahlins, Marshall. 1976. *The Use and Abuse of Biology: An Anthropological Critique of Sociobiology.* Ann Arbor: The University of Michigan Press.

Salmon, Wesley C. 1988. "Introduction." In *The Limitations of Deductivism*, ed. Adolf Grünbaum and Wesley C. Salmon, 1–18. Berkeley: University of California.

Sarkar, Jadunath. 1963. *Mughal Administration.* 5th ed. Calcutta: M. C. Sarkar and Sons.

Schegloff, Emanuel A. 2006. "Interaction: The Infrastructure for Social Institutions, the Natural Ecological Niche for Language and the Arena in Which Culture Is Enacted." In *Roots of Human Sociality: Culture, Cognition, and Interaction*, ed. N. J. Enfield and Steven Levinson, 70–96. Berg, Oxford.

Schmidt, Karen L., and Jeffrey F. Cohn. 2001. "Human Facial Expressions as Adaptations: Evolutionary Questions in Facial Expression Research." *Yearbook of Physical Anthropology* 44 (S33): 3–24. http://dx.doi.org/10.1002/ajpa.20001.

Sewell, William H., Jr. 1992. "A. Theory of Structure: Duality, Agency, and Transformation." *American Journal of Sociology* 98 (1): 1–29. http://dx.doi.org/10.1086/229967.

Simpson, Jeffrey A., and Lane Beckes. 2010. "Evolutionary Perspectives on Prosocial Behavior." In *Prosocial Motives, Emotions, and Behavior: The Better Angels of Our Nature*, ed. Mario Mikulincer and Philip R. Sharer, 35–53. Washington, D. C.: American Psychological Association. http://dx.doi.org/10.1037/12061-002

Sinnett, Peter F. 1977. "Nutritional Adaptation Among the Enga." In *Subsistence and Survival: Rural Ecology in the Pacific*, ed. Timothy Bayliss-Smith and Richard G. Feachem, 63–90. London: Academic Press.

Sinopoli, Carla M., and Kathleen D. Morrison. 1995. "Dimensions of Imperial Control: The Vijayanagara Capital." *American Anthropologist* 97 (1): 83–96. http://dx.doi.org/10.1525/aa.1995.97.1.02a00110.

Smith, Eric A. 2003. "Human Cooperation: Perspectives from Behavioral Ecology." In *Genetic and Cultural Evolution of Cooperation*, ed. Peter Hammerstein, 401–427. Cambridge: MIT Press.

Smith, Peter K. 1996. "Language and the Evolution of Mind-Reading." In *Theories of Theories of Mind.*, 344–354. Cambridge: Cambridge University Press. http://dx.doi.org/10.1017/CBO9780511597985.021

Snodgrass, Anthony. 1981. *Archaic Greece: The Age of Experiment.* Berkeley: University of California Press.

Sosis, Richard, and Candace Alcorta. 2003. "Signaling, Solidarity, and the Sacred: The Evolution of Religious Behavior." *Evolutionary Anthropology* 12 (6): 264–74. http://dx.doi.org/10.1002/evan.10120.

Sperber, Dan. 2006. "Why a Deep Understanding of Cultural Evolution Is Incompatible with Shallow Psychology." In *Roots of Human Sociality: Culture, Cognition, and Interaction*, ed. N. J. Enfield and Steven Levinson, 431–49. Berg, Oxford.

Stone, Valerie E. 2006. "Theory of Mind and the Evolution of Social Intelligence." In *Social Neuroscience: People Thinking About Thinking People*, ed. John T. Cacioppo, Penny S. Visser, and Cynthia L. Pickett, 103–129. Cambridge: MIT Press.

Sung-hsing, Wang. 1985. "On the Household and Family in Chinese Society." In *The Chinese Family and Its Ritual Behavior*, ed. H. Jih-Chang and C. Ying-Chang, 50–58. Academica Sinica, Institute of Ethnology, Taipei.

Sussman, Robert W., and Audrey R. Chapman. 2004. "The Nature and Evolution of Sociality." In *The Origins and Nature of Sociality*, ed. Robert W. Sussman and Audrey R. Chapman, 3–19. New York: Aldine de Gruyter.

Sussman, Robert W., and Paul A. Garber. 2004. "Rethinking Sociality: Cooperation and Aggression Among Primates." In *The Origins and Nature of Sociality*, ed. Robert W. Sussman and Audrey R. Chapman, 161–190. New York: Aldine de Gruyter.

Tambiah, Stanley J. 1976. *World Conqueror and World Renouncer: A Study of Buddhism and Polity in Thailand Against a Historical Background.* Cambridge: Cambridge University Press. http://dx.doi.org/10.1017/CBO9780511558184

Taylor, Romeyn. 1998. "Official Religion in the Ming." In *The Cambridge History of China*, Volume 8, *The Ming Dynasty, 1368–1644, Part 2*, ed. Dennis Twitchett and F. W. Mote, 840–92. Cambridge: Cambridge University Press.

Tomasello, Michael. 2008. *Origins of Human Communication.* Cambridge, MA: MIT Press.

Tooby, John, and Leda Cosmides. 1992. "The Psychological Foundations of Culture." In *The Adapted Mind: Evolutionary Psychology and the Generation of Culture*, ed. Jerome H. Barkow, Leda Cosmides, and John Tooby, 19–136. New York: Oxford University Press.

Tooby, John, and Leda Cosmides. 1997. "Think Again." In *Human Nature: A Critical Reader*, ed. Laura Betzig, 292–294. New York: Oxford University Press.

Trivers, Robert L. 1971. "The Evolution of Reciprocal Altruism." *Quarterly Review of Biology* 46 (1): 35–57. http://dx.doi.org/10.1086/406755.

van den Bergh, Pierre, and David P. Barash. 1977. "Inclusive Fitness and Human Family Structure." *American Anthropologist* 79 (4): 809–23. http://dx.doi.org/10.1525/aa.1977.79.4.02a00030.

Vanhaeren, Marian, Francesco d'Errico, Chris Stringer, Sarah L. James, Jonathan A. Todd, and Henk K. Mienis. 2006. "Middle Paleolithic Shell Beads in Israel and Algeria." *Science* 312 (5781): 1785–8. http://dx.doi.org/10.1126/science.1128139.

Vansina, Jan. 1973. *The Tio Kingdom of the Middle Congo, 1880–1892.* London: International African Institute and Oxford University Press.

Vansina, Jan. 1978. *The Children of Woot: A History of the Kuba Peoples.* Madison: University of Wisconsin Press.

van Zantwijk, Rudolf A. M. 1985. *The Aztec Arrangement: The Social History of Pre-Spanish Mexico.* Norman: University of Oklahoma Press.

Whitley, James. 2001. *The Archaeology of Ancient Greece.* Cambridge: Cambridge University Press.

Wilks, Ivor. 1993. *Forests of Gold: Essays on the Akan and the Kingdom of Asante.* Athens, OH: Ohio University Press.

Wilson, David Sloan. 2002. *Darwin's Cathedral: Evolution of Religion, and the Nature of Society.* Chicago: University of Chicago Press.

Wilson, Edward O. 1978. *On Human Nature.* Cambridge: Harvard University Press.

Winterhalder, Bruce. 1986. "Diet, Choice, Risk, and Food Sharing in a Stochastic Environment." *Journal of Anthropological Archaeology* 5 (4): 369–92. http://dx.doi.org/10.1016/0278-4165(86)90017-6.

Wolf, Eric R. 1999. *Envisioning Power: Ideologies of Dominance and Crisis.* Berkeley: University of California Press.

Wright, Walter L., Jr. 1935. *Ottoman Statecraft: The Book of Counsel for Vesirs and Governors.* Westport, CT: Greenwood Press.

6

Agency and Collective Action

Insights from North American Historical Archaeology

Dean J. Saitta

The concept of human agency has been widely used in archaeology over the past twenty years, and especially in the last decade (for reviews see Barrett 2001; Dobres and Robb 2000; Dornan 2002; Johnson 1989; Knapp and van Dommelen 2008). Agency theories in archaeology developed, in part, as a corrective to the often bloodless models of social life and change produced by various systems-theoretical and other processual approaches. Their development has been a good thing for the discipline. Agency theories have put people back into culture along with the cognitive factors—for instance, the frameworks of meaning by which people assign significance to events and things—that inform and motivate their actions. They have moved us to think about the freedom or "relative autonomy" that individuals have to maneuver within cultural systems and structures of social power. They have reunited society with history. In so doing, agency theories have rediscovered a key insight of the older Boasian, culture history approach that dominated archaeological thinking before the advent of processual archaeology: that the particulars of local historical context are worth investigating for their own sake, rather than simply serving as fodder for sweeping evolutionary narratives driven by cultural laws.

Several scholars have emphasized that *individual* agency is just one particular form of agency, and that the autonomous individual exercising rational choice and free will is a relatively recent invention specific to modernity (e.g., Thomas 2000; Hodder and Hutson 2003). Thomas (2000), for example, argues that humans always carry out their projects in the context of a concrete material world that includes other people. Thus, it is inadequate to consider human beings apart from the relationships in which they find themselves. Barrett (2001) agrees, noting that agency must include the operation of social collectives that extend beyond the individual's own body and lifespan. Indeed, Johannes Fabian (1994) has noted that human acting is always acting in company. Hodder (2004) helpfully suggests that agency, like power, is less a thing we possess than a capacity that we exercise. With Thomas, he sees the group as forming part of the resources used for individual agency, and thus views group behavior as another form of individual agency.

McGuire and Wurst (2002) push the critique of agency theory the farthest, from the standpoint of an explicitly activist archaeology that seeks to engage with the political present. They argue that theories of individual agency in postprocessual archaeology are as ideological as the cultural systems theories that preceded them. They identify the focus on the individual agent as a sustaining belief of modern capitalism. Capitalism depends for its survival on cultural processes that constitute people as free and unfettered individuals. Thus it works, through its cultural forms, to universalize this historically contingent idea. Where this ideology is internalized and taken for granted, it obscures the oppositional nature of class groupings and exploitation in society. It also produces the kind of self-serving "identity politics" that can fragment and debilitate collective movements for change. Thus, McGuire and Wurst find advocacy of individual agency models by scholars intending to use their research to challenge class, gender, and racial inequalities in the modern world to be misguided and contradictory. By embracing the logic, language, and symbolism of individual agency, activist scholars are in fact reinforcing that which they wish to critique. By projecting and universalizing that which is contingent, they help to propagate existing social relations. This notion of agency lacks transformative, emancipatory, and revolutionary potential (Harvey 1973).

These critiques are clear in suggesting that individuals are always and everywhere thoroughly enmeshed in a web of social relations. Collective action results from the shared consciousness or solidarity that defines a community of individuals. Such consciousness may be based in class, gender, ethnicity, race, age, physical ability, or some combination of these (and other) identities. People make history as members of social groups whose common consciousness derives from shared existential anxieties, political interests, and social relations. This perspective is evident in the current volume. Citing theorists from Marx through Giddens and Bourdieu, Roscoe (chapter 3) notes that humans are not just self-interested. Rather, they have multiple, specific interests. Such interests also have "lifetimes": some are situational, and others more enduring. Carballo (chapter 1) notes that the structure of collective action is contextual and "segmentary": groups of individuals who cooperate on the basis of certain interests in some settings are adversarial in others. This chapter, and the work reported within, respects the arguments of Carballo, Roscoe, and others. Further, to the extent that particular interests and actions are traceable to larger forces like global capitalism, and to the extent that community is always a delicate relation between fluid processes of self-identification and relatively permanent associations like that between person and nation-state (Harvey 2000: 240), an archaeology of collective action needs grand narratives of the structural and long term as well as small narratives of lived moments (Hodder 1999: 147).

This chapter considers cases of collective action that have been investigated by North American historical archaeologists. The focus is on "bottom-up" efforts by politically and economically oppressed groups to resist the forces that produce their oppression. The touchstone is historical archaeology's great triumvirate of race, class, and gender—the key identities that, depending on circumstances, the particular social interests at stake, and the "durability" of those interests either integrate or divide groups of individuals in society. In all cases material culture is understood as playing an active role in such efforts. Objects are considered key elements of the strategies that humans use to engage with their world; that is, as political and tactical weapons that themselves have agency (Gell 1998). The interest is in distilling insights relevant for developing an archaeology of cooperation, and evaluating the relative merits of different perspectives on the topic. For example, can the organizing epistemologies and theories of historical archaeology be usefully squared with those that inform the evolutionary archaeologies? Or does something important get lost in the bargain?

HISTORICAL ARCHAEOLOGIES OF COLLECTIVE ACTION

Historical archaeologists have made important breakthroughs in our understanding of cooperative behavior in the past. Paynter (2000) offers a comprehensive review of the existing literature. Race, gender, and class-based forms of collective action are also considered by contributors to Leone and Potter (1988, 1999), McGuire and Paynter (1991), Scott (1994), Delle, Mrozowski, and Paynter (2000), Van Bueren (2002), Hall and Silliman (2006), and others.

Several assumptions about race, gender, and class identity tend to guide collective action studies in historical archaeology. Identities are understood to be multiple, fluid, and situational. Orser (2010) notes that historical archaeologists today are more inclined to speak in terms of "vectors of inequality" than to focus on fixed notions of status. Understood in this way, identities are seen to be intertwined and thus difficult to study in isolation from each other. In other words, identities are constituted relationally (Meskell and Preucel 2004). Brubaker and Cooper (2000) have critiqued this "soft," constructivist view of identity, arguing that it can allow any number of putative identities to proliferate, empty the term of meaning, and thereby lose "analytical purchase" on the world. They are equally critical of stronger, categorical views that fix and essentialize identity and thus inform the sort of identity politics critiqued by McGuire and Wurst (2002). Brubaker and Cooper instead argue for the use of alternative terms like "identification" and "self-understanding." Here, I stick with the relational understanding

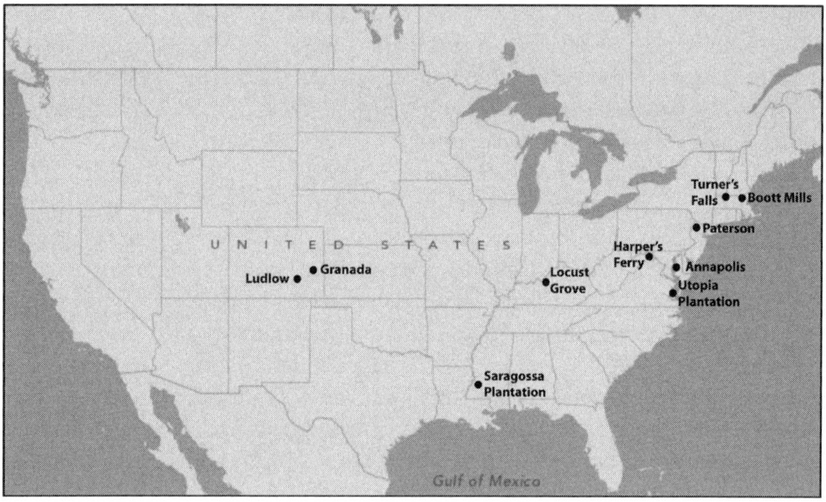

FIGURE 6.1 *Archaeological sites or nearby towns mentioned in text*

of identity while remaining cognizant of the fact that *all* conceptions of the world have merits and liabilities as entry points for critical analysis and social change (Saitta 2005).

The following review is necessarily selective. The studies described successfully demonstrate, or show great potential to demonstrate, how shared existential anxiety and identity can produce specific collective strategies for achieving change (see Figure 6.1 for a map of archaeological sites or nearby towns mentioned in the text). Because of the interpenetrability of race, class, and gender, my assignment of a study to one or another of these organizing categories is in some cases arbitrary. All of these studies, however, are illustrative of what is possible with an archaeology attuned to collective action.

Race

African diaspora studies provide a rich source of insights about race-based collective action in the past. Much discussion and debate has swirled around the existence and meaning of "Africanisms"—objects that either have a clear connection to African cultural practice or show significant commonalities among African diaspora communities—in the New World (Mullins 2004). There is a spreading recognition that a search for Africanisms is unproductive if it invests objects with a static identity, or reinforces a monolithic view of African culture (Orser 1998). The same can be said of the search for any other objects presumed to be associated with ethnic identity (Upton 1996). Alternatively, material objects are best viewed relationally—as having fluid meanings dependent on context that conceivably reference something in addition to, and even other than, racial or ethnic culture. That is, they are best seen as Hodderian "symbols in action"—

as active representations of otherness manipulated by individuals and groups within power relations (Hodder 1982; Leone 2005; Orser 1998; Singleton 1995).

Singleton (2005) summarizes important work by Lorena Walsh and Patricia Samford that implicates slave collective agency in the Chesapeake region. Walsh shows that at Utopia Plantation in Virginia, slaves built housing using Anglo-Virginian carpentry techniques but used African ideas of domestic space in placing houses in a square formation around an open courtyard. These courtyards would have provided central places for cooking and socializing. Singleton also reviews interesting studies of the rectangular and square subfloor pits that were dug within slave houses. Samford resists functional interpretations that relate pits to storage or to the concealment of pilfered items and, instead, favors a ritual interpretation. Using accounts of West African Igbo and Yoruba religious practices, Samford suggests that these pits served as household shrines used to bury religious items. Singleton notes that the existence of these pits often produced conflict between slaveholders and slave laborers to the extent that they served to challenge slaveholder control over living spaces.

The most famous examples of slave collective agency are associated with colonoware pottery studies. Colonoware is a low-fired, unglazed, handmade, locally produced earthenware found on African American sites in the eighteenth century. Colonoware vessels were used for preparing, serving, and storing food. They are found in shapes that resemble both European and African forms. A long debate about who made colonoware has been resolved in favor of production by a number of groups, including Native Americans (Orser 1996: 117–123). The colonoware vessel is an "intercultural artifact" (Singleton and Bograd 2000). Thus, interpretation needs to respect not only the form of these objects, but also the geographical area where they are found and the relational context in which they are used.

Working in the South Carolina Lowcountry, Leland Ferguson (1991, 1992) offers the most compelling case for colonoware vessels as instruments of slave agency geared toward collective resistance. Colonoware is found in particular abundance on Lowcountry sites, especially those associated with slaves. Ferguson documents, via quantitative and qualitative analysis, that colonoware in this region connected slave foodways to West African precedents. He convincingly shows that the forms of colonoware vessels recall West African patterns. A high frequency of bowls and a bimodal size distribution of jars reflect the West African tradition of serving starches in larger vessels and sauces or relishes in smaller ones. Bowls and jars both have rounded bases, distinguishing them from Anglo-European flat and tripodal bases. Another contrast with European dining practices of the time lies in the fact that the vast majority of colonoware containers—98 percent of the sample studied by Ferguson—lack cutlery marks (Ferguson 1991: 35).

Thus, Lowcountry slaves were apparently eating like their African ancestors rather than their European masters, and by extension using foodways to build community. Additional support for an African ethos comes from evidence indicating that colonoware pots—like Samford's Chesapeake pits—functioned in slave religious practices. A small number of colonoware bowls have features that recall a generalized West African "Bakongo" religious iconography. Bakongo refers to a "generalized cultural expression" that crosscuts ethnic differences in the Congo-Angola region of Africa, where about 40 percent of South Carolina slaves originated (Ferguson 1999: 118). The iconographic features or "cosmograms" include rounded ring bases and cross and circle designs incised into the pot's surface. In Bakongo culture clay pots are used in renewal rituals as containers for medicines and charms, and the cross and circle symbolize harmony with the universe and the continuity of life. Interestingly, in the South Carolina Lowcountry, colonoware pots are often excavated in streamside and river bottom contexts. In Bakongo cosmology water is associated with the separation between the living and spirit worlds. The water context association combined with their form and markings reinforces the interpretation of certain colonoware pots as "magic bowls" employed in community ritual.

Several lines of material evidence, along with historical analysis of Bakongo cosmology and oral testimony from a twentieth-century Georgia healer (see Ferguson 1999) thus converge to make a compelling case that the production and distribution of colonoware pottery served slave collective agency. Such agency is also evident in Lowcountry house forms, even more so than in the Chesapeake (Singleton 2005). Slaves having different ethnic roots in Africa used material objects to help build a "creolized" subculture that blended African cultural elements with other elements and, at the same time, distanced this subaltern culture from the dominant Anglo-European rationalizations that supported the planter social order. To the extent that no status differences or other boundaries are reflected within the colonoware assemblage, slaves were nurturing reciprocity and community. In short, material culture was used to build and support a pan-African sense of syncretic culture among the diverse peoples enslaved in the South Carolina Lowcountry (Ferguson 1999; Orser 1998).

Finally, work by Paul Mullins (1999) on African American use of material goods after emancipation in Annapolis explores change over time in how segments of this population expressed their collective identity by *reinterpreting* artifacts associated with genteel white consumer culture. Between 1850 and 1930 emancipated African Americans acquired previously inaccessible mass-produced parlor goods that were symbolically charged representations of American abundance and nationalism, signaling their owner's affluence and belonging (Orser 1998). These "knickknacks" were used by whites to materialize and naturalize white privilege, and to justify discrimination against blacks (Brumfiel 2003). On Mullins's view, emancipated blacks procured these items in order to articulate their aspirations for full citizenship in a capitalist, consumer-oriented society. These objects do not indicate a desire to assimilate. Blacks gave the objects new meanings in the interest of combating old racist notions of black material inferiority, distancing themselves from old racist caricatures generally, and negotiating expanded space for themselves in a new national order (Orser 1998).

Of course, historical archaeology's contributions to studies of collective action geared toward identity maintenance and/or political resistance do not stop with analyses of African American material culture. Brighton (2004) shows how smoking pipes bearing the symbol of the "Red Hand" galvanized Irish American identity and working-class solidarity in late nineteenth-century Paterson, New Jersey. The Red Hand was associated with the Ireland Home Rule movement in the 1880s, and its use by working-class Irish Americans in Paterson signified both a connection to their homeland and a sense of place and empowerment in the United States. Shackel (2010) shows how a particular set of consumer goods from a cross-section of African American and European American households in New Philadelphia, Illinois, produced a sense of shared group consciousness in a rural community shaped by racial hostilities and strife. The work of Bonnie Clark and her students at the World War II Japanese American internment camp of Granada (Amache) in southeastern Colorado implicates several dimensions of collective action under conditions of institutional confinement. Slaughter (2006) notes that the brewing of sake was against camp regulations, but at least one surviving internee remembers sake fermenting in the wash house boiler room in her housing block. Anyone having legitimate access to the boiler room would not have had legitimate access to leftover rice. Brewing rice probably came from the mess halls, so cooks in the camp were complicit along with, perhaps, many more service workers. Internees also created small gardens in the public areas of the camp between housing blocks. These gardens likely had practical functions, such as providing shade and some relief from the stark military landscape. But some likely articulated with the reinforcement of group identity given the evidence of overt Japanese landscaping techniques. Moreover, Amache gardens often use official camp construction materials (e.g., wire, concrete block, concrete) that were probably "liberated" from War Relocation Authority stockpiles in much the same way that leftover rice was liberated from internment camp kitchens.

Gender

Scholars researching gender have long been at the forefront of efforts to produce more nuanced understandings of social power relationships and organizational change. Paralleling historical archaeology's initial interest in documenting the slave presence through the search for Africanisms, early work in the archaeology of gender was dedicated to making women's lives more visible—"finding women" in the archaeological record. Later work turned more fully relational, studying how women and men interacted in divisions of labor and other social arrangements (e.g., contributors to Gero and Conkey 1991). Currently there is an impressive diversity of theoretical standpoints and research questions among archaeologists concerned with gender (Nelson 2006). This has led to important breakthroughs in our understanding of gender roles and strategies in the past.

Spencer-Wood's (1991, 1994, 2003) research on nineteenth- and early twentieth-century "domestic reform" sites in Boston and elsewhere explores strategies by which female domestic reformers sought to improve the conditions of women's lives by expanding their roles in both private and public spaces. She illustrates how reformers used the material world to accomplish this goal. Reformers employed a variety of material strategies to "invade" public space, or blur the boundaries between public and private, in ways conducive to expanding women's presence and influence. Institutions dedicated to domestic reform—various women's clubs, cooperative homes, YWCAs, and other voluntary organizations—were made visually dominant parts of landscape. In some instances they were purposely built as the tallest or largest building in the neighborhood. Domestic reformers also played a central role in the emerging City Beautiful Movement. Women physically shaped and exercised control over public landscapes by introducing playgrounds, children's gardens, and green spaces.

Similar "little tactics of the habitat" (Foucault 1980: 149) were applied by reformers at smaller scales. Reform activists in Boston sought to move women out of poverty by experimenting with communal built spaces and socialized housekeeping in new cooperative women's homes. Archaeological excavations at the Magdalen Society Asylum in Philadelphia indicate that mid-nineteenth-century reformers used plain and edged white ceramics with the intention of instilling in their "fallen" women residents the moral values of modesty, frugality, simplicity, and conservativism (Spencer-Wood 1994: 194). More draconian measures like massive brick walls were used to separate and protect the Magdalens from worldly temptations and other undesirable elements. But reformers were not always so heavy handed. Archaeological evidence also shows that the Magdalen Society reformers loosened up over time as evidenced by an increase in decorated ceramics in asylum assemblages and evidence for the relaxing of other rules. Reformers were also capable of yielding to reformees who themselves exercised collective agency; witness the successful lobbying of working-class women to enhance their personal privacy through the creation of more single rooms at the Chicago YWCA (Spencer-Wood 1994: 195). Spencer-Wood's work clearly shows the archaeological potential of domestic reform sites to inform about women's collective agency, and the negotiations between reformers and working-class women over how to construct women-friendly built environments.

Diana Wall's (1991, 1994, 1999) work in New York City also focuses on the dynamics of gender, class, and materiality. Her study of ceramic assemblages of working- and middle-class households in nineteenth-century Greenwich Village illuminates class-based differences in consumer patterns in ways that disclose female collective agency (Wall 1999). Wall interprets middle- and upper-class use of Gothic twelve-sided ironstone plates as related to the perceived role of women as guardians of a family's and society's morals. An Italianate style that paralleled the genteel style of middle-class architecture is interpreted in the same way; the style created good moral character and good people. In contrast, working-class households used a whole array of molded designs absent from middle- and upper-class assemblages. While the meaning of this variation is not entirely clear, it is certain that working-class people were not emulating middle- and upper-class understandings of women as moral guardians of the home.

Wall (1991) also compared the teaware from a working-class family to the teaware from a middle-class family. Both households had plain, paneled "Gothic" wares that were similar to their tableware. The two households differed in that the middle-class family had a second set of decorated porcelain teaware. Wall associates the two kinds of teaware with use in different social settings: morning and afternoon tea. Morning tea was a family affair, while afternoon tea was a venue for socializing with community members. She suggests that middle-class women had greater investment in displaying their status as way to impress upon friends the refinement and gentility of their families, and elevate their family's position in the class structure. Lower-class women lacked this interest. Instead, sharing tea may have been a way to create and affirm cooperative social relations. Rather than asserting their status through decorated porcelain teaware, working-class women created community by using plain wares that did not elicit competition (1991: 79).

Margaret Wood's (2002) study of working-class women in the Colorado Fuel and Iron Company coal mining town of Berwind in southeastern Colorado illustrates how women contribute to household economies in ways that make collective action possible; in this case the great Coal Field Strike of 1913–1914 (McGovern and Guttridge 1972). It is well known that labor strikes are hatched as much at the kitchen table as they are at the points of industrial production (i.e., on assembly lines and in the shafts). They are family affairs. Domestic trash at Berwind dating before the 1913 strike contains high frequencies of tin cans, large cooking pots, and big serving vessels. Mass-produced tin cans—especially large ones—represent 52 percent of all metal vessels recovered. In contrast, food storage vessels such as home canning jars represent only 1 percent of all metal artifacts. At this time it is known that coal town households routinely took in single male miners as boarders to make ends meet, given the very low wages paid by the coal company. Census records indicate that at Berwind in 1910, 53 percent of all nuclear families had one or more unrelated persons boarding in their homes (Wood 2002: 73). On average there were three boarders per household. Thus, archaeological evidence suggests that before the strike, women used store-bought canned foods to make stews and soups to feed the household. Wood calculates that through this activity women accounted for about 25 percent of the household's total income. This activity also likely provided more variety in fruits and vegetables for the woman's own family.

After the strike the Colorado Fuel and Iron Company strongly discouraged—or, in Wood's (2002: 77) words, "waged a quiet war on"—boarding as way to reduce worker opportunities for building collective solidarity. The company established and operated its own boardinghouses so that the behavior of single male miners could be more tightly controlled. Census records indicate that in 1920 the number of families taking in boarders had shrunk to 6 percent. Mining families no longer had income from boarders, and wages continued to remain very low. This forced some new strategizing by women on the homefront. Excavation in poststrike contexts revealed significant differences in household artifact assemblages that reflect changed strategies. Big pots and cans decrease in the trash and glass canning jars and lids increase. Mass-produced tin cans decrease to 38 percent of the total, while home canning jars increase to 29 percent. There is a significant increase—a doubling and tripling from prestrike levels—of glass food preparation bottles, such as catsup, mustard, and pepper sauce. These numbers indicate that women were doing much more home food production after the strike in order to provide for their families. Poststrike deposits also show an increase in the bones of rabbits and chickens, as well as an increase in fencing wire. The latter likely reflects more gardening related to the home production of canned foods.

Wood's analysis thus opens a window onto the shared existential realities and anxieties of women that were likely instrumental in creating interfamily ties of mutual support and assistance. These alliances would have paralleled those formed among men in the mine shafts. Both kinds of solidarity would have been required for organizing and sustaining the strike of 1913–1914 (see also Long 1985).

Finally, Amy Young's (2003) analyses of antebellum plantation landscapes show how African American women and men used different strategies to provide for their families and build community solidarity. Women at Locust Grove Plantation near Louisville, Kentucky, worked the spaces between slave houses and the communal yard between rows of houses. They conducted generalized reciprocal exchanges of items such as decorated ceramics, glass tableware, buttons, and other objects. Archaeological recovery of matched ceramic items and other artifacts from different houses indicate that they were shared out or given as gifts among the slave families. These reciprocal relations established bonds of kinship that helped the community cope with the predations and deprivations of slavery. They ensured the future of children whose parents were sold away, provided emotional support during periods of sickness and solace upon the death of a family member, and reached out to new slaves entering the community.

Young (2003) also considered male roles at Saragossa Plantation in Adams County, Mississippi, just outside Natchez (see also Young, Tuma, and Jenkins 2001). Here ethnographic, historic, and archaeological evidence converge to indicate the strategic importance of male hunting in slave communities. At Saragossa males worked the fields, forests, and streams beyond the slave quarters and the communal yard. Male hunting of small game (squirrel, raccoon, rabbit) and some deer provided sustenance for the community. This was likely accomplished through clandestine night hunting, as predicted by Paynter and McGuire (1991). But the hunting also had social and psychological purposes. It served to integrate newcomers into the slave community under conditions of a constantly fluctuating population. And it reinforced male self-worth (i.e., male as "breadwinner") in a deeply emasculating slave system. Together, these different female and male activities strengthened the entire slave community.

Class

Several important studies have shown how workers struggle with industrial capitalists over the conditions under which their labor is appropriated and compensated and its products distributed. Paynter and McGuire (1991) is a key source for much interpretive theory in this area. They note how collective resistance by workers in an industrial setting can take many forms including malingering, sabotage of machinery, and destruction of products, strategies that can all have archaeological correlates.

Nassaney and Abel (1993, 2000) investigated such strategies in the Connecticut River Valley of western Massachusetts. They analyzed material remains at the John Russell Cutlery Company in Turner's Falls, one of world's leading nineteenth-century knife manufacturers. Relocated from Greenfield and opened in 1870, the Turner's Falls plant was a prototype modern cutlery factory. Major modernization in the 1880s was informed by new techniques of managing work that separated product conception and production, subdivided the process of production, and standardized production tasks. These techniques degraded human labor by deskilling the work force (Braverman 1974). Archaeologists found a large quantity of artifacts related to primary production along the factory's riverbank. Discarded materials included inferior and imperfectly manufactured parts from various stages of the production process. Nassaney and Abel interpret this material as the residue of worker contempt toward, and defiance of, the new system of closely regulated work discipline. Workers may have intentionally spoiled knives—a kind of industrial sabotage—as way to assert some degree of autonomy on the shop floor. Documentary evidence suggests that the historical context was exactly right for expecting such action. Declining real wages, deteriorating work conditions, and layoffs produced frequent disputes between managers and workers in the late nineteenth century.

Shackel (2000, 2004) offers similar sorts of insights in his study of nineteenth-century sites in Harpers Ferry, West Virginia. Here, renovation of the local beer bottling works revealed hundreds of bottles accumulated in the factory walls and in the basement of the building's elevator shaft. All bottles date between 1893 and 1909. Working conditions at this time were deplorable: workers suffered fourteen-hour days and exposure to dramatic temperature swings and noxious acids. Accident rates were 30 percent higher than in other trades. Evidence from the walls and shaft suggests that workers intentionally and covertly *consumed* the products of their labor, and concealed their subversive behavior by disposing otherwise reusable bottles out of the view of their supervisors. These workers were, in effect, defying industrial discipline by drinking the owner's profits.

Shackel (2000) also compared household assemblages of managers and workers employed at the local armory during the mid-nineteenth-century transition from piecework to wage labor. Archaeological excavations revealed differences between managers and wage laborers in the consumption of tablewares. The houses of managers displayed the latest goods including pearlwares, whitewares, and ceramics with shell and transfer print designs. Managers were thus fully embracing the consumer culture associated with industrialization. On the other hand, houses of wage laborers contained unfashionable, out-of-date goods like creamwares and shell-edged ceramics. Shackel suggests that this working-class purchasing behavior was purposeful, motivated by a nostalgic longing for the "good old days" when family members had more control over their everyday lives. The assemblages recall a time when husbands were craftsmen, and when wives had better access to markets. Working-class men and women thus exercised agency in a way that critiqued the new industrial system.

The work of Beaudry, Cook, and Mrozowski (1991; Beaudry and Mrozowski 2002) at Boott Mills in Lowell, Massachusetts, explores how nineteenth- and twentieth-century workers expressed class identity and personal aspirations in a tightly managed environment. Lowell was the nation's first mass industrial city, and corporate paternalism loomed large. Lowell is the archetypal example of town planning for social control, and it provided a model that was emulated elsewhere. Industrialists in Lowell incorporated landscape as an active element in the reinforcement of social class distinctions. They located the textile mill, worker housing, and manager housing close together as a way to maximize surveillance and control and accentuate hierarchical structure. The construction of standardized worker housing with rooms of uniform size and shape would have sent a message of worker expendability and interchangeability, thereby producing compliance with the status quo. In contrast, managers' houses were distinguished by higher-quality facing materials and fashionable interiors (Mrozowski 1991).

Excavations in the backlots of typical boardinghouses, however, produced abundant evidence of worker noncompliance with the strict social order. Despite their limited power and economic means, workers were apparently creating their own identities and building up a "subculture" of resistance. An abundance of medicine bottles suggests consumption for alcohol content, as way to defy company discouragement of drinking and other efforts to control workers' leisure. Workers also created another distinctive category of pipes—short-stemmed white clay pipes—to express membership and pride in the working class. But workers were not entirely rejecting the notion of upward class mobility. Aspirations in this direction are indicated by ceramics suggesting middle-class dining habits and inexpensive costume jewelry that imitates costlier "class-conscious" items.

Our own work at the Ludlow Tent Colony in southeastern Colorado shows class-based collective agency manifested in a number of different ways (Saitta 2007; Larkin and McGuire 2009). The Ludlow Colony was occupied by the families of striking coal miners during Colorado's 1913–1914 coal field troubles. Many of Ludlow's occupants likely came from the coal camp of Berwind, discussed above. On April 20, 1914, the Ludlow Colony—numbering over 100 tents and about 200 people—was burned, and a couple dozen occupants killed, by an armed force of company gunmen and hired mercenaries. The attack appeared intended to break the long and acrimonious coal strike, and came to be known as the Ludlow Massacre. Archaeological work has aimed to clarify the everyday strategies of survival, social integration, and public image making crucial to the success of collective labor action. For example, the layout of the colony on the open Colorado prairie appears to have been strategic (Jacobson 2002, 2006). Family tents were laid out at a 45-degree angle to the east–west section road, running southwest–northeast rather than parallel. This diagonal arrangement would have restricted a passer-by's ability to peer into the colony, essentially terminating their view at the perimeter line of tents. Such concern for privacy is not surprising given the colony's exposed location in a larger landscape and the fact that it was subject to search by the state militia and other local authorities looking to keep the peace between striking miners and armed coal company operatives.

A collective concern to present an image of order and solidarity to an outside, "Progressive Era" world that often disparaged immigrant miners as volatile, uncivilized foreigners was also paramount. The colony contained numbered tents and named streets and featured a prominently located communal meeting place and medical facility. A baseball field for playing America's pastime was laid out directly across the section road. Within the colony a significant number of excavated artifacts reflect strong ethnic affiliations, including buttons inscribed with Habsburg eagles, embossed bottles from Italian and Croatian cities, and a suspender part bearing, in Italian, the inscription "Society of Tyrolean Alpinists." However, there is nothing in the distribution of these objects to suggest that the colony had ethnically distinct precincts. The public image presented was one of social order, unity, and solidarity.

Tent artifact assemblages at Ludlow offer insight into other strategies for building collective unity and solidarity out of social and cultural differences. Like the workers at Boott Mills discussed above and immigrant workers generally, mining families striking at Ludlow may have been expressing their aspirations for upward mobility in their new country with material culture. Ludlow colonists were aware of American middle-class values that prescribed elaborate matched table settings and formal teawares (Gray 2005). We have found in Ludlow deposits matched or near-matched sets of teaware having floral designs with gilded accents and embossed pieces. A child's tea set has also been excavated, and was likely used for teaching these middle-class values. But while the occupants of Ludlow's tents possessed the material culture that symbolized and transmitted traditions of tea taking, they did not necessarily fully embrace this tradition. A set of demitasse cups was excavated from one tent cellar, suggesting that the occupants also consumed espresso or coffee. According to Mary Thomas, a survivor of the massacre, she and her neighbors regularly shared coffee (O'Neal 1971).

Ludlow strikers thus may have sought to convey civility by using finely decorated vessels, but they did so on their own terms. They used their fine teawares to convey a message of gentility and upward mobility while perhaps maintaining a cultural preference for coffee. Through their daily practice, they negotiated a balance between traditional cultural values and those attached to American middle-class status. The stratigraphic context of the decorated and undecorated wares in one excavated tent cellar also suggests conscious strategizing to build class solidarity out of social and cultural difference. Most of the decorated vessels were recovered from *below* the charred floorboards in the feature fill, in the deepest part of the cellar. During the final excavation of the cellar it was noted that many of the vessels, including the decorated teaware, were associated with metal hardware and wood fragments. This suggests that they were stored in a piece of furniture for safekeeping. The demitasse set was also excavated in this context. In contrast, most of the plain ware was removed from the strata *above* the floorboards. This stratigraphic positioning suggests that the household used plainware most frequently in their daily practice, while reserving decorated vessels and the demitasse set for special occasions. These practices would parallel those reconstructed by Wall (1991) for her working-class families in Greenwich Village. In both social contexts the use of plainware would not have elicited the sort of comparison and competition that could threaten community solidarity. That the Ludlow strikers chose to store their decorated and loosely matched teaware—as well as the demitasse set, an object perhaps most loaded with an ethnic "charge"—reflects both the value they placed on those objects and their commitment to building community solidarity. They were not totally rejecting Americanizing influences, but rather negotiating a careful balance between American and Old World identities that would serve the cause of collective action.

CONCLUSIONS FOR AN ARCHAEOLOGY OF COOPERATIVE BEHAVIOR

Historical archaeologists have done fine work illuminating race, gender, and class-based forms of collective action in the past. They have theorized the social and economic conditions under which collective action, in both slave-based and capitalist modes of production, is expected to occur. They have documented specific strategies that disenfranchised and marginalized people in various political and economic circumstances used to cope with social inequality and oppression. They have identified specific material cultures that helped to galvanize group cooperation and symbolize group identity in particular times and places. Their work is rich in theoretical and methodological implications for studying not only the modern capitalist world, but also organizational variation and change in the ancient, prehistoric world.

The question remains: Can the organizing epistemologies and theories of historical archaeology—or what we might more broadly term historical anthropology—be usefully squared with those that inform evolutionary anthropology? Or does something important get lost in the bargain? Certainly there are a number of contentious issues that divide these paradigms (see O'Brien and Lyman 2004; Pauketat 2004). On the specific issue of cooperative behavior, evolutionary approaches constitute a diverse lot (Shennan 2008). Still, some squaring of theoretical commitments is possible (e.g., O'Brien and Lyman 2000). Evolutionary and historical anthropologies both recognize that humans have evolved capacities for cooperative behavior. Both recognize that material conditions (e.g., a shared experience of economic misery and deprivation) can be powerful spurs to collective action. Such parallels and convergences between evolutionary and historical approaches to understanding human social life and change go back, arguably, to the Boasians. Boas is often presented in histories of anthropology as a severe critic of all forms of evolutionary thought. Alternatively, Lewis (2001) persuasively argues that Boas was a "historicist" in the same sense as Darwin. That is, he was aware of, and sympathetic to, a Darwinian model of change recognizing that the world is open, diverse, undetermined, and shaped by historical contingency as well as human agency (see also O'Brien and Lyman 2000).

Recognizing these convergences and overlaps may help explain why substantive inferences about the past produced by many evolutionary archaeological frameworks are consistent with those produced by any number of other, nonevolutionary archaeological frameworks (Saitta 2002). Neiman's (2008) evolutionary interpretation of subfloor pits in slave houses at Monticello and throughout the wider Chesapeake as "safe deposit boxes" used by enslaved people to increase the security of their food supply strikes me as fully consistent with historical and "agentic" approaches to understanding the past. So too is Galle's (2010) interpretation of the metal buttons and refined ceramic wares used by eighteenth-century Chesapeake slaves as "signals" that communicated to potential allies the owner's personal skills, purchasing power, social mobility, and knowledge of the outside world. Galle recognizes that evolutionary theory and agency theory can speak to and even harmonize with each other. Regrettably, however, she caricatures the latter as relativist, subjectivist, and too often disinterested in archaeological data, when quite the opposite is the case.

There can be significant differences, however, with respect to the larger ambitions of those anthropologies geared to producing knowledge of cooperative action today and in the past. There is a much greater likelihood that practitioners of historical anthropology will orient their work toward using knowledge of world to intervene in the world; that is, to accomplish not only *explanatory* but also *emancipatory* work (Saitta 2008). Orser (2010) notes that "giving voice to the voiceless" is a major strength of historical archaeology, and that an increasing number of scholars are recognizing the political nature of their work. McGuire (2008) describes this critical, engaged approach as turning on the Marxist notion of "praxis": a commitment to know, critique, and change the world. Preucel and Mrozowski (2010) describe it as constituting a "New Pragmatism" in archaeological inquiry, one informed by the work of Boas's contemporaries William James and John Dewey, among others. However described, this approach to inquiry frames and justifies research questions and theories based on their relevance to society today. It prioritizes their accessibility to public as well as scholarly audiences. It understands that evaluation of competing ways of knowing the past must be made on pragmatic grounds; that is, on the extent to which theories and interpretations of the past serve perceived human need. This critical, activist edge is much less apparent in the evolutionary anthropologies than in the historical anthropologies even if some practitioners of the former are sympathetic to the cause (e.g., Galle 2010: 21). Interestingly, Lewis (2001) sees Deweyian pragmatism to be as much of an organizing influence on Boasian anthropology as Darwinian evolutionism. It stands to reason that such an orientation would require a distinctive—and perhaps incommensurable—set of organizing concepts, metaphors, analogues, and heuristics.

Historical anthropologists have no illusions that their work will change the world. As McGuire (2008) notes, there are better ways of accomplishing social change than by doing archaeology. But at the same time we should not minimize the potential of public scholarship for producing critical thought about how the contemporary world came to be and how alternative arrangements for organizing human social life have different consequences and effects in the world. Orser (1998: 76) asserts that the results of historical archaeology such as those described in this chapter "have potential meaning for all people seeking to understand how the social inequalities of today were materially expressed in the past." Our scholarly and public outreach work with the descendant community of coal miners and trade unionists living in towns around the Ludlow Massacre Memorial in southeastern Colorado amply illustrates the truth of Orser's claim (McGuire 2004; Saitta 2007, especially chapter 7; see also Shackel 2009). Turning critical thought into collective action that seriously challenges and eliminates the various social and institutional inequalities that bedevil us is another matter. History tells us that the potential for group cooperation and solidarity that springs from a common experience of class can all too easily be eroded by the lived experiences of race, ethnicity, nationality, gender, and sexuality (and vice versa). Thus, better understanding of how the intersection of these and other potentially divisive social identities complicates collective action remains the major challenge facing an engaged historical anthropology.

REFERENCES

Barrett, John. 2001. "Agency, the Duality of Structure, and the Problem of the Archaeological Record." In *Archaeological Theory Today*, ed. Ian Hodder, 141–164. Cambridge: Polity Press.

Beaudry, Mary, Lauren Cook, and Stephen Mrozowski. 1991. "Artifacts and Active Voices: Material Culture as Social Discourse." In *The Archaeology of Inequality*, ed. Randall McGuire and Robert Paynter, 150–191. Oxford: Basil Blackwell.

Beaudry, Mary, and Stephen Mrozowski. 2002. "Cultural Space and Worker Identity in the Company City: Nineteenth Century Lowell, Massachusetts." In *The Archaeology of Urban Landscapes: Explorations in Slumland*, ed. A. Mayne and Tim Murray, 118–131. Cambridge: Cambridge University Press.

Braverman, Harry. 1974. *Labor and Monopoly Capital.* New York: Monthly Review Press.

Brighton, Stephen. 2004. "Symbols, Myth-Making, and Identity: The Red Hand of Ulster in late Nineteenth-Century Paterson, New Jersey." *International Journal of Historical Archaeology* 8 (2): 149–64. http://dx.doi.org/10.1023/B:IJHA.0000043699.88179.13.

Brubaker, Rogers, and Frederick Cooper. 2000. "Beyond "Identity"." *Theory and Society* 29 (1): 1–47. http://dx.doi.org/10.1023/A:1007068714468.

Brumfiel, Elizabeth. 2003. "It's a Material World: History, Artifacts, and Anthropology." *Annual Review of Anthropology* 32 (1): 205–23. http://dx.doi.org/10.1146/annurev.anthro.32.061002.093335.

Delle, James, Stephen Mrozowski, and Robert Paynter, eds. 2000. "Introduction." In *Lines That Divide: Historical Archaeologies of Race, Class, and Gender*, ed., xi–xxxi. Knoxville: University of Tennessee Press.
Dobres, Marcia-Ann, and John Robb, eds. 2000. *Agency in Archaeology*. London: Routledge.
Dornan, Jennifer. 2002. "Agency and Archaeology: Past, Present, and Future Directions." *Journal of Archaeological Method and Theory* 9 (4): 303–29. http://dx.doi.org/10.1023/A:1021318432161.
Fabian, Johannes. 1994. "Ethnographic Objectivity Revisited: From Rigor to Vigor." In *Rethinking Objectivity*, ed. Alan Megill, 81–108. Durham: Duke University Press.
Ferguson, Leland. 1991. "Struggling with Pots in South Carolina." In *The Archaeology of Inequality*, ed. R. Paynter and R. McGuire, 28–39. London: Blackwell.
Ferguson, Leland. 1992. *Uncommon Ground: Archaeology and Early African America*. Washington, D.C.: Smithsonian Institution Press.
Ferguson, Leland. 1999. "The Cross is a Magic Sign": Marks on Eighteenth-Century Bowls from South Carolina. In *"I Too Am America": Archaeological Studies of African-American Life*, ed. Teresa Singleton, 116–131. Charlottesville: University Press of Virginia.
Foucault, Michel. 1980. *Power and Knowledge*. New York: Pantheon.
Galle, Jillian. 2010. "Costly Signaling and Gendered Social Strategies Among Slaves in the Eighteenth-Century Chesapeake: An Archaeological Perspective." *American Antiquity* 75:19–43.
Gell, Alfred. 1998. *Art and Agency: An Anthropological Theory*. Oxford: Clarendon Press.
Gero, Joan, and Margaret Conkey, eds. 1991. *Engendering archaeology*. Oxford: Basil Blackwell.
Gray, Amie. 2005. *Contested Ideals: Cultural Citizenship at the Ludlow Tent Colony*. Unpublished Master's Thesis, Department of Anthropology, University of Denver.
Hall, Martin, and Stephen W. Silliman, eds. 2006. *Historical Archaeology*. Oxford: Blackwell Publishing.
Harvey, David. 1973. *Social Justice and the City*. London: Arnold.
Harvey, David. 2000. *Spaces of Hope*. Berkeley: University of California Press.
Hodder, Ian. 1982. *Symbols in Action*. Cambridge: Cambridge University Press.
Hodder, Ian. 1999. *The Archaeological Process*. Oxford: Blackwell.
Hodder, Ian. 2004. "The "Social" in Archaeological Theory: An Historical and Contemporary Perspective." In *A Companion to Social Archaeology*, ed. Lynn Meskell and Robert Preucel, 23–42. Oxford: Blackwell.
Hodder, Ian, and Scott Hutson. 2003. *Reading the Past*. 3rd ed. Cambridge: Cambridge University Press.
Jacobson, Michael. 2002. Ideological Clash: A Study of Experience in the Colorado Coalfield War 1913–1914. Unpublished Master's Thesis, Department of Anthropology, Binghamton University, New York.
Jacobson, Michael. 2006. The Rise and Fall of Place: The Development of a Sense of Place and Community in Colorado's Southern Coalfields, 1890–1930. Unpublished Doctoral Dissertation, Department of Anthropology, Binghamton University, New York.
Johnson, Matthew. 1989. "Conceptions of Agency in Archaeological Interpretation." *Journal of Anthropological Archaeology* 8 (2): 189–211. http://dx.doi.org/10.1016/0278-4165(89)90024-X.
Knapp, A. Bernard, and Peter van Dommelen. 2008. "Past Practices: Rethinking Individuals and Agents in Archaeology." *Cambridge Archaeological Journal* 18 (01): 15–34. http://dx.doi.org/10.1017/S0959774308000024.

Larkin, Karen, and Randall McGuire, eds. 2009. *The Archaeology of Class War.* Boulder: University Press of Colorado.
Leone, Mark. 2005. *The Archaeology of Liberty in an American Capital: Excavations in Annapolis.* Berkeley: University of California Press.
Leone, Mark, and Parker Potter, Jr., eds. 1988. *The Recovery of Meaning: Historical Archaeology in the Eastern United States.* Washington, D.C.: Smithsonian Institution Press.
Leone, Mark, and Parker Potter, Jr. 1999. *Historical Archaeologies of Capitalism.* New York: Plenum Press.
Lewis, Herbert S. 2001. "Boas, Darwin, Science, and Anthropology." *Current Anthropology* 42 (3): 381–406. http://dx.doi.org/10.1086/320474.
Long, Priscilla. 1985. "The Women of the Colorado Fuel and Iron Strike." In *Women, Work, and Protest: A Century of U.S. Women's Labor History*, ed. R. Milkman, 62–85. London: Routledge and Kegan Paul.
McGovern, George, and Leonard Guttridge. 1972. *The Great Coalfield War.* Boston: Houghton Mifflin.
McGuire, Randall. 2004. "Colorado Coalfield Massacre." *Archaeology* 57 (6): 62–70.
McGuire, Randall. 2008. *Archaeology as Political Action.* Berkeley: University of California Press.
McGuire, Randall, and Robert Paynter, eds. 1991. *The Archaeology of Inequality.* London: Basil Blackwell.
McGuire, Randall, and LouAnn Wurst. 2002. "Struggling With the Past." *International Journal of Historical Archaeology* 6 (2): 85–94. http://dx.doi.org/10.1023/A:1016080520656.
Meskell, Lynn, and Robert Preucel. 2004. "Identities." In *A Companion to Social Archaeology*, ed. Lynn Meskell and Robert Preucel, 121–141. Oxford: Blackwell Publishing.
Mrozowski, Stephen. 1991. "Landscapes of Inequality." In *The Archaeology of Inequality*, ed. Randall McGuire and Robert Paynter, 79–101. Oxford: Basil Blackwell.
Mullins, Paul. 1999. *Race and Affluence: An Archaeology of African America and Consumer Culture.* New York: Plenum.
Mullins, Paul. 2004. "Ideology, Power, and Capitalism: The Historical Archaeology of Consumption." In *A Companion to Social Archaeology*, ed. Lynn Meskell and Robert Preucel, 195–211. Oxford: Blackwell. http://dx.doi.org/10.1002/9780470693605.ch9.
Nassaney, Michael, and Marjorie Abel. 1993. "The Political and Social Contexts of Cutlery Production in the Connecticut Valley." *Dialectical Anthropology* 18 (3-4): 247–89. http://dx.doi.org/10.1007/BF01303678.
Nassaney, Michael, and Marjorie Abel. 2000. "Urban Spaces, Labor Organization, and Social Control: Lessons from New England's Nineteenth-Century Cutlery Industry." In *Lines That Divide: Historical Archaeologies of Race, Class, and Gender*, ed. James Delle, Stephen Mrozowski, and Robert Paynter, 239–275. Knoxville: University of Tennessee Press.
Neiman, Frazier. 2008. "The Lost World of Monticello: An Evolutionary Perspective." *Journal of Anthropological Research* 64:161–93.
Nelson, Sarah, ed. 2006. *Handbook of Gender in Archaeology.* Lanham, MD: Altamira.
O'Brien, Michael, and R. Lee Lyman. 2000. "Darwinian Evolutionism is Applicable to Historical Archaeology." *International Journal of Historical Archaeology* 4 (1): 71–112. http://dx.doi.org/10.1023/A:1009556427520.
O'Brien, Michael, and R. Lee Lyman. 2004. "History and Explanation in Archaeology." *Anthropological Theory* 4 (2): 173–97. http://dx.doi.org/10.1177/1463499604042813.
O'Neal, Mary T. 1971. *Those Damn Foreigners.* Minerva, CA: Hollywood Press.

Orser, Charles. 1996. *A Historical Archaeology of the Modern World*. New York: Plenum Press.
Orser, Charles, Jr. 1998. "The Archaeology of the African Diaspora." *Annual Review of Anthropology* 27 (1): 63–82. http://dx.doi.org/10.1146/annurev.anthro.27.1.63.
Orser, Charles. 2010. "Twenty-First-Century Historical Archaeology." *Journal of Archaeological Research* 18 (2): 111–50. http://dx.doi.org/10.1007/s10814-009-9035-9.
Pauketat, Timothy. 2004. "Archaeology without Alternatives." *Anthropological Theory* 4 (2): 199–203. http://dx.doi.org/10.1177/1463499604042814.
Paynter, Robert. 2000. "Historical Archaeology and the Post-Columbian World of North America." *Journal of Archaeological Research* 8 (3): 169–217. http://dx.doi.org/10.1023/A:1009455104130.
Paynter, Robert, and Randall McGuire. 1991. "The Archaeology of Inequality: Material Culture, Domination, and Resistance." In *The Archaeology of Inequality*, ed. Randall McGuire and Robert Paynter, 1–27. Cambridge: Basil Blackwell.
Preucel, Robert, and Stephen Mrozowski eds. 2010. *Contemporary Archaeology in Theory: The New Pragmatism*. Oxford: Wiley-Blackwell.
Saitta, Dean J. 2002. Review of Archaeological Theory Today, ed. Ian Hodder. *Journal of Anthropological Research* 58: 419–420.
Saitta, Dean J. 2005. "Dialoguing with the Ghost of Marx: Mode of Production in Archaeological Theory." *Critique of Anthropology* 25 (1): 27–35. http://dx.doi.org/10.1177/0308275X05048611.
Saitta, Dean J. 2007. *The Archaeology of Collective Action*. Gainesville: University Press of Florida.
Saitta, Dean J. 2008. "Ethics, Objectivity, and Emancipatory Archaeology." In *Archaeology and Capitalism: From Ethics to Politics*, ed. Philip Duke and Yannis Hamilakis, 267–80. Walnut Creek, CA: Left Coast Press.
Scott, Elizabeth, ed. 1994. *Those of Little Note: Gender, Race, and Class in Historical Archaeology*. Tucson: University of Arizona Press.
Shackel, Paul. 2000. "Craft to Wage Labor: Agency and Resistance in American Historical Archaeology." In *Agency in Archaeology*, ed. Marcia-Ann Dobres and John Robb, 232–246. London: Routledge.
Shackel, Paul. 2004. "Labor's Heritage: Remembering the American Industrial Landscape." *Historical Archaeology* 38:44–58.
Shackel Paul. 2009. *The Archaeology of American Labor and Working Class Life*. Gainesville: University Press of Florida.
Shackel, Paul. 2010. "Identity and Collective Action in a Multiracial Community." *Historical Archaeology* 44:58–71.
Shennan, Stephen. 2008. "Evolution in Archaeology." *Annual Review of Anthropology* 37 (1): 75–91. http://dx.doi.org/10.1146/annurev.anthro.37.081407.085153.
Singleton, Teresa. 1995. "The Archaeology of Slavery in North America." *Annual Review of Anthropology* 24 (1): 119–40. http://dx.doi.org/10.1146/annurev.an.24.100195.001003.
Singleton, Teresa. 2005. "Before the Revolution: Archaeology and the African Diaspora on the Atlantic Seaboard." In *North American Archaeology*, ed. Timothy Pauketat and Diana Loren, 319–336. Oxford: Blackwell.
Singleton, Teresa, and Mark Bograd. 2000. "Breaking Typological Barriers: Looking for the Colono in Colonoware." In *Lines That Divide: Historical Archaeologies of Race, Class, and Gender*, ed. James Delle, Stephen Mrozowski, and Robert Paynter, 3–21. Knoxville: University of Tennessee Press.
Slaughter, Michelle Ann. 2006. An Archaeological and Ethnographic Examination of the Presence, Acquisition, and Consumption of Sake at Camp Amache, a World

War II Japanese Internment Camp. Unpublished Master's Thesis, Department of Anthropology, University of Colorado at Denver.
Spencer-Wood, Suzanne. 1991. "Toward an Historical Archaeology of Materialistic Domestic Reform." In *The Archaeology of Inequality*, ed. R. McGuire and R. Paynter, 231–286. Oxford: Blackwell.
Spencer-Wood, Suzanne. 1994. "Diversity and Nineteenth-Century Domestic Reform: Relationships among Classes and Ethnic Groups." In *Those of Little Note: Gender, Race, and Class in Historical Archaeology*, ed. Elizabeth Scott, 175–208. Tucson: University of Arizona Press.
Spencer-Wood, Suzanne. 2003. "Gendering the Creation of Green Urban Landscapes in America at the Turn of the Century." In *Shared Spaces and Divided Places*, ed. Deborah Rotman and Ellen Savulis, 24–61. Knoxville: University of Tennessee Press.
Thomas, Julian. 2000. "Reconfiguring the Social, Reconfiguring the Material." In *Social Theory in Archaeology*, ed. Michael Schiffer, 143–155. Salt Lake City: University of Utah Press.
Upton, Dell. 1996. Ethnicity, Authenticity, and Invented Traditions. *Historical Archaeology* 30: 1–7.
Van Bueren, Thad, ed. 2002. *Communities Defined By Work: Life in Western Work Camps. Historical Archaeology* 36(3).
Wall, Diana d.Z. 1991. "Sacred Dinners and Secular Teas: Constructing Domesticity in Mid-19th-Century New York." *Gender in Historical Archaeology*, ed. Donna Seifert. *Historical Archaeology* 25(4): 69–81.
Wall, Diana d.Z. 1994. *The Archaeology of Gender: Separating the Spheres in Urban America*. New York: Plenum.
Wall, Diana d.Z. 1999. "Examining Gender, Class, and Ethnicity in Nineteenth-Century New York City." *Historical Archaeology* 33:102–18.
Wood, Margaret. 2002. "A House Divided: Changes in Women's Power Within and Outside the Household, 1900–1930." In *The Dynamics of Power*, ed. Maria Donovan, 66–87. Carbondale, Illinois: Center for Archaeological Investigations.
Young, Amy. 2003. "Gender and Landscape: A View from the Plantation Slave Community." In *Shared Spaces and Divided Places*, ed. Deborah Rotman and Ellen Savulis, 104–134. Knoxville: University of Tennessee Press.
Young, Amy, Michael Tuma, and Cliff Jenkins. 2001. "The Role of Hunting to Cope with Risk at Saragossa Plantation, Natchez, Mississippi." *American Anthropologist* 103 (3): 692–704. http://dx.doi.org/10.1525/aa.2001.103.3.692.

PART II

Case Studies

7

Free-Riding, Cooperation, and Population Growth

The Evolution of Privatization and Leaders in Owens Valley, California

Jelmer W. Eerkens

As discussed in the opening chapter (Carballo, chapter 1), the evolution of cooperation among humans is a topic that continues to receive intense research by social and evolutionary scientists alike. Many economic, biological, and political science models examine the evolution of cooperation from a theoretical viewpoint (e.g., Axelrod 1997; Axelrod and Hamilton 1981; Bird 1999; Bowles and Gintis 2004; Boyd, Gintis, and Bowles 2010; Boyd and Richerson 2009; Dawkins 1976; Gardner and West 2010; Henrich and Boyd 2001; Nowak 2006; Trivers 1971; Winterhalder 1986, 1997). These models provide testable hypotheses for social scientists involved in lab experiments and ethnographic research, to examine short-term evolutionary processes in cooperation (e.g., Borgerhoff Mulder et al. 2010; Bowles, Smith, and Borgerhoff Mulder 2010; Gurven and Winking 2008; Henrich et al. 2004, 2010; Shariff and Norenzayan 2007). However, only archaeology can provide the data to test hypotheses about the long-term evolutionary consequences of cooperation in specific social, environmental, and temporal settings. Strangely, the evolution of cooperation, by comparison, has not been extensively studied by archaeologists.

All of the chapters in this volume show that cooperation is a complex notion that varies across many dimensions. Two individuals may cooperate in one activity, for example, to increase the average per-person yield in a cooperative hunt, but compete in another, for example, in inviting guests to feasts to partake in the spoils of such a hunt (and any resulting prestige or social debt that such feasts might bring). Humans are also incredibly complex in the social networks they form, and hence, the context in which cooperation occurs (e.g., Feinman, chapter 1; Saitta, chapter 6; Smith, chapter 12). Cooperation can occur along many different but related dimensions, including biological, political, religious, and economic, among others. Archaeologists are particularly well equipped to examine patterns in economic cooperation, and I focus on that topic in this chapter. Theory suggests that the political and especially religious dimensions of cooperation ought to be at the forefront of such research as well. Unfortunately, the archaeological record is often less complete and accessible for these topics, especially in the region used as a case study here, the Owens Valley of eastern California.

The varying biological, social, political, and religious interests of people can sometimes be satisfied through individual actions. However, as discussed in all the chapters in this volume, humans are cooperative creatures, and many of our interests are more often advanced by cooperation with other social entities, for different amounts of time and at different social scales. Some cooperation (e.g., between a male and female) occurs over years to decades and results in biological reproduction and the formation of nuclear families. Other cooperation, across all members of those nuclear families (e.g., the intrahousehold level) but at the same temporal scale, facilitates the transmission of knowledge and/or the extraction of resources from the environment, commonly referred to as *sharing*. In other cases, different households within a village may cooperate for shorter amounts of time to provide defense from, or raids on, other villages (Roscoe, chapter 3; Spencer, chapter 9) or to construct water-management systems (Chabot-Hanowell and Lucero, chapter 10). As well, larger intervillage or interregional levels (which would include what archaeologists identify as exchange or trade) of cooperation can result in the formation of complex castes, states, and empires (e.g., Carballo, chapter 11; Feinman, chapter 2; Smith, chapter 12).

Because decision-making and evolutionary processes are ultimately manifested at the level of the individual, I stick as close to that scale as is archaeologically possible by examining households (see also Pluckhahn, chapter 8, for such a household-based approach). Thus, I examine differing spatial scales in economic cooperation among households, and examine data from the prehistoric Owens Valley to test hypotheses about the long-term evolution of cooperative behavior.

COOPERATION AND FREE-RIDING

Much experimental and modeling research in a range of disciplines shows that cooperative enterprises are extremely vulnerable to cheating, or free-riding (e.g., Gurven and Winking 2008; Hardin 1968; Henrich and Boyd 2001; Ostrom 1990; Pruitt and Riechert 2009; Ratnieks and Wenseleers 2005). When cooperation leads to the production of nonexcludable resources, free-riders can gain all the benefits of cooperative ventures without incurring any of the costs. Rational choices on the part of self-interested individuals who aim to gain maximum benefits at the lowest possible costs encourage free-riding in such settings. The behavior is irrational at a higher group level, where cooperation would lead to higher payoffs for the group as a whole, but makes sense to individuals seeking to maximize personal fitness. This result is also sometimes known as the "Tragedy of the Commons" problem, described by Garrett Hardin (1968), and has spurred decades of research on cooperation and individual action among social and evolutionary scientists.

In some situations, humans tolerate free-riding by non-kin. In particular, when resource package size is larger than can be used by an individual in a short amount of time, there is little benefit to hoarding. Similarly, when the costs of defending a resource outweigh the benefits of consuming it in the near (or distant) future, people will tolerate some theft (also known as *tolerated scrounging* (Bliege Bird and Bird 1997; Blurton-Jones 1984; Gurven 2006; Gurven et al. 2000; Winterhalder 1996). Likewise, free-riding may also be tolerated if the supplier is paid in an alternative currency, such as prestige or mating opportunities.

While sometimes tolerated, in most cases free-riding is a drain on energy expenditure and offers little incentive for individuals to increase production beyond the bare minimum required to survive. As a result, individuals are not inspired to undertake cooperative ventures that would otherwise increase their net production of food (or prestige or any other currency) and/or biological fitness. In these cases, free-riding is not tolerated, and individuals will seek means to limit its effects. In short, to foster larger-scale cooperation and gain the potential benefits of cooperative ventures (e.g., higher returns per unit of time, space, or energy invested), people must resolve the free-rider issue. Much theorizing and empirical research on how humans go about eliminating the effects of free-riders suggests that there are three types of solutions to the problem. Note that these types are not necessarily mutually exclusive.

The first strategy is to somehow transform nonexcludable resources into excludable ones. A common means to achieve this end is to restrict access to the resources subject to free-riding. For example, governments typically issue permits, at a small cost, to certain individuals to graze land or mine minerals. With competitors, there are incentives for free-riders to overgraze, since all users pay the environmental costs while one person gains the benefits. Likewise, with mining, there is little incentive on the part of individuals to build and maintain significant infrastructure, such as access roads or electricity lines, if all users gain the benefits. By limiting the number of competitors the number of free-riders is minimized, and there is less incentive on the part of individuals to overgraze; or, in the case of mines, there is more incentive to build and maintain the infrastructure necessary to exploit resources.

Transferring ownership over resources to individuals eliminates free-riding, in which case free-riding becomes outright theft from another. At the same time, privatization also restricts the scale of non-kin cooperation. As a result, the construction of pyramids, freeway systems, and other large-scale ventures that require the coordination and cooperation of hundreds to tens of thousands of individuals are difficult to realize under this solution.

A second solution to the free-rider problem is similar to the first, but restricts the spatial or social scale of cooperation to a level that is above the individual, but below all potential individuals who might want to access the resource. Often this is accomplished by restricting access to the resource, and cooperation in exploiting it, only among close kin. Biologists explain such cooperation and tolerance of free-riding among related individuals through reference to kin-selection theory (or inclusive fitness). In these cases, and from a genetic evolutionary perspective, individuals can support free-riders because they share many of the same genes, which will be differentially passed on by such behavior (e.g., Hamilton 1964), and free-riders who do not promote the lineage incur extra genetic costs when cheating blood relatives. More recently some have argued that the same effect can be realized among non-kin when individuals restrict cooperation to only include trustworthy individuals, ones who are unlikely to cheat. For example, costly signaling be may used to show trustworthiness (e.g., Bliege Bird, Smith, and Bird 2001; Gintis, Smith, and Bowles 2001; Hawkes 1991; Sosis 2000) and may promote reciprocal altruism, or cooperation, among non-kin. Others have questioned the validity of such models on theoretical grounds, but there is some empirical support for reciprocal altruism in ethnographic field research (e.g., Gurven 2004; Gurven et al. 2000; Trivers 1971). In any case, as with privatization, this solution severely limits the scale of cooperation and does not foster large-scale cooperative projects.

The third major solution involves increasing the costs to those who free-ride. Simulation and mathematical modeling shows that some type of physical or social punishment, or often just the threat of punishment, is usually enough to deter potential free-riders (Boyd, Gintis, and Bowles 2010; Henrich and Boyd 2001; Johnson and Bering 2006). Thus, governments can use the threat of fines, jail, or even death to enforce conformance to cooperative rules (e.g., paying taxes to construct bridges and other public infrastructure). Likewise, small-scale groups can use gossip, public humiliation, ostracization, or physical violence to punish cheaters. Punishment can even involve the threat of supernatural punishment, which may explain the important role of religion and morality in fostering cooperative behaviors (Johnson and Bering 2006; Henrich et al. 2010; Richardson and McBride 2009; Sosis 2003). The deliverance of punishment, or threat thereof, generally requires some degree of organization or coordination among cooperators. Such organization usually necessitates the presence of one or more recognized leaders with the authority and/or power to mete out punishment (O'Gorman, Henrich, and Van Vugt 2009). A potential threat to the benefits of cooperation, of course, are leaders who abuse such powers and cheat the system themselves (e.g., by taking a greater percentage of the product) or abuse their ability to punish.

The solutions that individuals and societies collectively settle upon to solve free-riding problems will vary with social, economic, and demographic conditions. One arena where archaeology can make enormous contributions to the understanding of cooperation, and evolutionary change therein, is to examine how and under what circumstances societies employ one or another strategy to cope with free-riders, and how societies transition between different production strategies. In this chapter I explore one such transition in a particular setting. However, I believe that the conditions that fostered this transition are present in the evolution of many societies. In particular, I focus on the effects of more sedentary settlement patterns and higher population densities on societies, and the choices that individuals make to cope with these conditions.

OWENS VALLEY AS A CASE STUDY

The Owens Valley is a long north–south trending valley in eastern California (Figure 7.1). Separated from western (or cismontane) California by the Sierra Nevada Mountains, which form the western boundary of the Owens Valley, this region is part of the Great Basin geographic province in North America. The valley basin is notable for the relatively large and permanent Owens River, which is fed by a dozen or more perennial creeks draining the eastern Sierra Nevada. These creeks, and the Owens River, flow through an otherwise dry high desert environment. The Owens River ends at the large, shallow, and saline Owens Lake, which has no further outlet.

The Owens Valley has a number of advantages for studying long-term cultural evolutionary processes. First, the archaeological record is relatively accessible, being neither deeply buried nor destroyed by recent urban sprawl. Second, a rich and detailed ethnographic record collected by Julian Steward (1933, 1938) and others anchors Paiute lifeways at or shortly after the time of contact with European American settlers. Third, a long history of scientific archaeological research beginning in the 1950s and 1960s (Lanning 1963; Riddell 1951; Riddell and Riddell 1956) and continuing through to the present day (e.g., Basgall 2008; Basgall and McGuire 1988; Bettinger 1975, 1999; Delacorte 1999; Gilreath and Hildebrandt 1997; Yohe 1998), provides a wealth of comparative data from a range of sites, environments, and time periods.

As in many regions, the Owens Valley witnessed two broad but related trends over the course of prehistory. First, all data indicate that population levels within the Owens Valley generally increased over time (Bettinger 1999; Meyer, Young, and Rosenthal 2009). Although there were certainly brief reversals (i.e., periods when populations decreased, or perhaps even disappeared altogether), for example, during the warmer and drier Medieval Climatic Anomaly (Jones et al. 1999; though see Basgall 2008), the general trend indicates slowly increasing numbers of inhabitants in the valley. Second, the scale of residential mobility appears to have decreased over time. Again, a gradual process of sedentarization is unlikely, but there is little doubt that the degree of residential mobility decreased markedly over time (e.g., Basgall 1989; Bettinger 1999; Eerkens, Spurling, and Gras 2008). It appears that this transition to full sedentism began about 1,500 years ago and was clearly in place by 700 years ago (Basgall 1989; Eerkens 2003).

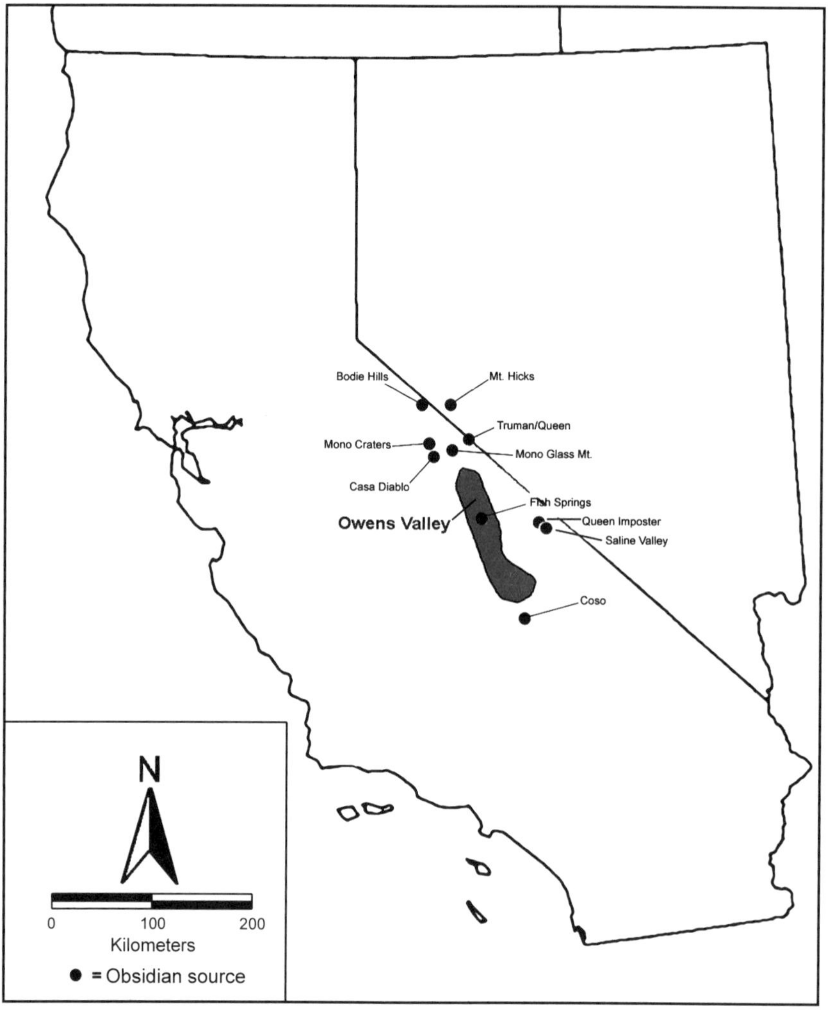

FIGURE 7.1 *Map of study area, centering on the Owens Valley and showing regional obsidian sources and surrounding geographic features.*

Why or how these two processes were correlated is irrelevant to the focus of this paper, but several archaeologists, myself included, have argued that these concurrent processes set the stage for a fundamental shift in social organization in the valley (Bettinger 1999; Delacorte 1999; Eerkens 2004; Eerkens and Spurling 2008). Prior to the shift (pre–1500 BP), it appears that hunter-gatherers traveled about in small bands, likely a small number of related households (e.g., one to four households). Although we do not currently know the kinship composition of such households in the ancient Owens Valley, based on analogies with other mobile hunting and gathering societies, they were probably composed of close kin (nuclear and extended families). In other words, these close-knit, egalitarian social groups had a high genetic coefficient of relatedness. Further, they saw each other on a daily basis, increasing trust between individuals and households, and facilitating a high degree of cooperation, cooperation that increased production and food yields compared to individuals or households working alone.

In such close-knit groups, it appears that most resources, especially food, were widely and freely shared. Thus, most goods were publicly owned and nonexcludable to other members of the group. As a result, households appear as redundant economic units, where each household has the same range of tools and subsistence remains. Strong kin-based social rules, a common world view and religion, and the threat of ostracism were likely all that was needed to discourage potential free-riders from cheating others in the social group.

As societies grew larger and settled down after 1500 BP, the coefficient of relatedness within groups would have decreased. In a growing but widely cooperative population, people would have been interacting with more and more individuals, especially unrelated ones, and interactions with any one person would have been more sporadic. Cooperating with such individuals would have been more risky for several reasons. First, because these interactions occurred *between* rather than *within* kin groups, it may have been harder for individuals to enforce social cooperative rules and/or punish cheaters. Lacking centralized leadership, the power to punish may not have extended outside local kin groups (e.g., Fried 1967; Service 1962; Steward 1938). Second, because people saw each other less often and they interacted less intensively, there was less opportunity to build trust between potential cooperators. Third, a world view encouraging free and widespread sharing may have encouraged some to free-ride at the expense of others.

Lacking solutions to the social conditions developing after 1500 BP, free-riding would have been an effective means to increase individual returns. As a result, cooperative ventures may have been destabilized and average return rates on cooperative hunting or gathering ventures may have decreased for cooperators. This situation would have lead cooperators to abandon traditional economic strategies. As discussed above, one strategy to deal with free-riding is to transform nonexcludable resources into excludable ones and restrict access to only certain members of society. This is what has been proposed in Owens Valley across the 1500 to 700 BP temporal frame as part of a "privatization" model (Bettinger 1999; Delacorte 1999; Eerkens 2004). In this sense, "privatization" refers to the process of restricting ownership of resources and the distribution thereof, especially postharvesting, to smaller numbers of individuals. In particular, these smaller numbers of individuals are assumed to reflect nuclear to extended families.

TESTING THE PRIVATIZATION MODEL

The privatization model makes some specific predictions about human behavior that can be tested against the archaeological record. Each independent and failed attempt to falsify the model should lend it greater validity. Three of these predictions are reviewed below and compared to data from the archaeological record in Owens Valley.

New Subsistence Pursuits

As populations and free-riders increase in number, cooperation in various subsistence pursuits should decrease. Because social norms about the sharing of certain foods are often well established, it may be difficult to simply assert exclusive ownership rights over traditional foods. Instead, individuals should turn increasingly to new food resources, especially those that can be individually collected, processed, and stored. The gross caloric return rates on such resources may be lower, but the net consumed return rates might be higher because foods are not lost to free-riders.

I have tested this prediction in the Owens Valley (Eerkens 2004) using data generated over the last thirty years from a number of excavations (e.g., Basgall and McGuire 1988; Bettinger 1989; Delacorte 1999; as well as still-unpublished work of my own). Flotation remains from households show that the exploitation of seeds increased around 700 BP (see Figure 7.2). A *t*-test comparing assemblages before and after 700 BP is difficult because of one very dense sample (a density over 2,000 seeds per liter), resulting in unequal and very high standard deviations between the pre– and post–700 BP samples. If this anomalous sample is removed, a one-tailed *t*-test comparing pre– and post–700 BP assemblages is significant at the 0.05 level ($p = 0.47$). However, I believe the variation within the two time periods is also significant, being much higher after 700 BP. An F-test comparing assemblage variation before and after 700 BP supports this notion and is highly significant ($p < 0.0000001$). Even removing the anomalous high-density sample, the variances are different though the probability is less significant ($p = 0.07$).

The seed data are consistent with artifacts recovered from sites in the region. At 700 BP new technologies appear, such as pottery and certain types of more portable milling stones, which facilitated the processing and storage of mass-collected seeds. All of these technologies can be produced, operated, and maintained by individuals. Likewise, the resulting products can be stored by individuals within houses, where they are also conveniently out of sight from others. Patterns in the distribution of some artifacts, especially ceramic sherds, also indicate preferential use within houses, either for cooking or storage.

The collection, processing, and storage of small seeds is unlike many other foods, where cooperation greatly increases yields. Lacking a mill powered by moving water or horses, there are no economies of scale with seed collection or processing. Moreover, while learning when and where to collect seeds is an acquired skill, the production of seed-gathering tools and the actual activity of seed harvesting and grinding are not especially learning- or skill-intensive. In other words, gathering and processing seeds as parts of cooperative groups is

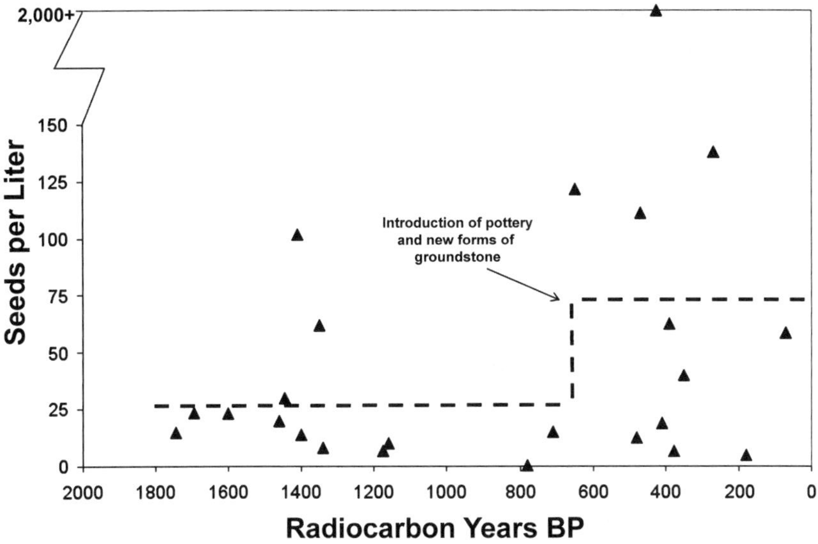

FIGURE 7.2 *Density of seeds per liter of soil from house floor assemblages in the Owens Valley, showing an increase in seed density beginning around 700 years ago, which coincides with the introduction of seed-processing material technologies. Note also the greater variation among houses after 700 BP.*

unlikely to increase per-individual yields. Thus, the transition to more intensive seed use cannot be explained by increased cooperation among individuals or households. Nor is there evidence for any marked changes in climate that would increase seed availability or decrease other higher-ranked resources. Instead, it is suggestive of decreased sharing between spatially associated household units and the increased production of "private" or nuclear-family-owned food resources.

On the other hand, cooperative drives of antelope, rabbits, or even fish can result in the capture of massive numbers of animals, far more per person on average than individual efforts to capture these animals. We do not have evidence of driving lanes in the Owens Valley, but they have been recorded in other parts of the Great Basin (e.g., Arkush 1995; Hockett and Murphy 2009; Lubinksi 1999; Raymond 1982; Thomas 1988). A careful study of the distribution of dates from such driving lanes relative to habitation sites has not been undertaken and is needed. However, it is clear that while approximately half of the driving lanes date to pre–1500 BP, the vast majority of habitation sites postdate this time frame. This suggests that while some cooperative hunting continued into late prehistoric times, the importance of cooperative hunting was much greater prior to 1500 BP.

Indeed, Bettinger (1999) has argued that the introduction of the bow and arrow around 1500 BP ushered in a new style of hunting in the Great Basin. The greater accuracy of the bow and the ability of hunters to stay nearly still during the release of a projectile facilitated individual hunting. Return rates of these hunters, while hunting alone, may have been commensurate with return rates of atlatl-using hunters working in cooperative groups. It is further possible that increased hunting proficiency led to a collapse in the large game population around 1000 BP, causing the disappearance of the well-known Coso style hunting-related rock art (e.g., Garfinkel, Young, and Yohe 2010; Gilreath and Hildebrandt 2008; Hildebrandt and McGuire 2002). Such a collapse may have made free-riding an even more attractive option to hungry bow hunters.

Of course, a focus on small seeds, or even individually hunted prey taken by bow and arrow, does not mean that people could not occasionally participate in game drives or other cooperative subsistence activities. My argument is only that the *focus* of subsistence changed markedly toward the exploitation of individually gathered foods, and that this trend reflects a concern on the part of households to gain the majority of their calories via these means rather than through extra-familial cooperative ventures.

Interhousehold Heterogeneity

As people focus more on collecting privately owned resources, the level of cooperation and resource redistribution between households should, on average, decrease. A reflection of this should be a shift from interhousehold redundancy during times of high cooperation and sharing of nonexcludable goods, to greater interhousehold heterogeneity under a strategy focused on exploiting private and excludable goods. Ideally, we can examine houses that are contemporaneously occupied in the archaeological record. However, most dating techniques (e.g., radiocarbon, obsidian hydration, luminescence) do not provide the temporal precision needed to demonstrate contemporaneity, and we must examine houses that date to roughly the same time period.

I have tested this prediction in the Owens Valley (Eerkens and Spurling 2008) with artifact composition and imported goods found in households. Obsidian was an essential raw material in the region for producing a wide range of tools and is ubiquitous at archaeological sites. However, a source of obsidian is not present in the southern Owens Valley, and it had to be imported from sources to the south, east, and north. Geochemical analyses, such as X-ray fluorescence, neutron activation analysis, or inductively coupled plasma-mass spectrometry, allow archaeologists to determine the geographic origin of obsidian artifacts with a high degree of accuracy. Figure 7.3 shows the diversity of obsidian geochemical sources recorded in household assemblages in the southern Owens Valley, near Owens Lake. The data are standardized for sample size, and show the number of geochemical sources present in houses relative to samples size. It also groups houses by culture-historical period, where Marana assemblages correlate with dates after 700 years ago, and Haiwee (700–1500 BP) and Newberry (1500–3500 BP) assemblages predate this point in time.

The figure suggests an earlier period, between 2000 and 1500 BP, where source diversity was higher, but interhousehold differences were minimal (i.e.,

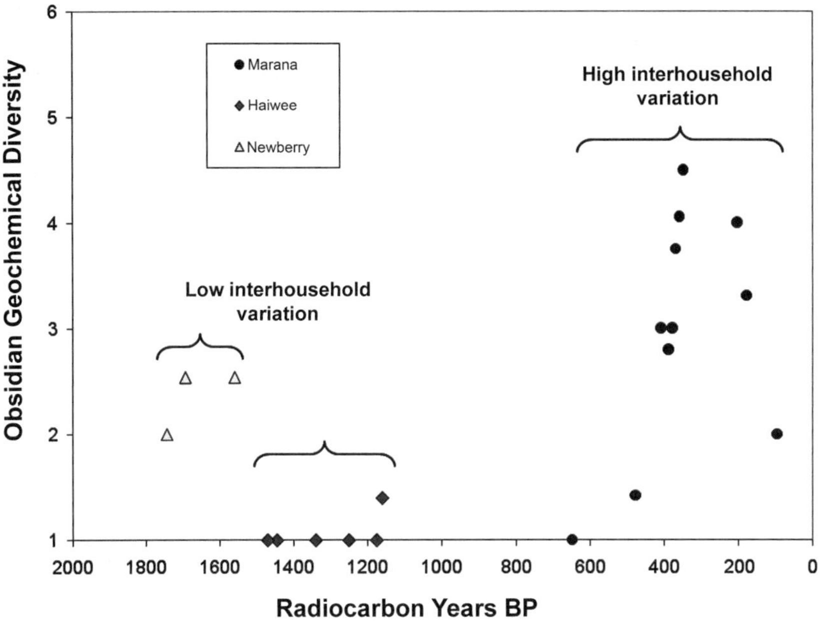

FIGURE 7.3 *Obsidian geochemical diversity in households over time in the southern Owens Valley, showing high redundancy before 650 BP and high interhousehold variation after then.*

redundancy is high). Obsidians from long distances away are not uncommon in these assemblages, likely reflecting a high degree of mobility (Eerkens, Spurling, and Gras 2008), but all households had access to the same set of obsidians. Likewise, there are no notable differences in other exotic goods, such as marine shell beads, or even more locally produced items such as groundstone. I attribute these patterns in the distribution of various goods to widespread sharing of raw materials and artifacts between household units (i.e., a public goods system with little private ownership), resulting in greater household redundancy.

Between 1500 and 1000 BP, source diversity decreases markedly to essentially one, or occasionally two, obsidian sources (the closest sources only), but interhousehold differences remain minimal. The decrease in source diversity is a reflection of a marked shift in settlement patterns toward sedentism, and corresponds to a number of other changes in local lifeways, such as the construction of more substantial and permanent domiciles (Eerkens 2004). Yet, homogeneity across households in their access to exotic and local goods, I argue, represents the maintenance of a public-goods system where resources are widely shared between economic units.

Beginning around 700 BP, there is a notable shift in interhousehold variation. As shown in Figure 7.3, some houses after this time period still have access to only one or two geochemical types, while others have access to four or five types. I have argued that access to these different sources is a reflection of access to trading networks and social connections (Eerkens and Spurling 2008; Eerkens, Spurling, and Gras 2008). In other words, some household units had access to trading partners living in a diverse range of habitats to the south, east, and north, while others were poorly connected and could only access the closest source (or had to scavenge existing sites). Far-flung social connections may have provided some families with a better safety net against local resource shortfall, or perhaps, access to religiously, socially, or politically valuable goods and services (Eerkens 2012).

The data from Figure 7.3 are consistent with those in Figure 7.2. Both figures suggest increasing differentiation of household units over time. I argue that this pattern is a reflection of lower levels of local cooperation as families tried to solve the free-rider problem. At the same time, the greater diversity of obsidian geochemical types, and greater numbers of other imported goods such as marine shell beads (see Eerkens and Spurling 2008), indicate greater extralocal cooperation. In other words, while less effort was expended on cooperating with people in the immediate surroundings, more seems to have been expended on cooperating with long-distance contacts.

Intrahousehold Activity Areas

Third, as household groups privatize resources and focus on defining and increasing the use of excludable goods, greater effort should be spent on hoarding and defending those goods. One simple way to do this is to remove goods from public view and process and store them indoors, within domiciles. This should be reflected by more intensive use of space within houses for such resources (e.g., Wiessner 1982). In particular, space within the house should be devoted to preparing and/or storing particular economic resources, specifically, those that were formerly nonexcludable. Archaeologically, the signature of this should be a more partitioned use of interior space and the formation of distinctive "activity areas."

Again, I have tested this hypothesis in the southern Owens Valley, albeit on a more limited scale than above. Prehistoric houses in the Owens Valley occasionally have internal features that are physically visible during excavation, such as artifact caches, hearths, and subfloor pits. However, such internal features are not ubiquitous and in any case are unlikely to represent the full range of activities that take place within houses. Many activities do not physically alter the inside of a house at a macrolevel and will not be discovered during the course of field excavation. In this respect, organic residues left behind on floors are likely to represent a broader range of activities and may better indicate the interior division of space in ancient times.

Over the last five years, my graduate students and I have excavated several prehistoric houses where we systematically laid a grid over a section of an exposed house floor and took small soil samples at regular intervals. These soil samples were then sorted for microartifacts and bone, and subjected to a range of

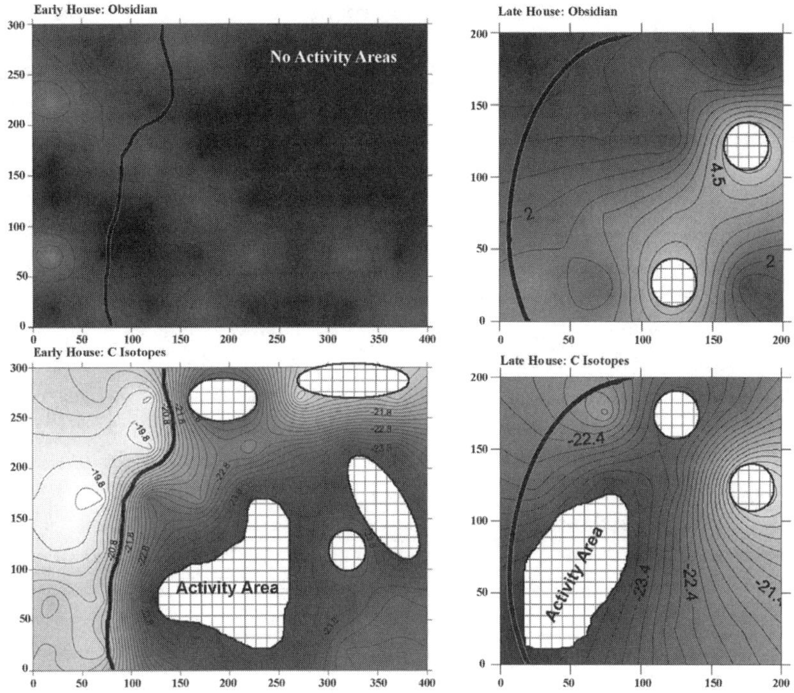

FIGURE 7.4. *Distribution of microflakes of obsidian and carbon isotope ratios across two house floors, one dated to 1450 BP (left) and the other to 650 (right). Interior sections are to the right of the line.*

chemical analyses, including measurement of pH, carbon and nitrogen isotopic composition, and parts-per-million (ppm) content of a range of elements, including aluminum, calcium, carbon, copper, iron, potassium, magnesium, manganese, nitrogen, sodium, phosphorus, sulfur, strontium, and zinc (see Eerkens and Santy n.d.; Santy and Eerkens 2010).

The chemical-based analyses of sediments are still ongoing, but preliminary results support the privatization model. For example, Figure 7.4 compares two houses, one radiocarbon dated to 1450 BP (with 12 m^2 of sediment systematically sampled, including 9 m^2 of actual floor and 3 m^2 of floor exterior) and the other to 650 BP (with 4 m^2 sampled, including 3.5 m^2 of actual floor and 0.5 m^2 of floor exterior). Note that the spatial area of the earlier house is over twice as much as the later house. Samples were taken every 50 cm. In all figures the actual floor occurs on the right side of the bold black line, which demarcates the edge of the floor as recorded in the field (i.e., based on physical features). Neither house was completely exposed, although more of the floor was exposed than from which systematic samples were collected. The earlier house (left side of Figure 7.4) was larger and appears to have had a more square to rectangular shape. The later house (right side of Figure 7.4) was smaller and had a more circular shape.

I defined "activity areas" as spatially contiguous areas where values for a particular measure (e.g., pH, total C, C isotope ratios) are noticeably higher or lower than the immediately surrounding floor. In the figures below, these show up as concentrations of topographic lines or areas of dark or white, and are highlighted by cross-hatching. I then calculated an average number of "activity areas" per square meter of house floor as an indication of the intensity of subdivision of space within houses.

The top left panel in Figure 7.4 shows that obsidian in the earlier house is nearly absent. No sample within the house had more than one flake (most had none), though some samples from outside the house had two and three flakes. As a result, no activity areas there are definable based on the density of obsidian. By contrast, the later house has no obsidian on the exterior (though only one sample comes from this context), and all but one sample had at least one flake. Furthermore, at least two spatially isolated concentrations or activity areas are evident, of seven and six flakes, respectively. These activity areas could represent locations where obsidian was used to cut something, leading to the deposition of microflakes, or areas where obsidian tools were finished or resharpened. Patterns in the distribution of bone largely mirror these results, where the earlier house is relatively clean and the later house has more noticeable concentrations of bone.

The carbon isotopes are less dramatically different between the early and late houses, but suggest a similar pattern. The earlier house has a noticeably different carbon isotope ratio signature within than outside, making it possible to define the edge of the house using this measure only. Within the house, a large activity area is visible in the lower center part of the map defined by highly negative carbon isotope ratios, an area where excavators recorded a small hearth. Four other activity areas are also present, including another less negative area near the top right where an ash deposit was recorded by excavators. The later house shows three activity areas within the house. One of these, the larger area in the lower left with more negative isotopic readings, overlies an area where excavators recorded a hearth (note also that one of the obsidian concentrations is directly adjacent to this hearth).

When we standardize the number of activity areas by total floor area examined, both the density of obsidian and carbon isotope ratios suggest a higher density of activities in the later house (per unit area). The same general pattern is true of most other measures (e.g., pH, N isotopes, total K). While the increased use of excludable versus nonexcludable goods is not the only process by which interior space becomes more partitioned (e.g., Brooks and Yellen 1987; Fletcher 1995; Kent 1987), this result is consistent with the privatization model.

Discussion

The data from the Owens Valley suggest a marked shift in cooperation behavior around 700 years ago. When evaluating economic data, such as faunal and floral remains, and access to imported goods, the archaeological record suggests a high degree of household redundancy before this date and heterogeneity afterwards. This heterogeneity is consistent with a model of increasing family-level independence and less intravillage cooperation. The archaeological record is less clear regarding cooperation in other realms, such as religious and/or political activities.

While there appears to be a reduction in intravillage economic cooperation, there is evidence for increasing cooperation beyond the scale of the village. As mentioned, an increase in imported goods, such as obsidian and shell beads, indicates higher levels of trade after 700 years ago than before it. This is especially true among certain household units that appear to have been well connected with other individuals living in nearly all cardinal directions. I have argued elsewhere (Eerkens 2012; Eerkens, Neff, and Glascock 2002) that such social connections were essential in providing access to more distant foraging territories in times of local resource shortfall. The distribution of some conveyed artifacts, pottery vessels in particular, follows clines that maximize climatic difference. That is, people tended to establish and maintain contact with other people from regions where precipitation was most likely to be different. This maximized the chances that, when local resources failed due to too little rainfall, regions where people maintained contacts would not have simultaneously failed. Access to foraging territories appears to have been reciprocal (Eerkens 2012).

The net effect of this transition was that the spatial level of cooperation shifted in the Owens Valley. Prior to 700 BP cooperation focused on the local scale, where other related families and immediate kin provided a safety net to food shortage. After 700 BP people shifted cooperative ventures to the regional scale, where more distant kin or non-kin provided such a safety net. In this particular situation, cooperators at the regional scale may have been less apt to cheat or free-ride, for example, by not granting access to their foraging territory after previously gaining access to someone else's. This is because climate is unpredictable, and the potential punishment of denying access to others currently in need (when you have a surplus), is a similar denial in the future when you yourself are in need (and are running a deficit).

The finding that families were largely independent of other families within a community is consistent with economic patterns recorded shortly after the time of contact with Anglo-American settlers (i.e., ethnographically). Steward (1933, 1938) was adamant that Paiute family groups had complete autonomy from other families, and did not regularly share basic staples such as seed resources. Even hunting grounds and piñon groves were owned by particular families, and access to them was carefully guarded, with physical punishment possible for violators.

At the same time, the ethnographic record does suggest some degree of intravillage cooperation in certain realms. For example, village headmen would help coordinate and oversee certain activities such as the construction of irrigation ditches (which fed small plots where wild plants were grown; see also Lawton et al. 1976), annual fandangos (feasts), war parties, and communal rabbit drives. Village headmen could also approve or veto witch killings (i.e., people suspected of practicing witchcraft). Yet, Steward is clear that the influence of such leaders was not extensive and they usually had little power to enforce their will.

At issue in understanding the evolution of cooperative behavior from prehistoric into ethnographic times, is whether such village headmen are a product of postcontact phenomena or are indigenous developments. We do not have direct archaeological evidence on when such leaders appeared. It is conceivable, for example, that families had to join forces and cooperate together, by necessity, to cope with displacement from traditional hunting and gathering lands and/or the increase in mortality from exposure to disease after contact. Alternatively, leaders may have appeared in late precontact times as a means to facilitate small-scale cooperative ventures that were otherwise difficult or impossible given the autonomy of family units and the effects of free-riders. This is the position that Bettinger (1983; Bettinger and King 1971) took in an evaluation of existing data in the early 1980s.

My own evaluation of the data, though still scant, is that such leaders did evolve prior to contact, but sometime well after 700 BP. Testing this hypothesis will require collecting new types of archaeological information in the region. For example, indirect evidence could come from the excavation and careful seriation of individual burials, which may indicate the presence and/or evolution of small-scale leaders. To date only a handful of burials have been excavated and analyzed and are equivocal about the presence of leaders (Eerkens 2010). As well, the excavation and dating of clearly cooperative features such as irrigation systems, feasting areas, or V-wing hunting traps that date to the last 700 years may show an increase in frequency around the time that leaders emerge. Clearly, more work is needed to address this interesting issue.

THE EVOLUTION OF PRIVATIZED, EXCLUDABLE GOODS AND SMALL-SCALE LEADERS

I have argued that changes in the demographics of societies cause changes in the costs and benefits of sharing and free-riding. Increases in the proportion of non-kin in a community could be caused by simple population growth, which should be a gradual process and less noticeable on a day-to-day basis. However, other factors can accentuate or cause similar demographic changes, such as changes in marriage rules favoring exogamy and/or in-migration of refugees into a community. Such changes may be more sudden and noticeable to individuals, eliciting different types of responses.

In societies where individuals have little or no authority to punish non-kin and widespread sharing and cooperation is the norm, an increase in the absolute number of non-kin in a community, in particular, will make free-riding an attractive option. As more and more individuals free-ride, people will be loath to engage in cooperative ventures outside the family unit. That is, there is little incentive to invest time and labor into such activities if the fruits of such investments are consistently lost. For certain types of goods, especially those that are nonexcludable, a "breakdown" in sharing and a shift to increasing privatization should ensue. In most small-scale societies, the easiest solution, not requiring extensive coordination and a broad societal shift in social norms, is for individuals to focus on exploiting different types of goods for which sharing rules are not well established. By doing so, individuals are not violating social norms by not sharing, and can possibly establish norms that make such goods excludable. Alternatively, individuals can attempt to hoard and protect nonexcludable goods within the household, resulting in a "closed" site layout (per Wiessner 1982). The latter solution, in particular, carries the risk of public skirmishes and/or ostracism from the community, when individuals demand a share of "public" (i.e., nonexcludable) goods.

Alternatively, societies may elect to establish formalized offices with individuals who can enforce rules or norms, including cooperation among non-kin. The presence of leaders who have some ability to mete out punishment within a community could hold in check the increase in free-riding expected with increasing numbers of non-kin. Of course, small-scale societies generally lack formal positions with that type of power (Bird and Bliege Bird 2010; Eerkens, Vaughn, and Kantner 2010). In any case, it is unlikely that individuals could create and assign themselves to such positions of leadership without the support of others in the group (Kantner 2010; Wiessner 2010). However, communities may opt to create such positions and elect such leaders if the benefits are obvious. In this respect, the increased decision-making power given to such individuals and reduction in free-riding behavior may outweigh the loss of autonomy of family units. In other words, a mutualistic or "win-win" situation ensues, where leaders gain personal powers (and likely personal wealth) and communities of people gain a means to control free-riding behavior (and an increase in production due to more cooperative behavior).

An alternative or complementary means to limit the effects of free-riding is for a community of unrelated individuals to adopt a common religion or world view that incorporates the threat of punishment for noncooperators. For example, if mythical or spiritual beings within such a religion are given the power to distribute some type of punishment (e.g., causing sickness or bad luck or limiting access to an afterlife), potential free-riders may opt to cooperate instead of cheat. Of course, there is a coordination issue here in getting individuals within a community to opt-in to a common religion or world view. But if such a religion is already in place, people born into the community or those that migrate in through marriage or other means may be persuaded or forced to adopt the dominant view. Further, influential leaders may be instrumental in bringing about the widespread adoption of such a religion and may use religion as an indirect means to threaten punishment to potential cheaters, rather than having to take on that responsibility directly. The role of religion in fostering group-wide cooperation and conformity has been discussed extensively elsewhere (e.g., Evans et al. 1995;

Sosis and Ruffle 2004; Tan and Vogel 2008; Wilson 2002), but its role in limiting free-riding has received less attention (for some exceptions see Iannaccone 1992; Richardson and McBride 2009; Sosis 2003).

In the Owens Valley case, there is evidence for the evolution of small-scale leaders by the late 1800s to early 1900s. There is less evidence for a widespread religion with mythical beings that would punish potential cheaters, though a generalized notion that bad deeds would bring about evil was present (Steward 1933). On the other hand, shamans could practice witchcraft, with the power to cause illness or death to other individuals. Such witchcrafting power, and the ability for any individual to hire a shaman to undertake such activities (Steward 1933: 314), may have been a deterrent to potential cheaters. Unfortunately, as mentioned above, the archaeological record is currently lacking with regards to the origins and evolution of leaders, shaman, and religious practices (though see Eerkens 2010).

In sum, the archaeological record in the Owens Valley indicates a marked change in cooperative behaviors over time. As a working model, I propose that as societies grew, the incentives for free-riding behavior increased, for example, by more individuals demanding shares of public or nonexcludable goods that they did not themselves procure. Individuals who did not participate in hunting and gathering activities may nevertheless have been able to procure shares of food once brought back to the village, because such foods were nonexcludable. To combat this effect, more effort was given to establishing and maintaining extralocal contacts, an important safety net against local resource shortfall, and less effort was expended on developing intravillage cooperative ventures. For various reasons, these extralocal contacts were less apt to free-ride or cheat than local within-village actors, and reciprocal altruism may have guided such long-distance cooperative interactions.

The archaeological data do not indicate that this process was gradual. Instead, a dramatic change around 700 BP marks this development. At some later point, perhaps in the latest of precontact times (ca. 300 BP), small-scale leaders evolved as a means to help solve the local free-riding issue. At that point, local cooperative activities, such as the construction of irrigation ditches and communal feasting, began to increase again in scope, leading to the situation recorded by Steward (1933) at the time of contact. I hope that future archaeological research will seek to collect the right types of data to test and refine this model.

REFERENCES

Arkush, Brooke S. 1995. *The Archaeology of CA-Mno-2122: A Study of Pre-Contact and Post-Contact Lifeways Among the Mono Basin Paiute.* University of California Anthropological Records No. 31, Berkley.

Axelrod, Robert. 1997. *The Complexity of Cooperation: Agent-Based Models of Competition and Collaboration.* Princeton, NJ: Princeton University Press.

Axelrod, Robert, and W. D. Hamilton. 1981. "The Evolution of Cooperation." *Science* 211 (4489): 1390–6. http://dx.doi.org/10.1126/science.7466396.

Basgall, Mark E. 1989. "Obsidian Acquisition and Use in Prehistoric Central-eastern California." In *Current Directions in California Obsidian Studies*, ed. R. E. Hughes, 111–26. Contributions of the University of California Archaeological Research Facility, No. 45.

Basgall, Mark E. 2008. "An Archaeological Assessment of Late-Holocene Environmental Change in the Southwestern Great Basin." In *From Avocados to Millingstones: Papers in Honor of Delbert L. True*, ed. M. E. Basgall and G. Waugh, 137–153. Monographs in California and Great Basin Anthropology No. 5. California State University, Sacramento.

Basgall, Mark E., and Kelly McGuire. 1988. *The Archaeology of CA-INY–30: Prehistoric Culture Change in the Southern Owens Valley, California*. Report submitted to California Department of Transportation, District 9, Bishop.

Bettinger, Robert L. 1975. "The Surface Archaeology of Owens Valley, Eastern California: Prehistoric Man-land Relationships in the Great Basin." Ph.D. dissertation, University of California, Riverside.

Bettinger, Robert L. 1983. "Aboriginal Sociopolitical Organization in Owens Valley: Beyond the Family Band." In *Development of Political Organization in Native North America*, ed. E. Tooker and H. M. Fried, 45–88. Proceedings of the American Ethnological Society, 1979. Washington, DC: American Ethnological Society.

Bettinger, Robert L. 1989. *The Archaeology of Pinyon House, Two Eagles, and Crater Midden: Three Residential Sites in Owens Valley, Inyo County, California*. Anthropological Papers of the American Museum of Natural History, no. 67. New York: American Museum of Natural History.

Bettinger, Robert L. 1999. "What Happened in the Medithermal." In *Models for the Millennium*, ed. C. Beck, 62–74. Salt Lake City: University of Utah Press.

Bettinger, Robert L., and Thomas F. King. 1971. "Interaction and Political Organization: A Theoretical Frameword for Archaeology in Owens Valley, California." *University of California Archaeological Survey Annual Report* 13: 187–95.

Bird, Doug W., and Rebecca Bliege-Bird. 2010. "Competing to Be Leaderless: Food Sharing and Magnanimity among Martu Aborigines. In *The Evolution of Leadership: Transitions in Decision Making from Small-Scale to Middle-Range Societies*, ed. J. Janter, K. Vahn, and J. Earkins, 21–49. Santa Fe: SAR Press.

Bird, R. 1999. "Cooperation and Conflict: The Behavioral Ecology of the Sexual Division of Labor." *Evolutionary Anthropology* 8 (2): 65–75. http:/dx.doi.org/10.1002/(SICI)1520-6505(1999)8:2<65::AID-EVAN5>3.0.CO;2-3.

Bliege Bird, R., and D. W. Bird. 1997. "Delayed Reciprocity and Tolerated Theft: The Behavioral Ecology of Food-Sharing Strategies." *Current Anthropology* 38 (1): 49–78. http://dx.doi.org/10.1086/204581.

Bliege Bird, Rebecca, Eric Alden Smith, and Douglas W. Bird. 2001. "The Hunting Handicap: Costly Signaling in Human Foraging Strategies." *Behavioral Ecology and Sociobiology* 50 (1): 9–19. http://dx.doi.org/10.1007/s002650100338.

Blurton Jones, N. G. 1984. "A Selfish Origin for Human Food Sharing." *Ethology and Sociobiology* 5 (1): 1–3. http://dx.doi.org/10.1016/0162-3095(84)90030-X.

Borgerhoff Mulder, Monique, Ila Fazzio, William Irons, Richard L. McElreath, Samuel Bowles, Adrian Bell, Tom Hertz, and Leela Hazzah. 2010. "Pastoralism and Wealth Inequality: Revisiting an Old Question." *Current Anthropology* 51 (1): 35–48. http://dx.doi.org/10.1086/648561.

Bowles, S., and H. Gintis. 2004. "The Evolution of Strong Reciprocity: Cooperation in Heterogeneous Populations." *Theoretical Population Biology* 65 (1): 17–28. http://dx.doi.org/10.1016/j.tpb.2003.07.001.

Bowles, Samuel, Eric Alden Smith, and Monique Borgerhoff Mulder. 2010. "The Emergence and Persistence of Inequality in Premodern Societies: Introduction to the Special Section." *Current Anthropology* 51 (1): 7–17. http://dx.doi.org/10.1086/649206.

Boyd, R., H. Gintis, and S. Bowles. 2010. "Coordinated Punishment of Defectors Sustains Cooperation and can Proliferate when Rare." *Science* 328 (5978): 617–20. http://dx.doi.org/10.1126/science.1183665.

Boyd, R., and P. J. Richerson. 2009. "Culture and the Evolution of Human Cooperation." *Philosophical Transactions of the Royal Society of London. Series B, Biological Sciences* 364 (1533): 3281–8. http://dx.doi.org/10.1098/rstb.2009.0134.

Brooks, A., and J. Yellen. 1987. "The Preservation of Activity Areas in the Archaeological Record: Ethnoarchaeological and Archaeological Work in Northwest Ngamiland, Botswana." In *Method and Theory for Activity Area Research: An Ethnoarchaeological Approach*, ed. S. Kent, 63–106. New York: Columbia University Press.

Dawkins, Richard. 1976. *The Selfish Gene*. Oxford: Oxford University Press.

Delacorte, Michael G. 1999. *The Changing Role of Riverine Environments in the Prehistory of the Central-Western Great Basin: Data Recovery Excavations at Six Prehistoric Sites in Owen's Valley, California*. Report submitted to California Department of Transportation, District 9, Bishop, California.

Eerkens, Jelmer W. 2003. "Sedentism, Storage, and the Intensification of Small Seeds: Prehistoric Developments in Owens Valley, California." *North American Archaeologist* 24 (4): 281–309. http://dx.doi.org/10.2190/GHAT-YV8A-M1BW-0HPH.

Eerkens, Jelmer W. 2004. "Privatization, Small-Seed Intensification, and the Origins of Pottery in the Western Great Basin." *American Antiquity* 69 (4): 653–70. http://dx.doi.org/10.2307/4128442.

Eerkens, Jelmer W. 2010. "Privatization of Resources and the Evolution of Prehistoric Leadership Strategies." In *The Evolution of Leadership: Transitions in Decision Making from Small-Scale to Middle-Range Societies*, ed. K. J. Vaughn, J. W. Eerkens, and J. Kantner, 73–94. Santa Fe, NM: SAR Press.

Eerkens, Jelmer W. 2012. "A Model for Predicting Economic Interaction in Arid Lands and an Evaluation in Eastern California Based on Brownware Ceramics." In *Meeting at the Margins: Prehistoric Cultural Interactions in the Intermountain West*, ed. David Rhode, 229–245. Salt Lake City: University of Utah Press.

Eerkens, Jelmer W., Hector Neff, and Michael Glascock. 2002. "Ceramic Production among Small-Scale and Mobile Hunters and Gatherers: A Case Study from the Southwestern Great Basin." *Journal of Antropological Archaeology* 21: 200–229.

Eerkens, Jelmer W., and Jenna Santy. n.d. Soil Geochemistry and the Identification of Activity Areas on Ancient House Floors. Manuscript in possession of the author.

Eerkens, Jelmer W., and Amy M. Spurling. 2008. "Obsidian Acquisition and Exchange Networks: A Diachronic Perspective on Households in the Owens Valley." *Journal of California and Great Basin Anthropology* 28:111–26.

Eerkens, Jelmer W., Amy M. Spurling, and Michelle A. Gras. 2008. "Measuring Prehistoric Mobility Strategies based on Obsidian Geochemical and Technological Signatures in the Owens Valley, California." *Journal of Archaeological Science* 35 (3): 668–80. http://dx.doi.org/10.1016/j.jas.2007.05.016.

Eerkens, Jelmer W., Kevin J. Vaughn, and John Kantner. 2010. "Introduction: The Evolution of Leadership." In *The Evolution of Leadership: Transitions in Decision Making from Small-Scale to Middle-Range Societies*, ed. K. J. Vaughn, J. W. Eerkens, and J. Kantner, 3–17. Santa Fe, NM: SAR Press.

Evans, T. D., F. T. Cullen, R. G. Dunaway, and V. S. Burton, Jr. 1995. "Religion and crime reexamined: the impact of religion, secular controls, and social ecology on adult criminality." *Criminology* 33 (2): 195–224. http://dx.doi.org/10.1111/j.1745-9125.1995.tb01176.x.

Fletcher, Roland. 1995. *The Limits of Settlement Growth: A Theoretical Outline*. New York: Cambridge University Press.

Fried, Morton. 1967. *The Evolution of Political Society: An Essay in Political Anthropology*. New York: Random House.

Gardner, Andy, and Stuart A. West. 2010. "Greenbeards." *Evolution; International Journal of Organic Evolution* 64 (1): 25–38. http://dx.doi.org/10.1111/j.1558-5646.2009.00842.x.

Garfinkel, Alan P., David A. Young, and Robert M. Yohe, II. 2010. "Bighorn Hunting, Resource Depression, and Rock Art in the Coso Range, Eastern California: A Computer Simulation Model." *Journal of Archaeological Science* 37 (1): 42–51. http://dx.doi.org/10.1016/j.jas.2009.08.010.

Gilreath, A. J., and W. R. Hildebrandt. 1997. *Prehistoric Use of the Coso Volcanic Field*. Contributions of the University of California Archaeological Research Facility, Berkeley.

Gilreath, A. J., and W. R. Hildebrandt. 2008. "Coso Rock Art within its Archaeological Context." *Journal of California and Great Basin Anthropology* 28 (1): 1–22.

Gintis, H., E. A. Smith, and S. Bowles. 2001. "Costly Signaling and Cooperation." *Journal of Theoretical Biology* 213 (1): 103–19. http://dx.doi.org/10.1006/jtbi.2001.2406.

Gurven, Michael. 2004. "To Give and to Give Not: The Behavioral Ecology of Human Food Transfers." *Behavioral and Brain Sciences* 27 (04): 543–83. http://dx.doi.org/10.1017/S0140525X04000123.

Gurven, Michael. 2006. "The Evolution of Contingent Cooperation." *Current Anthropology* 47 (1): 185–92. http://dx.doi.org/10.1086/499552.

Gurven, Michael, Kim Hill, Hillard Kaplan, Ana Hurtado, and Richard Lyles. 2000. "Food Transfers among Hiwi Foragers of Venezuela: Tests of Reciprocity." *Human Ecology* 28 (2): 171–218. http://dx.doi.org/10.1023/A:1007067919982.

Gurven, M., and J. Winking. 2008. "Collective Action in Action: Pro-Social Behavior In and Out of the Laboratory." *American Anthropologist* 110 (2): 179–90. http://dx.doi.org/10.1111/j.1548-1433.2008.00024.x.

Hamilton, W. D. 1964. "The Genetical Evolution of Social Behaviour." *Journal of Theoretical Biology* 7 (1): 1–16, 17–52. http://dx.doi.org/10.1016/0022-5193(64)90038-4.

Hardin, Garrett. 1968. "The Tragedy of the Commons." *Science* 162 (3859): 1243–8. http://dx.doi.org/10.1126/science.162.3859.1243.

Hawkes, K. 1991. "Showing Off: Tests of Another Hypothesis about Men's Foraging Goals." *Ethology and Sociobiology* 12 (1): 29–54. http://dx.doi.org/10.1016/0162-3095(91)90011-E.

Henrich, J., and R. Boyd. 2001. "Why People Punish Defectors: Weak Conformist Transmission can Stabilize Costly Enforcement of Norms in Cooperative Dilemmas." *Journal of Theoretical Biology* 208 (1): 79–89. http://dx.doi.org/10.1006/jtbi.2000.2202.

Henrich, J., R. Boyd, S. Bowles, C. Camerer, E. Fehr, and H. Gintis, eds. 2004. *Foundations of Human Sociality: Economic Experiments and Ethnographic Evidence from Fifteen Small-Scale Societies*. Oxford: Oxford University Press.

Henrich, J., J. Ensminger, R. McElreath, A. Barr, C. Barrett, A. Bolyanatz, J. Cardenas, M. Gurven, E. Gwako, N. Henrich, et al. 2010. "Markets, Religion, Community Size, and the Evolution of Fairness and Punishment." *Science* 327 (5972): 1480–4. http://dx.doi.org/10.1126/science.1182238.

Hildebrandt, W. R., and K. R. McGuire. 2002. "The Ascendance of Hunting during the California Middle Archaic: An Evolutionary Perspective." *American Antiquity* 67 (2): 231–56. http://dx.doi.org/10.2307/2694565.

Hockett, Bryan, and Timothy W. Murphy. 2009. "Antiquity of Communal Pronghorn Hunting in the North-Central Great Basin." *American Antiquity* 74:708–34.

Iannaccone, Laurence R. 1992. "Sacrifice and Stigma: Reducing Free-Riding in Cults, Communes, and other Collectives." *Journal of Political Economy* 100 (2): 271–91. http://dx.doi.org/10.1086/261818.

Johnson, Dominic, and Jesse Bering. 2006. "Hand of God, Mind of Man: Punishment and Cognition in the Evolution of Cooperation." *Evolutionary Psychology* 4:219–33.

Jones, Terry L., Gary M. Brown, L. Mark Raab, Janet L. McVickar, W. Geoffrey Spaulding, Douglas J. Kennett, Andrew York, and Phillip L. Walker. 1999. "Environmental Imperatives Reconsidered: Demographic crises in Western North America during the Medieval Climatic Anomaly." *Current Anthropology* 40 (2): 137–70. http://dx.doi.org/10.1086/200002.

Kantner, John. 2010. "Identifying the Pathways to Permanent Leadership." In *The Evolution of Leadership: Transitions in Decision Making from Small-Scale to Middle-Range Societies*, ed. K. J. Vaughn, J. W. Eerkens, and J. Kantner, 249–281. Santa Fe: SAR Press.

Kent, Susan. 1987. "Understanding the Use of Space: An Ethnoarchaeological Approach." In *Method and Theory for Activity Area Research: An Ethnoarchaeological Approach*, ed. S. Kent, 1–60. New York: Columbia University Press.

Lanning, Edward P. 1963. "Archaeology of the Rose Spring Site INY–372." *University of California Publications American Archaeology and Ethnology* 49 (3): 237–336.

Lawton, Harry W., Pilip J. Wilke, Mary DeDecker, and William M. Mason. 1976. "Agriculture among the Paiute of Owens Valley." *Journal of California Anthropology* 3 (1): 13–50.

Lubinksi, Patrick M. 1999. "The Communal Pronghorn Hunt: A Review of the Ethnographic and Archaeological Evidence." *Journal of California and Great Basin Anthropology* 21:158–81.

Meyer, Jack, D. Craig Young, and Jeff Rosenthal. 2009. Geoarchaeological Overview and Assessment of Caltrans Districts 6 and 9. Far Western Anthropological Research Group, Inc. Submitted to the California Department of Transportation, District 6, Fresno, California.

Nowak, M. A. 2006. "Five Rules for the Evolution of Cooperation." *Science* 314 (5805): 1560–3. http://dx.doi.org/10.1126/science.1133755.

O'Gorman, Rick, Joseph Henrich, and Mark Van Vugt. 2009. "Constraining Free Riding in Public Goods Games: Designated Solitary Punishers can Sustain Human Cooperation." *Proceedings. Biological Sciences* 276 (1655): 323–9. http://dx.doi.org/10.1098/rspb.2008.1082.

Ostrom, E. 1990. *Governing the Commons: The Evolution of Institutions for Collective Action*. Cambridge: Cambridge University Press.

Pruitt, Jonathan N., and Susan E. Riechert. 2009. "Frequency-Dependent Success of Cheaters during Foraging Bouts might Limit their Spread within Colonies of a Socially Polymorphic Spider." *Evolution; International Journal of Organic Evolution* 63 (11): 2966–73. http://dx.doi.org/10.1111/j.1558-5646.2009.00771.x.

Ratnieks, F., and T. Wenseleers. 2005. "Policing Insect Societies." *Science* 307 (5706): 54–6. http://dx.doi.org/10.1126/science.1106934.

Raymond, Anan S. 1982. "Two Historic Aboriginal Game Drive Enclosures in the Eastern Great Basin." *Journal of California and Great Basin Anthropology* 4 (2): 23–33.

Richardson, Gary, and Michael McBride. 2009. "Religion, Longevity, and Cooperation: The Case of the Craft Guild." *Journal of Economic Behavior & Organization* 71 (2): 172–86. http://dx.doi.org/10.1016/j.jebo.2009.03.012.

Riddell, Harold S. 1951. "The Archaeology of a Paiute Village Site in Owens Valley." *Reports of the University of California Archaeological Survey* 12:1–28.

Riddell, Harold S., and Francis A. Riddell. 1956. "The Current Status of Archaeological Investigations in Owens Valley, California." *Reports of the University of California Archaeological Survey* 33:28–33.

Santy, Jenna, and Jelmer W. Eerkens. 2010. The Organization of Domestic Space in Late Prehistoric Owens Valley Households. *Proceedings of the Society for California Archaeology* 24. Available at http://www.scahome.org/publications/proceedings/Proceedings.24Santy.pdf.

Service, Elman R. 1962. *Primitive Social Organization: An Evolutionary Perspective*. New York: Random House.

Shariff, A. F., and A. Norenzayan. 2007. "God is Watching You: Priming God Concepts Increases Prosocial Behavior in an Anonymous Economic Game." *Psychological Science* 18 (9): 803–9. http://dx.doi.org/10.1111/j.1467-9280.2007.01983.x.

Sosis, Richard. 2000. "Costly Signaling and Torch Fishing on Ifaluk Atoll." *Evolution and Human Behavior* 21 (4): 223–44. http://dx.doi.org/10.1016/S1090-5138(00)00030-1.

Sosis, Richard. 2003. "Why Aren't we all Hutterites? Costly Signaling Theory and Religious Behavior." *Human Nature (Hawthorne, N.Y.)* 14 (2): 91–127. http://dx.doi.org/10.1007/s12110-003-1000-6.

Sosis, Richard, and Bradley J. Ruffle. 2004. "Ideology, Religion, and the Evolution of Cooperation: Field Experiments on Israeli Kibbutzim." *Research in Economic Anthropology* 23:89–117. http://dx.doi.org/10.1016/S0190-1281(04)23004-9.

Steward, Julian H. 1933. "Ethnography of the Owens Valley Paiute." *University of California Publications in American Archaeology and Ethnography*, Berkeley 33: 233–350.

Steward, Julian H. 1938. *Basin-Plateau Aboriginal Sociopolitical Groups*. Washington, DC: Smithsonian.

Tan, Jonathan H.W., and Claudia Vogel. 2008. "Religion and Trust: An Experimental Study." *Journal of Economic Psychology* 29 (6): 832–48. http://dx.doi.org/10.1016/j.joep.2008.03.002.

Thomas, David H. 1988. *The Archaeology of Monitor Valley: 3. Survey and Additional Excavations*. American Museum of Natural History Anthropological Papers 66(2), New York.

Trivers, Robert L. 1971. "The Evolution of Reciprocal Altruism." *Quarterly Review of Biology* 46 (1): 35–57. http://dx.doi.org/10.1086/406755.

Wiessner, Polly. 1982. "Beyond Willow Smoke and Dogs' Tails: A Comment on Binford's Analysis of Hunter-Gatherer Settlement Systems." *American Antiquity* 47 (1): 171–8. http://dx.doi.org/10.2307/280065.

Wiessner, Polly. 2010. "The Power of One? Big Men Revisited." In *The Evolution of Leadership: Transitions in Decision Making from Small-Scale to Middle-Range Societies*, ed. K. J. Vaughn, J. W. Eerkens, and J. Kantner, 195–222. Santa Fe: SAR Press.

Wilson, David Sloan. 2002. *Darwin's Cathedral: Evolution, Religion, and the Nature of Society*. Chicago: University of Chicago Press.

Winterhalder, Bruce. 1986. "Diet Choice, Risk, and Food Sharing in a Stochastic Environment." *Journal of Anthropological Archaeology* 5 (4): 369–92. http://dx.doi.org/10.1016/0278-4165(86)90017-6.

Winterhalder, Bruce. 1996. "A Marginal Model of Tolerated Theft." *Ethology and Sociobiology* 17 (1): 37–53. http://dx.doi.org/10.1016/0162-3095(95)00126-3.

Winterhalder, Bruce. 1997. "Gifts Given, Gifts Taken: The Behavioral Ecology of Non-Market Intragroup Exchange." *Journal of Archaeological Research* 5 (2): 121–68. http://dx.doi.org/10.1007/BF02229109.

Yohe, Robert M., II. 1998. "The Introduction of the Bow and Arrow and Lithic Resource Use at Rose Spring (CA-INY–372)." *Journal of California and Great Basin Anthropology* 20:26–52.

8

Cooperation and Competition among Late Woodland Households at Kolomoki, Georgia

THOMAS J. PLUCKHAHN

How do collective social groups form and persist in light of the obstacles posed by the pursuit of individual self-interests? The dynamic between competition and cooperation has emerged as a major topic of concern, as evidenced by its inclusion on a list of the "big questions" in contemporary science compiled by contributors to the journal *Science* (Pennisi 2005). It is one of only a few topics on that list that pertain specifically to the social sciences (Steckel 2007; see also Feinman, chapter 13).

Given the attention archaeologists have devoted to the development of complex societies, one might imagine that the tension between competition and cooperation would form a topic of concerted research; yet this has not been the case. As Pauketat (2009: xvi–xiii) recently noted, archaeologists have paid surprisingly little attention to conflict, particularly among the native prehistoric societies of North America. On the opposite side of the coin, as Blanton and Fargher (2008: 1) observe, anthropologists have largely abdicated the study of cooperation and collective action to scholars in other disciplines.

Nevertheless, the dynamic between cooperation and competition has been implicit in anthropological treatments of social complexity almost since their inception (see contributions to this volume by Carballo, chapter 1; Feinman, chapter 2; Roscoe, chapter 3). In the mid-twentieth century, neoevolutionist archaeologists searched for the triggers that stimulated evolutionary progression from one type or stage to the next, looking mainly to stimuli external to society (e.g., Sahlins 1958; Service 1958, 1962). Frequently, those external stimuli were framed in terms of intersocietal conflict (Carneiro 1970, 1978, 1981; Dye 2009).

Archaeologists prescribing to stricter evolutionary approaches (i.e., Darwinian archaeologists and behavioral ecologists) have modeled the dynamic between competition and cooperation in greater detail (e.g., Bonhage-Freund and Kurland 1994; Kantner 1996; Kohler 2004; Kohler, VanBuskirk, and Ruscavage-Barz 2004; Kohler, Van Pelt, and Yap 2000; Kohler and Van West 1996; Stanish 2004; Stanish and Haley 2005). Drawing insights from evolutionary ecology (e.g., Richerson, Boyd, and Henrich 2003; Smith 2003; Smith and Bird 2005) and game theory (e.g., Axelrod 1984, 1997), they assume—based on natural selection—that individuals act according to their self-interests (see extended discussion by Blanton and Farger, chapter 5). This presents an impediment to the development of larger and more complex social formations, in that individuals can reap the benefits of collective action without participating (thus negating the selective advantages of cooperation). External competition (Kohler 2004: 4) and internal communal ritual (Stanish and Haley 2005: 64–65) foster greater cooperation and the development of larger, more complex societies.

Another, more diverse group of archaeologists assert that decisions regarding cooperation are complex and historically contingent rather than determined by natural selection. Many of these are rooted in Marxist theories of internal relations (Gramsci 1971; Ollman 1976), looking to conflicts within societies as the sources of change (e.g., Bender 1990; Gilman 1981; McGuire 1992; Nassaney 1992, 2001; Sassaman 1993). Some also draw from theories of agency and practice (Bourdieu 1977, 1980; Giddens 1979; Saitta, chapter 6). In contrast with evolutionary models, Marxists generally see communal ritual and cooperative labor as features already well established in small-scale societies; inherent contradictions in these are manipulated by individuals and factions to suit their own interests, giving rise to greater complexity (Bender 1990; Gilman 1981; Lee 1990; Nassaney 1992, 2000; Pauketat 2000; Saitta and Keene 1990). Thus, in these works internal competition is generally assumed to become more pronounced as societies become more complex, although divisions may be masked by ideologies promoting a more egalitarian vision.

Dual-processualists (Blanton et al. 1996; see also Feinman, chapter 2) strike something of a balance between evolutionary and Marxist perspectives. Like the latter, they often draw from practice theory and emphasize strategic political action, albeit within the constraints imposed by culture (Blanton et al. 1996: 2). Like evolutionary archaeologists, they also emphasize regularities across cultures, positing a distinction between two categories of political economic strategies: network, wherein actors strive for prominence through the maintenance of exclusionary ties to people and groups outside their area, as manifested in patrimonial rhetoric and prestige goods exchange; and corporate, in which power is shared across subgroups and emphasis is placed on cooperative labor projects and rituals reinforcing broad themes such as fertility and renewal (Blanton et al.

1996: 6–7). Without assuming any inherent tendencies in human behavior, dual-processualists posit a scenario generally similar to evolutionary archaeologists with respect to the relationship between complexity and cooperation, associating exclusionary strategies with small-scale, autonomous villages and corporate systems with larger, more complex polities (Blanton et al. 1996: 2–3).

Thus, to briefly summarize, for neoevolutionary archaeologists, competition between or among societies is the force driving the development of larger and more complex social formations. Many neo-Marxists instead favor conflicts internal to society. For dual-processual and some evolutionary archaeologists, on the other hand, larger social groups develop in conjunction with more effective cooperative social mechanisms.

Clearly there are fundamental philosophical divisions among archaeologists regarding the relationship between cooperation and societal complexity. Still, as David Carballo (chapter 1) suggests, and as the contributors to this volume as a whole clearly demonstrate, the dynamics of cooperation may be profitably explored from multiple theoretical approaches. Further, as Saitta (chapter 6) observes, approaches that may be opposed in terms of their basic assumptions regarding inherent human tendencies in regard to cooperation may nevertheless share other insights, and may produce "converging and overlapping" explanations.

I submit that a focus on archaeological households[1] offers a potential point of articulation or convergence for diverse perspectives on cooperation, particularly if we view households as the historical constructs of agents situated within larger social and material landscapes and macrohistorical processes. Specifically, a focus on households allows archaeologists to narrate "smaller stories" that more closely express the lived experiences of the people of the past (Gerritsen 2004: 143), while also recognizing that households are embedded within larger social networks and constrained by larger structures and long-term processes (Hendon 1996; Souvatzi 2008).

This chapter considers cooperation and conflict among households of the Late Woodland period in the southeastern United States. This area and time period offer a productive venue for such research because, as previous archaeological studies have suggested (e.g., Dye 2009: 69–98), there appear to have been dramatic changes in the dynamics of cooperation. The Middle Woodland period (ca. 100 BC to AD 600) witnessed a fluorescence of mound construction and long-distance exchange, best known for the Hopewell societies of the Midwest but also true of contemporaneous sites in the Southeast. These developments, coincident with (in some cases) experimentations with horticulture and larger and more permanent villages, have been interpreted as evidence of cooperative relationships between households and communities, perhaps to mitigate the uncertainties of food production (Braun 1986, 1987; Brose and Percy 1974; Muller 1997: 122, 128–129; Percy and Brose 1974; Seeman 1979) or the social stresses of life in larger communities (Pluckhahn 2003, 2010a, 2010b; but see also Seeman 1988 and Dye 2009: 69–98).

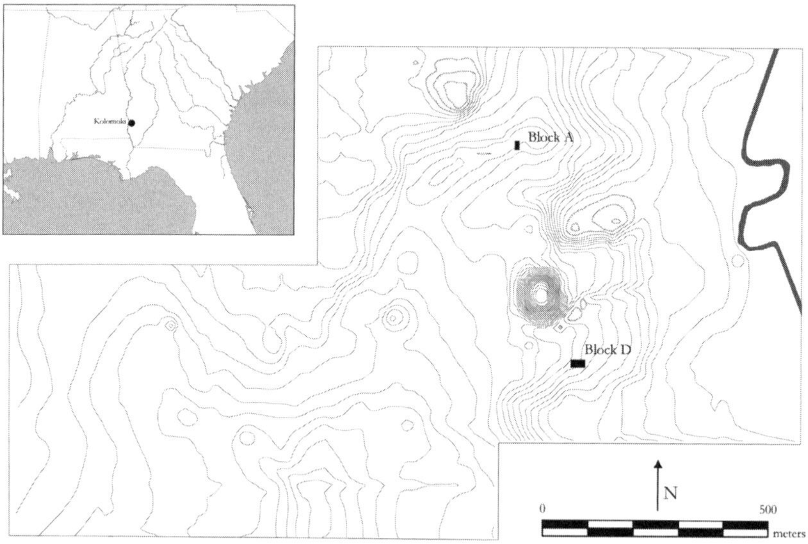

FIGURE 8.1 *The Kolomoki site and the locations of Blocks A and D.*

In contrast with the Midwest, where Hopewell ceremonialism declined precipitously by around AD 400 (McElrath, Emerson, and Fortier 2000), mound construction and long-distance exchange continued into the early Late Woodland (AD 600–800) in some parts of the Southeast, such as the Gulf Coast (Muller 1997: 123; Nassaney and Cobb 1991: 314). However, by the terminal Late Woodland (AD 800–1000) many of the trends initiated during the Middle Woodland were reversed here as well; long-distance exchange of prestige goods became more infrequent, settlements became more dispersed, and mound building declined. These changes have been attributed to increased competition and conflict among households, assumed by some archaeologists to have resulted from the introduction of the bow and arrow (Muller 1997: 129–130), and by others from a greater investment in swidden agriculture (Kohler 1991).

I will evaluate this generalized trajectory against the archaeological record for the Middle and Late Woodland societies of the Gulf Coast and adjacent interior portions of Alabama, Georgia, and Florida, broadly defined by the Swift Creek and Weeden Island ceramic traditions. I focus specifically on one of the largest settlements from these periods, the Kolomoki site (9ER1) in the lower Chattahoochee Valley of southwestern Georgia (Figure 8.1). Recent excavations at Kolomoki reveal fundamental changes in households over the course of the Late Woodland, between approximately AD 600 and 850. I relate these changes to the waxing and waning of social institutions—particularly public ceremony—that encouraged cooperation among households. My analysis reveals consistency with the generalized sequence described above, reinforcing the utility of conceptualizing cooperation in terms of long-term structures and trends, whether these are interpreted from evolutionary or macrohistorical perspectives. However, there are also significant points of discrepancy, highlighting the need to examine the historical contingencies of particular cases and—perhaps more important—serving as a reminder that cooperation and competition are inevitably counterposed, overlapping, and highly nuanced (see also Roscoe, chapter 3; Saitta, chapter 6).

With this in mind, the discussion of competition and cooperation requires a perspective that is not predisposed toward explanations that are either strictly materialist or idealist, behavioral or social, external or internal, or that are rooted solely in either agency or structure. Recognizing the nuances inherent to competition and cooperation—particularly in the sort of early village societies that are the focus of this study—the approach taken here is grounded in theories of practice, agency, and structuration (Bourdieu 1977, 1980; Giddens 1979, 1984). More specifically, my interpretation draws from the refinement of these concepts by Sewell (2005). Sewell's approach offers a number of advantages over previous formulations of the relationship between agency and structure, as critiqued by several authors in recent years (e.g., Dornan 2002). First, Sewell (2005: 124–125) recognizes a greater reflexivity between agency and structure, thus avoiding the tendency toward explanations grounded in either unbridled agency or structural determinism. Extension of this perspective to the archaeological record permits a greater appreciation of temporal changes in competition and cooperation. Next, Sewell (2005: 145) recognizes that agency is "profoundly social or collective" even as exercised by individuals (see also Pauketat 2000, 2001a, 2001b; Saitta 1999, 2007), a point particularly relevant for understanding the social pressures inherent to small-scale, early village societies. Sewell's (2005: 145–151) recognition of a multiplicity of structures at various levels and of different types, including both mental schemas and material resources is also important, in that it facilitates the understanding of competition and cooperation at various social and spatial scales (see also Saitta and Keene 1990; Sassaman 2000: 151). Finally, the discussion here takes heed of Sewell's argument that historically based studies can be attuned to both events and macrohistorical patterns, mitigating a charge that historical approaches have tended to be too particularistic (O'Brien and Lyman 2004).

This case study generally follows the inclusive definition of cooperation offered by Mead (1937: 8), and discussed more fully by Carballo (chapter 1), as "the act of working together to one end." However, consistent with the theoretical perspective described above, I conceive of cooperative or collective action in terms of contemporary understandings of agency, structure, and practice. Thus, I would shift emphasis from "the act" to human actions and dispositions, especially those with meaningful political implications, with the understanding that such actions may be either rational or irrational, and may have consequences that are either intentional or unintentional or both.

THE MIDDLE WOODLAND AND EARLY LATE WOODLAND PERIODS

Although not all of the earthworks at Kolomoki have been adequately dated, sufficient work has been done to suggest that mound construction and ceremonialism reached a zenith during the Middle Woodland period and continued, to a lesser extent, into the early Late Woodland period. During the former period, at least four mounds were constructed along a central east–west axis, centered on a circular plaza. These included two elaborate burial mounds (D and E), one large platform mound (A), and one smaller mound of uncertain function (K) (Pluckhahn 2003: 193). Excavations in the two burial mounds by William Sears (1951, 1953, 1956) revealed a complicated series of mound construction episodes and a number of burials with nonlocal goods, consisting mainly of personal adornments of copper, mica, marine shell, and meteoric iron. There are differences in mortuary treatments indicative of status differentiation, but for the most part these appear to have been communal burial facilities open to large segments of the population. As with many Middle Woodland burial mounds along the Gulf Coast excavated by C. B. Moore (1900, 1901, 1902, 1903a, 1903b, 1905, 1907), the mortuary ceremonies culminated with the placement of caches of pottery on the eastern sides of mounds "for the dead in common" (Moore 1902: 161; see also Willey 1949: 405). These caches include a class of elaborate effigy vessels apparently produced as mortuary offerings (Milanich et al. 1997 [1984]; Sears 1956).

Middle Woodland ceremony was not limited to mortuary rituals. Mound A, the large flat-topped mound at the site, is poorly understood, but also appears to date to this period (Pluckhahn 2003: 56–58). Knight's (1990, 2001) examination of similar mounds in the region reveals that the summits of Middle Woodland platform mounds are frequently covered by seemingly random arrangements of posts, some very large and the others smaller. He suggests that the platforms served as stages for rituals and feasts, the former including the repetitive replacement of larger posts to emphasize renewal, the latter including the conspicuous display of meats on scaffolds comprised of small posts.

The formality of Middle Woodland ritual is paralleled in the highly structured nature of domestic contexts from this time period at Kolomoki. Systematic sampling of off-mound areas—including more than 1,300 shovel tests and controlled surface collections—revealed the existence of a discontinuous circular village surrounding a very large central plaza (Pluckhahn 2003: 120–125). Based on the size of the scatter, I have suggested that there may have been as many as 50 households and a total population of around 300 (Pluckhahn 2003: 191). Sampling suggests that these households were little differentiated by status (Pluckhahn 2003: 92–105); none of the elaborate ceramics or ornaments found with burials in the mounds have been identified in village middens.

Targeted excavations in the village culminated with the complete excavation of one house in an area designated Block A (Pluckhahn 2003: 148–165) (Figure 8.2). The house consisted of a square, semisubterranean, "keyhole" structure with a projecting entrance passage and a central hearth, similar to examples exca-

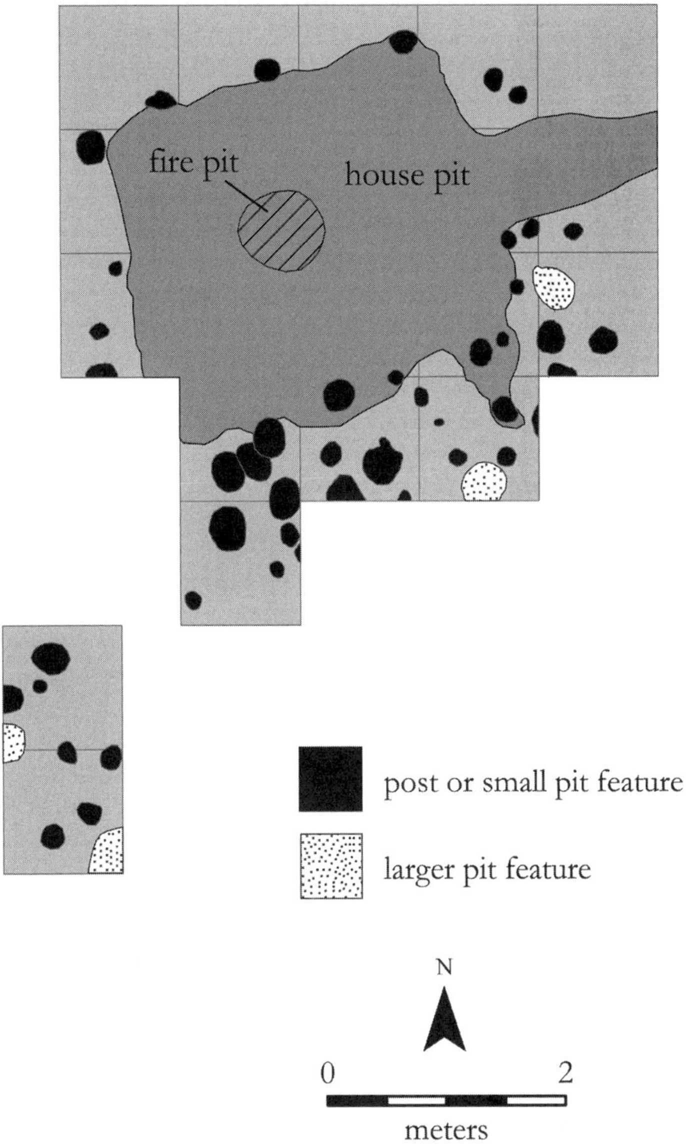

FIGURE 8.2 *Close-up of the pit structure in Block A.*

vated on Woodland sites in the Midwest (Binford et al. 1970; Kelly et al. 1987; Kelly, Ozuk, and Williams 1990). The living floor is quite small, measuring only about 2.5 × 3 m and with a total area of 7.9 m[2] (Pluckhahn 2003: 148–165). Five radiocarbon dates from the structure exhibit two sigma calibrated dates extending from cal AD 420 to 980 (Pluckhahn 2011).[2] Two more reliable dates from

Block A (one on a kernel of *Zea mays* and the other on *Carya* nutshell) are nearly identical, with two-sigma calibrated ranges overlapping between cal AD 570 and 650 (Pluckhahn 2003, 2011), dating the occupation here to the terminal Middle Woodland or early Late Woodland period.

Faunal and botanical samples from the fill of the house pit and from surrounding features suggest a year-round occupation (Pluckhahn 2003: 163–165). Presumably, warmer portions of the year would have been spent in a more lightly constructed pole structure, as might be indicated by the scatter of posts to the south of the pit house. The subsistence regime included limited cultivation of maize (with approximately 10 percent ubiquity) but, interestingly, few other cultigens. The presence of an interior hearth containing the remains of a young deer indicates some degree of production and consumption by the co-resident group. However, pit features were absent from the house and infrequent in the surrounding area with only seven pits (including storage, processing, and fire pits) in the 31 m^2 of excavated area in Block A. With the exception of a fire pit at the center of the house—which accounts for more than half of the total volume (132.5 liters) of all pit features in the block—the pits here were small, with a mean volume of just 29.3 liters. The paucity of storage may suggest that production was at least partially organized at a level above the domestic group (Flannery 2002; Wesson 1999), perhaps with extensive sharing by members of extended kin groups or the community at large. The fill from the house pit provides some additional corroboration of this, given that the number of Minimum Number of Individuals (MNI) for deer well exceeds the quantity that could reasonably be expected for the small number of people who could have resided in the house (Pluckhahn et al. 2006). This high MNI, coupled with an overrepresentation of meaty cuts of deer, suggests procurement of deer in quantity and at some distance from the village (Pluckhahn, Compton, and Bonhage-Freund 2006), perhaps along the lines of the communal deer drives documented in ethnohistoric accounts of the Southeast (Hudson 1976: 276; Swanton 1946: 317–321; Waselkov 1978).

True to the impressions formed on the basis of the sampling data, the ceramic assemblage from Block A contains few indications of high status (Table 8.1). Discounting sherds that are too small or eroded to identify with confidence, the remaining collection of 8,515 sherds is dominated by utilitarian plain and complicated stamped pottery, the vast majority of which was undoubtedly produced locally. More unusual, potentially higher status types—principally zoned red filmed, incised, and punctated wares (Milanich et al. 1997 [1984]: 120–139)—are either absent or present only in minute quantities.

An analysis of the Minimum Number of Vessels (MNV) presents a similar picture (Table 8.2). The fifty-five vessels identified in the Block A assemblage consist mainly of forms used primarily for cooking and storage (collared jars, neckless jars), or cooking and serving (open bowls, restricted bowls), or all of these tasks (simple bowls/unrestricted jars). Vessel forms associated exclusively with serving such as cups and shallow dishes/plates are exceedingly rare, with each category represented by a single, undecorated (plain) vessel.

TABLE 8.1 Relative frequencies of surface treatments in assemblages of identifiable Woodland pottery from Blocks A and D.

Surface Treatment	*Block A (n = 8,515)* *percent*	*Block D (n = 9,272)* *percent*
plain	59.92	77.18
complicated stamped	39.32	10.19
dentate stamped	0.03	0
incised	0.12	3.68
punctate	0.07	3.38
net marked	0	0.39
ridge pinched	0	0.08
red filmed	0.42	4.92
check stamped	0	0.14
cord marked	0	0.04
other	0.12	0

TABLE 8.2 Relative frequencies of vessel forms identified in MNV (Minimum Number of Vessels) analysis of Blocks A and D.

Vessel Form	*Block A (n = 55)* *percent*	*Block D (n = 57)* *percent*
simple bowl/unrestricted jar	23.64	24.56
restricted bowl	12.73	19.30
open bowl	18.18	17.54
collared jar	30.91	15.79
neckless jar	10.91	12.28
cup	1.82	1.75
dish/plate	1.82	8.77

Note: Analysis was conducted on larger rim sherds (> 5 percent of orifice diameter) and diagnostic body sherds. Vessel forms represent a compromise between the classifications recently devised by Hally (2009) and Wallis (2011) (the latter derived mainly from Willey [1949]).

In short, the assemblage is consistent with the notion that the possession and display of exotic ceramic wares and other prestige goods were limited largely to the contexts of public ceremony. While the Block A household represents only a small sample of the estimated number of households that were present during the Middle and early Late Woodland periods, the broader applicability of these observations is supported by the sampling data noted above.

THE TERMINAL LATE WOODLAND

During the Late Woodland, less investment was made in public ceremony at Kolomoki, consistent with the general models for this period described above.

Mound construction continued into the initial portion of the period, but at a much more modest scale. Two small dome-shaped mounds (Mounds B and C) were added to the south and north of Mound A, respectively (Pluckhahn 2003: 58–59). Although these were extensively excavated (Sears 1956: 10–11), their purpose remains enigmatic; Mound B consisted of nothing more than a collection of very large posts, while Mound C appears to have been constructed of sweepings from the plaza. Later in the Late Woodland, two low platform mounds (Mounds F and H) were added; these probably served as stages for ceremonies, but both appear to have been used for only brief intervals before they were ritually sealed (Larson 1952; Pluckhahn 2003: 67–69; Sears 1956: 13).

Previous work suggests that the decline in mound construction during the Late Woodland was accompanied by substantial changes in the village plan. The formal, circular village arrangement that held through the first two centuries of the site's existence broke down into a seemingly random scatter of occupations on the margins of the site (Pluckhahn 2003: 120–125). The reduced size of this component suggests a decline in population; I have suggested a total of 27 households and 135 people, or about one-half the levels of the Middle Woodland (Pluckhahn 2003: 209–211).

Notably, however, it is against this background of declining population and public works that greater distinctions may have emerged among households. Shovel tests and test units first suggested that ceramic types previously restricted mainly to mounds became more widely distributed in domestic contexts (Pluckhahn 2003: 208–211). More recent excavations have centered on one such context, in an area referred to as Block D located about 150 m south of Mound A (Pluckhahn 2011). Four radiocarbon dates from Block D have two-sigma calibrated ranges spanning the Late Woodland, from cal AD 650 to 1020. However, the two-sigma calibrated ranges for three of the four dates overlap between cal AD 780 and 880. This more restricted interval, corresponding with the late Late Woodland, is considered a more accurate approximation of the period of occupation. Thus, the households in Block A and D are likely separated by only one or two centuries.

Block D encompasses 52 m^2, of which 38 m^2 were contiguous 1 × 1 m units that together form a block about 8 m long (north–south) and 6 m wide (east–west) (Pluckhahn 2011) (Figure 8.3). The evidence for domestic architecture was less conclusive here than in Block A, but an arcing pattern of post features suggest the presence of an oval structure of single set posts measuring about 7.3 m long and 5.2 m wide. At the center of the oval pattern and presumed house, there a basin-shaped pit (Feature 171) with relatively high densities of fire-cracked rock and other stone, suggesting use as a hearth.

Several lines of evidence support the interpretation of a structure in Block D (Pluckhahn 2011). First, there is regularity to the spacing of exterior posts, as well to the spacing of interior support posts. Next, pit features appear to be nonrandomly positioned with respect to the structure. Specifically, larger pit features (particularly large, bell-shaped pits) are more common in the interior

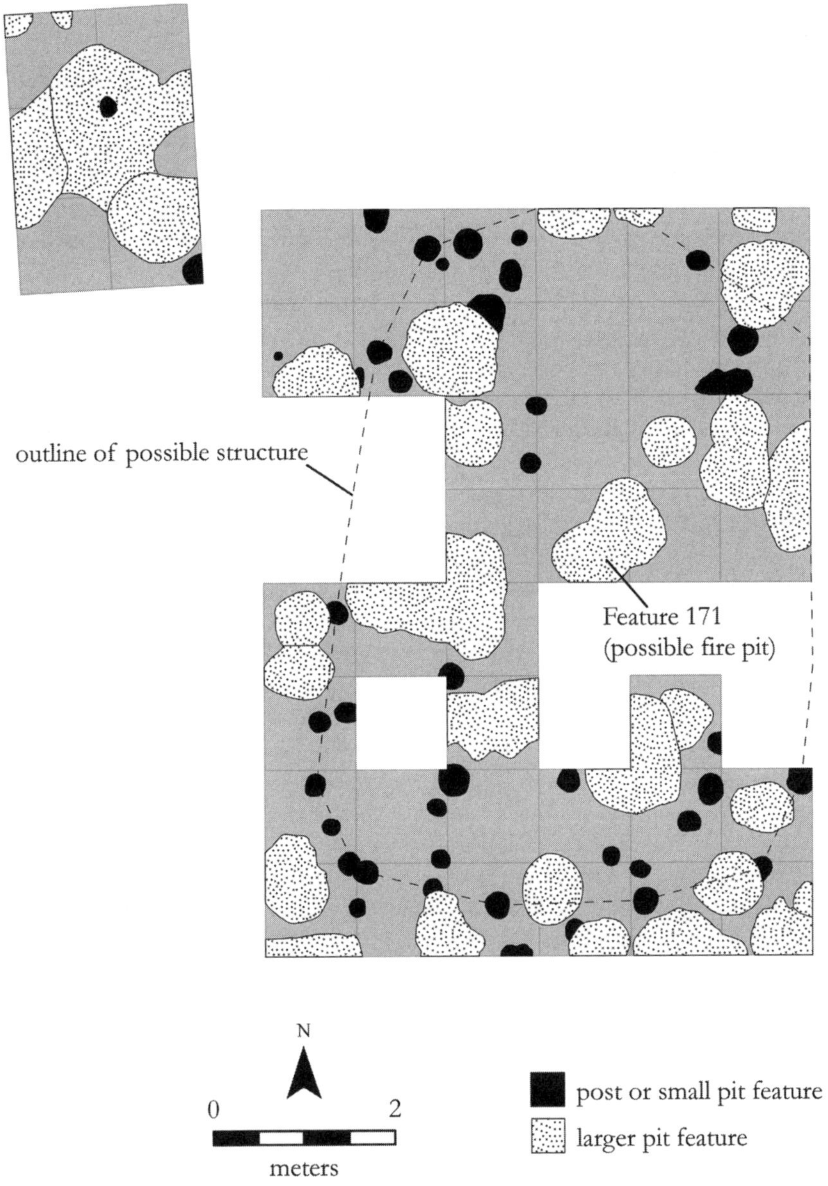

FIGURE 8.3 *Close-up of the structure in Block D.*

of the structure. The distributional patterns of several classes of artifacts lend additional support to the interpretation of a structure in Block D. Finally, the structure in Block D corresponds closely in size, shape, and method of construction with terminal Late Woodland structures excavated at the Sycamore

(Milanich 1974) and Woodland Terrace (Mickwee 2009) sites to the south of Kolomoki.

Comparison of Blocks A and D suggests that households were dramatically reorganized during the Late Woodland period (Pluckhahn 2011). First and most obvious, there was less investment in domestic architecture, with the probable abandonment of semisubterranean construction for set post architecture. On the other hand, houses—and perhaps households as well—appear to have become larger. The set post structure in Block D, with a floor area of 29.1 m^2, is more than three times the size of the pit house in Block A.

Storage, and by extension production and consumption, also appear to have been dramatically reorganized (Pluckhahn 2011). Storage pits are present within the presumed house in Block D and ubiquitous in the surrounding area, with a total of 38 pits in 52 m^2 of excavation. Several of these pits are very large, measuring as much as 2 m in diameter and up to 1.5 m deep. The 38 pits in Block D had a combined volume of 5175.5 liters and a mean volume of 136.2 liters. Equalizing for area of excavation, Block A had 6.6 liters storage/m^2 while the storage capacity in Block D was more than 10 times this, at 99.5 liters/m^2.

Maize cultivation does not appear to have been a factor in the increased importance of storage. No evidence of maize was identified in macrobotanical analysis of flotation samples from features in Block D (Bonhage-Freund 2010; Pluckhahn 2011). As a check on recovery, pollen and phytolith analysis was conducted on a smaller subset of soil samples from pit features from the block (Pluckhahn 2011; Yost and Cummings 2010). Although *Zea mays* pollen was identified in two of the four samples, maize phytoliths were not present, suggesting the pits were not used for maize storage. Given the prevalence of nutshell and oak pollen in Block D features, the storage of mast resources seems a more likely interpretation. Also notable was the recovery of phytoliths derived from the arrowroot family (*Marantaceae*); the starchy roots of several members of this family were used for food and medicine by native peoples of the Southeast and elsewhere in the Americas (Moerman 1998). More generally, however, the increase in storage may indicate less permanent occupation, assuming food was cached for later use (DeBoer 1988). Importantly, it may also indicate that households had greater autonomy over production and consumption (DeBoer 1988; Wesson 1999).

An increase in household autonomy would be consistent with other aspects of the domestic artifact assemblage. First, the ceramic assemblage from our Late Woodland excavation block reveals greater diversity in both decorative attributes and vessel forms, suggesting both lessened constraint in ceramic production and use and more intensive interactions with other households (Pluckhahn 2011). Decorative attributes include a number of varieties that are uncommon to the area, such as the net-marked and ridge-pinched ceramics more commonly associated with sites on the Gulf Coast (see Table 8.1). Also notable are the increases in incised, punctate, and red-filmed wares, from well less than 1 percent each in Block A to over 3 percent each in Block D. A chi-square test indicates very significant differences between the two assemblages with respect to decorative attributes ($\chi^2 = 1282.123, p < 0.001$).[3] Differences in vessel form are difficult to evaluate in statistical terms due to the low (< 5) expected values for several categories. Still, there are conspicuous changes, particularly in the increased relative frequencies of specialized serving vessels like dishes/plates (from 1.82 percent in Block A to 8.77 percent in Block D) (see Table 8.2). In the Block D assemblage, most of these plates and dishes are decorated with elaborate and labor-intensive red-filmed and incised decorations that would be well suited to conspicuous display.

Next, the Block D lithic assemblage is marked by the first appearance of small triangular bifaces (Pluckhahn 2011; Pluckhahn and Norman 2010), generally considered the earliest arrow points in the region (Blitz 1988; Nassaney and Pyle 1999). Five of these were recovered from Block D; none have been identified from Block A or any other contexts at Kolomoki (in an assemblage of more than 200 projectile points). The limited number and relative frequency of arrow points in Block D suggests that they were as yet only minor additions to the more traditional repertoire of spear points. However, it is not unreasonable to assume that the improved efficiency of bow hunting contributed to greater seasonal dispersal and increased household subsistence autonomy (Bettinger 1999; Muller 1997: 130).

DISCUSSION

To recapitulate, changes in households at Kolomoki appear to track the waxing and waning of public ceremony. During the Middle Woodland and early Late Woodland periods, when ritual and long-distance exchange were at their peak, houses were formally arranged in a large, circular village. Domestic architecture also appears to have been formalized. Households were small and invested little effort in storage, suggesting that at least some share of production and consumption may have been organized at a level above the co-resident group. As I have argued elsewhere (Pluckhahn 2003), these might have been facets of a deliberate strategy to reduce distinctions among households, as indicated also by relatively redundant and unexceptional domestic ceramic assemblages.

In the Late Woodland, as ceremony declined in importance, community patterns were less formalized. Domestic architecture was also less formal, but households appear to have increased in size. Storage capacity increased markedly, perhaps indicating greater autonomy of basic subsistence. This autonomy may also be manifested in the increased appearance of status-related ceramics vessels in domestic assemblages.

These general trends appear to square nicely with previous evolutionary models of the dynamics of cooperation among Middle and Late Woodland societies (Dye 2009; Muller 1997) (see Eerkens, chapter 7, regarding a similar trajectory in the Owens Valley, California). In the former period, experimentation with maize horticulture and life in larger villages may have been facilitated by greater cooperation among households, reinforced through community-wide rituals.

The florescence of Middle Woodland ceremony, as well as the conspicuous formality of the village plan and domestic architecture, could perhaps be profitably explained with reference to costly signaling theories—that is, these were markers that were costly to fake, thus discouraging free-riders and encouraging cooperation (Gintis, Smith, and Bowles 2001; Iannaccone 1992; Irons 2001). In the Late Woodland, the advent of new and potentially more efficient hunting techniques accompanying the introduction of the bow and arrow may have granted households greater autonomy over their own subsistence, providing less incentive for such formalized cooperation (Muller 1997: 129–130).

While the material record of Kolomoki displays consistency with this generalized schema, closer examination also reveals subtleties that are not so easily accommodated. First, although the Middle Woodland can be reasonably generalized as a period of greater cooperation among households within villages such as Kolomoki, there are indications of competition and conflict at larger social and spatial scales.

For example, I have argued elsewhere (Pluckhahn 2010a, 2010b) that leadership at Kolomoki during the Middle Woodland—rather than being vested in one or at least a very restricted number of individuals (and thus more competitive in nature)—may have taken the form of a more collective and crosscutting social group such as a sodality, a point developed in much greater detail by Carr (2008: 644–650) for contemporaneous Hopewell societies in Ohio. Yet assuming this is the case, there was undoubtedly still competition within and among such groups, as evidenced by the ethnographic record of native societies of the southwestern United States, where particular clans often "owned" the ceremonies, ceremonial facilities, and ritual paraphernalia (Kantner 2004: 252; Whiteley 1987) and where the leaders of sodalities were drawn from high-ranked clans and enjoyed privileged access to the esoteric knowledge necessary for the sodalities to survive (McGuire and Saitta 1996: 210; Whiteley 1987).

There was also likely competition among Middle Woodland communities, as evidenced most obviously by the disparities in size and elaboration between sites like Kolomoki, with large villages and multiple earthworks, and smaller and simpler mound and village sites. More subtle signs of competition among communities are common. For example, the circular village plan at Kolomoki, while perhaps minimizing variation within households, would have also made the community look larger than it really was, thus providing an imposing landscape for visitors to the site. The choice of domestic architecture also bears scrutiny. Keyhole structures are unknown elsewhere in the immediate region, and would seem somewhat poorly suited to the semitropical climate of the Deep South. The use of this style of domestic architecture at Kolomoki suggests an attempt to demonstrate connections to distant lands, again perhaps reflecting competition at the community level. The same can also be said of the presence of maize; Kolomoki appears to mark the southernmost latitude in which maize has been positively documented in macrobotanical remains from Middle Woodland contexts. Maize may have provided an advantage in competitive feasting among rival communities, because it could be easily stored (Reber 2006) and because—at least in this region and time period—it would have remained the sort of novelty foodstuff often associated with such occasions (Hayden 2001: 40–41).

Likewise, glossing the Late Woodland as a period of increased conflict, while perhaps true in a general sense, also masks important nuances. Formalized, ritually sanctioned cooperation among households at Kolomoki may have declined, but the increased diversity in domestic ceramic assemblages and the appearance of specialized serving vessels in domestic assemblages suggest continued, and perhaps even increased, interaction among households. It may be that seasonal dispersal actually fostered more contact and cooperation among households of different communities, as Sassaman (1993: 225) has suggested for the interlude following the decline of Poverty Point exchange in the Late Archaic. Consistent with this notion, the larger houses of the terminal Late Woodland could have also provided more space for entertaining members of other households (Muller 1997: 188). At a finer scale, larger houses may indicate greater intrahousehold cooperation if they are indicative of the incorporation of family groups into larger, co-resident households.

The observations should not be construed as mere nitpicking with models of macroscale historical and evolutionary trends in the development of cooperation and conflict among early village societies. Such narratives fill an important need because, as Gerritsen (2004: 144) has observed for household studies, "the focus on practices of daily life stimulates detailed, small-scale, and synchronic studies, but at the same time appears to stand in the way of a perspective combining the small social scale with broader diachronic developments."

Yet general trends may obscure potentially important oscillations and countertrends, and these exceptions are more than just evolutionary noise or drift. They demonstrate that cooperation and conflict, rather than alternating in neat evolutionary progressions, are frequently counterposed at different social and spatial scales (see also Roscoe, chapter 3; Saitta, chapter 6). Moreover, they remind us that individual motivations for cooperation and competition are complex, compound, and potentially even conflicting. As Sewell (2005: 110) notes, "Big and ponderous social processes are never entirely immune from being transformed by small alterations in local social processes."

ACKNOWLEDGMENTS

An earlier version of this paper was presented at the Meeting of the Society for American Archaeology in Atlanta in 2009. Collections from Blocks A and D are curated at the University of Georgia, Athens. Excavation of Block A at Kolomoki was supported by a grant from the National Geographic Society to Stephen A. Kowalewski. Excavation of Block D was supported by a grant from the National Geographic Society to the author. Additional support was provided by the University of Georgia, the University of Oklahoma, and the University of South Florida. I thank all of the field school students and volunteers who participated in the excavations. I also appreciate the support of the Georgia Department of Natural Resources and the staff of Kolomoki Mounds State Historic Park. Finally, thanks to David Carballo for his helpful comments.

NOTES

1. I utilize the term "archaeological household" as defined by Nash (2009: 224): a "coresidential group that used the occupation surface, features, and the artifact assemblage of a dwelling," with "dwelling" defined to include one or more closely related structures and both indoor and outdoor spaces.

2. Calibrated date ranges calculated using OxCal v4.1.7 (© Christopher Bronk Ramsey 2010) and the IntCal09 calibration curve (Reimer et al. 2009).

3. For the chi-square test, the ridge-pinched, cord-marked, check-stamped, and dentate categories were collapsed into "other," since the low expected values (< 5) violate the assumptions of this test.

REFERENCES

Axelrod, Robert. 1984. *The Evolution of Cooperation.* New York: Basic Books.

Axelrod, Robert. 1997. *The Complexity of Cooperation: Agent-Based Models of Competition and Collaboration.* Princeton: Princeton University Press.

Bender, Barbara. 1990. "The Dynamics of Nonhierarchical Societies." In *The Evolution of Political Systems: Sociopolitics in Small-Scale Sedentary Societies*, ed. S. Upham, 247–263. Cambridge: Cambridge University Press.

Bettinger, Robert L. 1999. "What Happened in the Medithermal." In *Models for the Millennium*, ed. C. Beck, 62–74. Salt Lake City: University of Utah Press.

Binford, Lewis R., Sally R. Binford, Robert Whallon, and Margaret Ann Hardin. 1970. *Archaeology at Hatchery West.* Memoirs of the Society for American Archaeology No. 24, Washington, D.C.

Blanton, Richard, and Lane Fargher. 2008. *Collective Action in the Formation of Pre-Modern States.* New York: Springer.

Blanton, Richard E., Gary M. Feinman, Stephen A. Kowalewski, and Peter N. Peregrine. 1996. "A Dual-Processual Theory for the Evolution of Mesoamerican Civilization." *Current Anthropology* 37 (1): 1–14. http://dx.doi.org/10.1086/204471.

Blitz, John H. 1988. "Adoption of the Bow in Prehistoric North America." *North American Archaeologist* 9 (2): 123–45. http://dx.doi.org/10.2190/HN64-P1UD-NM0A-J0LR.

Bonhage-Freund, Mary Theresa. 2010. Paleoethnobotanical Analysis of Flotation Samples, NGS II Kolomoki Project (9ER1). Intermet Associates, Alma, Michigan. Submitted to Thomas Pluckhahn, University of South Florida, Tampa.

Bonhage-Freund, Mary Theresa, and Jeffrey A. Kurland. 1994. "Tit-For-Tat among the Iroquois: A Game Theoretic Perspective on Inter-Tribal Political Organization." *Journal of Anthropological Archaeology* 13 (3): 278–305. http://dx.doi.org/10.1006/jaar.1994.1016.

Bourdieu, Pierre. 1977. *Outline of a Theory of Practice.* Cambridge: Cambridge University Press.

Bourdieu, Pierre. 1980. *The Logic of Practice.* Stanford: Stanford University Press.

Braun, David P. 1986. "Midwestern Hopewellian Exchange and Supralocal Interaction." In *Peer Polity Interaction and Socio-Political Change*, ed. C. Renfrew and C. F. Cherry, 117–126. Cambridge: Cambridge University Press.

Braun, David P. 1987. Coevolution of Sedentism, Pottery Technology, and Horticulture in the Central Midwest, 200 BC-AD 600. In *Emergent Horticultural Economies of the Eastern Woodlands*, ed. W. Keegan, 153–181. Occasional Paper 7. Center for Archaeological Investigations, Southern Illinois University, Carbondale.

Brose, David S., and George Percy. 1974. Weeden Island Ceremonialism: A Reappraisal. Paper presented at the 39th Annual Meeting of the Society for American Archaeology, Washington, DC.

Carneiro, Robert. 1970. "A Theory of the Origin of the State." *Science* 169:773–38.

Carneiro, Robert. 1978. "Political Expansion as an Expression of the Principle of Competitive Exclusion." In *Origins of the State: The Anthropology of Political Evolution*, ed. Ronald Cohen and Elman R. Service, 205–224. Philadelphia: Institute for the Study of Human Issues.

Carneiro, Robert. 1981. "The Chiefdom: Precursor of the State." In *The Transition to Statehood in the New World*, ed. G. D. Jones and R. R. Kautz, 37–79. New York: Cambridge University Press.

Carr, Christopher. 2008. "Coming to Know Ohio Hopewell Peoples Better: Topics for Future Research, Masters' Theses, and Doctoral Dissertations." In *The Scioto Hopewell and Their Neighbors: Bioarchaeological Documentation and Cultural Understanding*, ed. D. Troy Case and Christopher Carr, 603–690. New York: Springer. http://dx.doi.org/10.1007/978-0-387-77387-2_15

DeBoer, Warren. 1988. "Subterranean Storage and the Organization of Surplus: the View from Eastern North America." *Southeastern Archaeology* 7 (1): 1–20.

Dornan, Jennifer L. 2002. "Agency and Archaeology: Past, Present, and Future Directions." *Journal of Archaeological Method and Theory* 9 (4): 303–29. http://dx.doi.org/10.1023/A:1021318432161.

Dye, David H. 2009. *War Paths, Peace Paths: An Archaeology of Cooperation and Conflict in Native Eastern North America*. Lanham, Maryland: AltaMira Press.

Flannery, Kent V. 2002. "The Origins of the Village Revisited: From Nuclear to Extended Households." *American Antiquity* 67 (3): 417–33. http://dx.doi.org/10.2307/1593820.

Gerritsen, F. 2004. "Archaeological Perspectives on Local Communities." In *A Companion to Archaeology*, ed. J. Bintliff, 141–154. Malden, MA: Blackwell Publishing.

Giddens, Anthony. 1979. *Central Problems in Social Theory: Action, Structure, and Contradiction in Social Analysis*. Berkeley: University of California Press.

Giddens, Anthony. 1984. *The Constitution of Society: Outline of a Theory of Structuration*. Berkeley: University of California Press.

Gilman, Antonio. 1981. "The Development of Social Stratification in Bronze Age Europe." *Current Anthropology* 22 (1): 1–24. http://dx.doi.org/10.1086/202600.

Gintis, Herbert, Eric Alden Smith, and Samuel Bowles. 2001. "Costly Signaling and Cooperation." *Journal of Theoretical Biology* 213 (1): 103–19. http://dx.doi.org/10.1006/jtbi.2001.2406.

Gramsci, Antonio. 1971. *Selections from the Prison Notebooks*. New York: International Publishers.

Hally, David J. 2009. Late Woodland to Mississippian Vessel Assemblage Change as a Response to Changing Foodways. Paper presented at the Annual Meeting of the Society for American Archaeology, Atlanta.

Hayden, Brian. 2001. "Fabulous Feasts: A Prolegomenon to the Importance of Feasting." In *Feasts: Archaeological and Ethnographic Perspectives on Food, Politics, and Power*, ed. Michael Dietler and Brian Hayden, 23–64. Washington: Smithsonian Institution Press.

Hendon, Julia A. 1996. "Archaeological Approaches to the Organization of Domestic Labor: Household Practice and Domestic Relations." *Annual Review of Anthropology* 25 (1): 45–61. http://dx.doi.org/10.1146/annurev.anthro.25.1.45.

Hudson, Charles. 1976. *The Southeastern Indians.* Knoxville: University of Tennessee Press.

Iannaccone, L. R. 1992. "Sacrifice and Stigma: Reducing Free-Riding in Cults, Communes, and Other Collectives." *Journal of Political Economy* 100 (2): 271–91. http://dx.doi.org/10.1086/261818.

Irons, W. 2001. "Religion as a Hard-to-Fake Sign of Committment." In *Evolution and the Capacity for Commitment*, ed. R. Neese, 292–309. New York: Russell Sage Foundation.

Kantner, John. 1996. "Political Competition among the Chaco Anasazi of the American Southwest." *Journal of Anthropological Archaeology* 15 (1): 41–105. http://dx.doi.org/10.1006/jaar.1996.0003.

Kantner, John. 2004. *Ancient Puebloan Southwest.* Cambridge: Cambridge University Press.

Kelly, John E., Andrew C. Fortier, Stephen J. Ozuk, and Joyce A. Williams. 1987. *The Range Site: Archaic Through Late Woodland Occupations.* American Bottom Archaeology FAI–270 Site Reports, Volume 16. Urbana: University of Illinois Press.

Kelly, John E., Stephen J. Ozuk, and Joyce Williams. 1990. *The Range Site 2: The Emergent Mississippian Dohack and Range Phase Occupations.* American Bottom Archaeology FAI–270 Site Reports, Volume 20. Urbana: University of Illinois Press.

Knight, Vernon J., Jr. 1990. *Excavation of the Truncated Mound at the Walling Site: Middle Woodland Culture and Copena in the Tennessee Valley.* Report of Investigations 56. Division of Archaeology, Alabama State Museum of Natural History. Tuscaloosa: University of Alabama.

Knight, Vernon J., Jr. 2001. "Feasting and the Emergence of Platform Mound Ceremonialism in Eastern North America." In *Feasts: Archaeological and Ethnographic Perspectives on Food, Politics, and Power*, ed. Michael Dietler and Brian Hayden, 311–333. Washington: Smithsonian Institution Press.

Kohler, Timothy A. 1991. "The Demise of Weeden Island, and Post-Weeden Island Cultural Stability." In *Non-Mississippianized Northern Florida. In Stability, Transformation, and Variation: The Late Woodland Southeast*, ed. M. S. Nassaney and C. R. Cobb, 91–110. New York: Plenum Press.

Kohler, Timothy A. 2004. "Introduction." In *Archaeology of Bandelier National Monument: Village Formation on the Pajarito Plateau, New Mexico*, ed. Timothy A. Kohler, 1–17. Albuquerque: University of New Mexico Press.

Kohler, T. A., Stephanie VanBuskirk, and Samantha Ruscavage-Barz. 2004. "Vessels and Villages: Evidence for Conformist Transmission in Early Village Aggregations on the Pajarito Plateau, New Mexico." *Journal of Anthropological Archaeology* 23 (1): 100–18. http://dx.doi.org/10.1016/j.jaa.2003.12.003.

Kohler, Timothy A., Matthew W. Van Pelt, and Lorene Y.L. Yap. 2000. "Reciprocity and Its Limits: Considerations for a Study of the Prehispanic Pueblo World." In *Alternative Leadership Strategies in the Prehispanic Southwest*, ed. Barbara J. Mills, 180–206. Tucson: University of Arizona Press.

Kohler, Timothy A., and Carla Van West. 1996. The Calculus of Self-Interest in the Development of Cooperation: Sociopolitical Development and Risk among the Northern Anasazi. In *Evolving Complexity and Environmental Risk in the Prehistoric Southwest*, ed. Joseph A. Tainter and Bonnie Bagley Tainter, 169–196. Studies in the Sciences of Complexity Proceedings 24, Addison-Wesley, Reading, MA.

Larson, Lewis H. 1952. Unpublished report summarizing excavations in Mound H, Kolomoki. On file at the University of Georgia Laboratory of Archaeology, Athens.

Lee, Richard B. 1990. "Primitive Communism and the Origin of Social Inequality." In *The Evolution of Political Systems: Sociopolitics in Small-Scale Sedentary Societies*, ed. S. Upham, 225–246. Cambridge: Cambridge University Press.

McElrath, Dale L., Thomas E. Emerson, and Andrew C. Fortier. 2000. "Social Evolution or Social Response? A Fresh Look at the "Good Gray Cultures" After Four Decades of Midwest Research." In *Late Woodland Societies: Tradition and Transformation across the Midcontinent*, ed. T. E. Emerson, D. L. McElrath, and A. C. Fortier, 3–36. Lincoln: University of Nebraska Press.

McGuire, Randall H. 1992. *A Marxist Archaeology*. San Diego: Academic Press.

McGuire, Randall H., and Dean J. Saitta. 1996. "Although They Have Petty Captains, They Obey Them Badly: The Dialectics of Prehispanic Western Pueblo Social Organization." *American Antiquity* 61 (2): 197–216. http://dx.doi.org/10.2307/282418.

Mead, Margaret. 1937. "Interpretative Statement." In *Cooperation and Competition among Primitive Peoples*, ed. Margaret Mead, 458–515. New York: McGraw-Hill.

Mickwee, Christopher. 2009. Wakulla in the Sandhills: Analysis of a Late Weeden Island Occupation in the Northwest Florida Interior Uplands. Paper presented at the meeting of the Florida Anthropological Society, Pensacola, Florida.

Milanich, Jerald T. 1974. "Life in a 9th Century Indian Household: A Weeden Island Fall-Winter Site on the Upper Apalachicola River, Florida." *Florida Bureau of Historic Sites and Properties Bulletin* 4:1–44.

Milanich, Jerald T., Ann S. Cordell, Vernon J. Knight, Jr., Timothy A. Kohler, and Brenda J. Sigler-Lavelle. 1997 [1984]. *Archaeology of Northern Florida, AD 200–900*. New York: Academic Press.

Moerman, Daniel E. 1998. *Native American Ethnobotany*. Portland, OR: Timber Press.

Moore, Clarence Bloomfield. 1900. "Certain Antiquities of the Florida West-Coast." *Journal of the Academy of Natural Sciences of Philadelphia, Second Series* 11 (3): 350–94.

Moore, Clarence Bloomfield. 1901. "Certain Aboriginal Remains of the Northwest Florida Coast. Part I." *Journal of the Academy of Natural Sciences of Philadelphia, Second Series* 11 (4): 421–97.

Moore, Clarence Bloomfield. 1902. "Certain Aboriginal Remains of the Northwest Florida Coast. Part II." *Journal of the Academy of Natural Sciences of Philadelphia, Second Series* 12 (2): 127–358.

Moore, Clarence Bloomfield. 1903a. "Certain Aboriginal Mounds of the Central Florida West-Coast." *Journal of the Academy of Natural Sciences of Philadelphia* 12:361–438.

Moore, Clarence Bloomfield. 1903b. "Certain Aboriginal Remains of the Apalachicola River." *Journal of the Academy of Natural Sciences of Philadelphia, Second Series* 12 (3): 440–94.

Moore, Clarence Bloomfield. 1905. "Certain Aboriginal Remains of the Lower Tombigbee River." *Journal of the Academy of Natural Sciences of Philadelphia, Second Series* 13 (2): 246–78.

Moore, Clarence Bloomfield. 1907. "Mounds of the Lower Chattahoochee and Lower Flint Rivers. *Journal of the Academy of Natural Sciences of Philadelphia.*" *Second Series* 13 (3): 426–57.

Muller, Jon. 1997. *Mississippian Political Economy*. New York: Plenum Press.

Nash, D. J. 2009. "Household Archaeology in the Andes." *Journal of Archaeological Research* 17 (3): 205–61. http://dx.doi.org/10.1007/s10814-009-9029-7.

Nassaney, Michael S. 1992. Communal Societies and the Emergence of Elites in the Prehistoric American Southeast. In *Lords of the Southeast: Social Inequality and the Native*

Elites of Southeastern North America, ed. A. Barker and T. Pauketat, 111–143. Archaeological Papers of the American Anthropological Association Number 3, Washington, D.C.
Nassaney, Michael S. 2000. "The Late Woodland Southeast." In *Late Woodland Societies: Tradition and Transformation across the Midcontinent*, ed. T. E. Emerson, D. L. McElrath, and A. C. Fortier, 713–30. Lincoln: University of Nebraska Press.
Nassaney, Michael S. 2001. "A Historical-Processual Development of Late Woodland Societies." In *The Archaeology of Traditions: Agency and History Before and After Columbus*, ed. T. R. Pauketat, 157–173. Gainesville: University Press of Florida.
Nassaney, M. S., and C. R. Cobb. 1991. "Patterns and Processes of Late Woodland Development in the Greater Southeastern United States." In *Stability, Transformation, and Variation: The Late Woodland Southeast*, ed. M. S. Nassaney and C. R. Cobb, 285–322. New York: Plenum Press.
Nassaney, Michael S., and Kendra Pyle. 1999. "The Adoption of the Bow and Arrow in Eastern North America: A View from Central Arkansas." *American Antiquity* 64 (2): 243–63. http://dx.doi.org/10.2307/2694277.
O'Brien, Michael J., and R. Lee Lyman. 2004. "History and Explanation in Archaeology." *Anthropological Theory* 4 (2): 173–97. http://dx.doi.org/10.1177/1463499604042813.
Ollman, Bertell. 1976. *Alienation*. Cambridge: Cambridge University Press.
Pauketat, Timothy R. 2000. "Tragedy of the Commoners." In *Agency in Archaeology*, ed. Marcia-Anne Dobres and John Robb, 113–129. New York: Routledge.
Pauketat, Timothy R. 2001a. "A New Tradition in Archaeology." In *The Archaeology of Traditions: Agency and History Before and After Columbus*, ed. T. R. Pauketat, 1–16. Gainesville: University Press of Florida.
Pauketat, Timothy R. 2001b. "Practice and History in Archaeology: An Emerging Paradigm." *Anthropological Theory* 1 (1): 73–98.
Pauketat, Timothy R. 2009. "Foreword." In *War Paths, Peace Paths: An Archaeology of Cooperation and Conflict in Native Eastern North America*, by David H. Dye., xi–xiii. Lanham, Maryland: AltaMira Press.
Pennisi, Elizabeth. 2005. "How did Cooperative Behaviors Evolve?" *Science* 309 (5731): 93. http://dx.doi.org/10.1126/science.309.5731.93.
Percy, George, and David Brose. 1974. Weeden Island Ecology: Subsistence and Village Life in Northwest Florida. Paper presented at the 39th Annual Meeting of the Society for American Archaeology, Washington, DC.
Pluckhahn, Thomas J. 2003. *Kolomoki: Settlement, Ceremony, and Status in the Deep South, AD 350 to 750*. Tuscaloosa: University of Alabama Press.
Pluckhahn, Thomas J. 2010a. "Practicing Complexity (Past and Present) at Kolomoki." In *Ancient Complexities: New Perspectives in Precolumbian North America*, ed. Susan M. Alt, 52–72. Salt Lake City: University of Utah Press.
Pluckhahn, Thomas J. 2010b. "The Sacred and the Secular Revisited: The Essential Tensions of Early Village Societies in the Southeastern U.S." In *Becoming Villagers: Comparing Early Village Societies*, ed. Matthew Bandy and Jake Fox, 100–118. Tucson: University of Arizona Press.
Pluckhahn, Thomas J., ed. 2011. The Household and the Making of Prehistory: Household Change in the Late Woodland at Kolomoki (9ER1). Department of Anthropology, University of South Florida, Tampa. Submitted to Georgia Department of Natural Resources, Atlanta. In preparation.
Pluckhahn, Thomas J., J. Matthew Compton, and Mary Theresa Bonhage-Freund. 2006. "Archaeological Correlates of Small-Scale Feasting: Evidence from the Woodland

Period Site of Kolomoki in Georgia." *Journal of Field Archaeology* 31 (3): 263–84. http://dx.doi.org/10.1179/009346906791071882.

Pluckhahn, Thomas J., and Sean P. Norman. 2010. What's the Point? Standardizing Middle/Late Woodland Projectile Point Identification (with an Example from Kolomoki). Paper presented at the Annual Meeting of the Florida Anthropological Society, Fort Myers, Florida.

Reber, Eleanora A. 2006. "A Hard Row to Hoe: Changing Maize Use in the American Bottom and Surrounding Areas." In *Histories of Maize: Multidisciplinary Approaches to the Prehistory, Linguistics, Biogeography, Domestication, and Evolution of Maize*, ed. John E. Staller, Robert H. Tykot, and Bruce F. Benz, 235–274. Burlington, MA: Academic Press.

Reimer, P. J., M. G. L. Baillie, E. Bard, A. Bayliss, J. W. Beck, P. G. Blackwell, C. Bronk Ramsey, C. E. Buck, G. S. Burr, R. L. Edwards, et al. 2009. "IntCal09 and Marine09 Radiocarbon Age Calibration Curves, 0–50,000 years cal BP." *Radiocarbon* 51 (4): 1111–50.

Richerson, Peter J., Robert T. Boyd, and Joseph Henrich. 2003. "Cultural Evolution of Human Cooperation." In *Genetic and Cultural Evolution of Cooperation*, ed. Peter Hammerstein, 357–388. Cambridge, Berlin: MIT Press and Preie Universität.

Sahlins, Marshall. 1958. *Social Stratification in Polynesia. American Ethnological Society, Monograph 29.* Seattle: University of Washington Press.

Saitta, D. J. 1999. Prestige, Agency, and Change in Middle-Range Societies. In *Material Symbols: Culture and Economy in Prehistory*, ed. John E. Robb, 135–149. Occasional Paper No. 26, Center for Archaeological Investigations, Southern Illinois University, Carbondale.

Saitta, D. J. 2007. *The Archaeology of Collective Action.* Gainesville: University Press of Florida.

Saitta, D. J., and Arthur S. Keene. 1990. Politics and Surplus Flow in Prehistoric Communal Societies. In *The Evolution of Political Systems: Sociopolitics in Small-Scale Sedentary Societies*, ed. S. Upham, 203–224. Cambridge University Press, Cambridge.

Sassaman, Kenneth E. 1993. *Early Pottery in the Southeast: Tradition and Innovation in Cooking Technology.* Tuscaloosa: University of Alabama Press.

Sassaman, Kenneth E. 2000. "Agents of Change in Hunter-Gatherer Technology." In *Agency in Archaeology*, ed. Marcia-Anne Dobres and John Robb, 148–168. New York: Routledge.

Sears, William H. 1951. *Excavations at Kolomoki: Season II, 1950.* Athens: University of Georgia Press.

Sears, William H. 1953. *Excavations at Kolomoki: Seasons III and IV, Mound D.* Athens: University of Georgia Press.

Sears, William H. 1956. *Excavations at Kolomoki: Final Report.* Athens: University of Georgia Press.

Seeman, Mark F. 1979. Feasting with the Dead: Ohio Hopewell Charnel House Ritual as a Context for Redistribution. In *Hopewell Archaeology: The Chillicothe Conference*, ed. D.S. Brose and N. Greber, 39–46. Kent State University Press, Kent, OH.

Seeman, Mark F. 1988. "Ohio Hopewell Trophy-Skull Artifacts as Evidence for Competition in Middle Woodland Societies Circa 50 BC-AD 350." *American Antiquity* 53 (3): 565–77. http://dx.doi.org/10.2307/281217.

Service, Elman. 1958. *A Profile of Primitive Culture.* New York: Harper and Row.

Service, Elman. 1962. *Primitive Social Organization: An Evolutionary Perspective.* New York: Random House.

Sewell, William H., Jr. 2005. *Logics of History: Social Theory and Social Transformation*. Chicago: The University of Chicago Press.
Smith, Eric A. 2003. "Human Cooperation: Perspectives from Behavioral Ecology." In *Genetic and Cultural Evolution of Cooperation*, ed. Peter Hammerstein, 401–427. Cambridge, Berlin: MIT Press and Preie Universität.
Smith, Eric A., and Rebecca Bliege Bird. 2005. "Costly Signaling and Cooperative Behavior." In *Moral Sentiments and Material Interests: The Foundations of Cooperation in Economic Life*, ed. Herbert Gintis, Samuel Bowles, Robert Boyd, and Ernst Fehr, 115–148. Cambridge: MIT Press.
Souvatzi, Stella G. 2008. *A Social Archaeology of Households in Neolithic Greece: An Anthropological Approach*. Cambridge: Cambridge University.
Stanish, Charles. 2004. "The Evolution of Chiefdoms: An Economic Anthropological Model." In *Archaeological Perspectives on Political Economies*, ed. Gary M. Feinman and Linda M. Nicholas, 7–24. Salt Lake City: University of Utah.
Stanish, Charles, and Kevin J. Haley. 2005. Power, Fairness, and Architecture: Modeling Early Chiefdom Development in the Central Andes. In *Foundations of Power in the Prehispanic Andes*, ed. Kevin J. Vaughn, Dennis Ogburn, and Christina A. Conlee, 53–70. Archaeological Papers of the American Anthropological Association, No. 14. American Anthropological Association, Arlington.
Steckel, Richard H. 2007. "Big Social Science History." *Social Science History* 31 (1): 1–34. http://dx.doi.org/10.1215/01455532-2006-012.
Swanton, John R. 1946. *Indians of the Southeastern United States*. Washington, D.C.: United States Government Printing Office.
Wallis, Neill J. 2001. *The Swift Creek Gift: Vessel Exchange on the Atlantic Coast*. Tuscaloosa: University of Alabama Press.
Waselkov, Gregory A. 1978. "Evolution of Deer Hunting in the Eastern Woodlands." *Midcontinental Journal of Archaeology* 3:15–34.
Wesson, Cameron B. 1999. "Chiefly Power and Food Storage in Southeastern North America." *World Archaeology* 31 (1): 145–64. http://dx.doi.org/10.1080/00438243.1999.9980436.
Whiteley, Peter M. 1987. "The Interpretation of Politics: A Hopi Conundrum." *Man* 22 (4): 696–714. http://dx.doi.org/10.2307/2803359.
Willey, Gordon R. 1949. *Archeology of the Florida Gulf Coast. Smithsonian Miscellaneous Collections 113*. Washington, D.C.: Smithsonian Institution.
Yost, Chad, and Linda Scott Cummings. 2010. Phytolith and Pollen Analysis of Pit Features from the Kolomoki Mounds Site (9er1), Southwestern Georgia. PaleoResearch Institute, Golden, Colorado. Submitted to the Department of Anthropology, University of Southern Florida, Tampa, Florida.

9

The Competitive Context of Cooperation in Pre-Hispanic Barinas, Venezuela

A Multilevel-Selection Approach

Charles S. Spencer

"Perhaps the most remarkable aspect of evolution," Martin A. Nowak commented in a recent paper (2006b: 1563), "is its ability to generate cooperation in a competitive world." Although the operation of variation-generating mechanisms combined with natural selection would seem to reward only selfish behavior—with each gene, cell, or organism shaped to pursue its own success to the detriment of its competitors—there is nonetheless abundant evidence of cooperation throughout the biological and cultural spheres, including the cooperation of genes in genomes, the cooperation of cells in organisms, and the cooperation of organisms in groups. Cooperation plays a key role in human societies of varying political complexity. "Humans," Nowak noted, "are the champions of cooperation: From hunter-gatherer societies to nation-states, cooperation is the decisive organizing principle of human societies" (2006b: 1560). Understanding how cooperation evolves is a major challenge for both biological and social scientists.

In this chapter I address the role of cooperation in the political dynamics of the El Gaván polity, a chiefdom of modest size that appeared around AD 550 along the Río Canaguá in the Venezuelan state of Barinas. I propose that the successful operation of this society depended upon a cooperative relationship that arose through a process of negotiation between leaders and followers. Underlying my argument is the assumption that this cooperative process had both "top-down" and "bottom-up" dimensions, expressed as: (1) *coordination*, the strategies and tactics that leaders use to marshal the support of followers;

and (2) *collaboration*, the participatory strategies and tactics pursued by followers. Although these are my definitions, they are consistent with standard usage, as recognized by Webster's Ninth New Collegiate Dictionary, which provides the following definition of *coordinate*: "to bring into a common action, movement, or condition," connoting the active leadership required for successful coordination. Webster's definition of *collaborate*, "to cooperate with an agency or instrumentality with which one is not immediately connected," conveys the key point that the interests of followers are not completely identical to those of leaders, although they will most likely overlap. Both coordination and collaboration are associated with potential benefits and costs. Coordination can benefit leaders by giving them access to a larger pool of labor and potential surplus production, both crucial to the long-term security of their privileged position; the costs of coordination can include the energy required to support group projects, such as warfare or infrastructural improvement, as well as the personal costs (probably unpleasant) to the leadership in the case of failure. Collaboration can benefit followers by providing the security and sense of community that come with large-group membership; the costs of collaboration can include the loss of personal autonomy and the labor and/or goods that are the price of participation in the larger social entity. This bilateral process of negotiation becomes especially delicate in those cases where aspiring leaders lack a full array of coercive mechanisms to compel compliance and potential followers have greater latitude to choose whether to follow. In such situations, we must seek to identify the contextual factors that favor this process of negotiated cooperation and the consequent success of the social entity that it unites. In the case of the El Gaván chiefdom, I argue that interpolity competition provided a context that favored intrapolity cooperation.

MODELING THE EVOLUTION OF COOPERATION

Evolutionary biologists have long sought to account for the prevalence of altruism, through which individuals engage in certain cooperative behaviors that do not enhance their own individual adaptiveness or fitness, and yet do provide benefits to others in their group (Nowak 2006a; Sober and Wilson 1998; Williams 1966, 1992; D. S. Wilson and E. O. Wilson 2007; Wynne-Edwards 1962, 1986). It would seem obvious that members of a social group must cooperate and do things for each other if the group is to enjoy adaptive success. Nevertheless, those members who do act selfishly (e.g., defectors, cheaters) can potentially benefit from the altruism of others and out-compete the cooperators; eventually the cooperators might die out, an outcome that would surely bode ill for the group's continued success. If natural selection is assumed to be operating only on individuals and/or genes, and not at all on groups of individuals (see Williams 1966), then it becomes difficult to explain the persistence of altruism, precisely because "traits that are 'for the good of the group' are usually not favored by selection within groups—what we have called the fundamental problem of social life" (D. S. Wilson and E. O. Wilson 2007: 336).

David Sloan Wilson has been a tireless advocate of the view that the evolution of cooperation is more effectively understood if one views selection as operating in a multilevel fashion, simultaneously adjudicating the relative fitness among groups as well as among individuals (e.g., 1975, 1976, 1983, 1989, 2007). From this perspective, altruistic traits will persist within a group if those traits are helpful to the group's survival, especially when a condition of serious competition develops between that group and other groups—i.e., when there is significant between-group selection. Moreover, D. S. Wilson and E. O. Wilson (2007: 336–338) have recently argued that a number of other contemporary approaches to the evolution of cooperation do not necessarily contradict—and indeed can often be subsumed by—the multilevel-selection perspective.

Prominent among these alternatives is inclusive fitness, or kin-selection, theory (Hamilton 1963, 1964), which hypothesizes that an altruistic act would be favored if *r*, the coefficient of relatedness (the probability of sharing a gene) between the donor and the recipient exceeds *c*/*b*, the cost-to-benefit ratio of the act (Nowak 2006b: 1560). Since the initial formulation, inclusive fitness theorists have tended to broaden the interpretation of *r*, noting that the members of a social group are more likely to have some genetic relationship to one other than to the members of other social groups within the larger population (D. S. Wilson and E. O. Wilson 2007: 337). This interpretive expansion has inevitably led to the inclusion of individuals of very low relatedness within the posited networks of altruism, thus reducing the relative impact of genetic relatedness on altruism to near insignificance (Nowak 2006b: Table 1). Inclusive fitness theory was initially, and most successfully, applied to a rather narrow range of phenomena, most notably the occurrence of altruism in hymenopterans, in which haplodiploidy is the method of sex determination (fertilized eggs become females, making sisters more closely related to one another than daughters are to their mothers). Yet this approach did not account for the altruistic (eusocial) behaviors that were increasingly noted in other species that did not use haplodiploidy—or the lack of altruism in still other species that did, thus weakening the argument that genetic relatedness is the prime mover behind altruism (Nowak, Tarnita, and Wilson 2010). Instead, in recent years there has been growing support for the viewpoint that altruistic behavior on the individual level is related to selection on the level of groups, which can consist of individuals of widely varying degrees of relatedness.

Another approach to the evolution of cooperation is direct reciprocity, which considers the options of cooperation versus defection, notably through the game-theoretic framework called the Prisoner's Dilemma, for which the most influential winning strategy has been Axelrod's (1984, 1997) "tit-for-tat" gambit (after a first round of cooperation, each player does whatever the other player has done in the previous round, either cooperate or defect). Because the pure tit-for-tat strategy is intolerant of mistakes, more robust versions have been proposed in recent years, including "generous-tit-for-tat" and "win-stay, lose-shift," the latter strategy consisting of repeating your previous move when you are doing well, but shifting otherwise (Nowak and Sigmund 1993). A key problem with the direct reciprocity approach is its assumption of repeated, direct encounters between the same two individuals. Nowak (2006b: 1561) noted that direct reciprocity can promote the evolution of cooperation only if the probability of another encounter between the same two individuals is greater than the cost/benefit ratio of the altruistic act in question. Especially in the case of human beings, the direct reciprocity approach would seem to be overly restrictive in its applicability.

Indirect reciprocity is a more robust approach than direct reciprocity, though it is analytically more complicated. An altruistic act, observed by others, enhances the reputation of the cooperator, who eventually receives a reward for his/her admirable behavior, even from those who did not directly observe the behavior but learned about it through information sharing (Nowak and Sigmund 1998). The dissemination of information is required for indirect reciprocity to promote widespread cooperation. It has been suggested that the evolution of language (a group-level trait) and associated human intelligence were promoted through successful indirect reciprocity (Nowak and Sigmund 2005). The evolution of cooperation will be promoted by indirect reciprocity, as Nowak (2006b: 1561) has hypothesized, if the probability of knowing an individual's reputation is greater than the cost/benefit ratio of the altruistic act. The persistence of a given altruistic act through indirect reciprocity necessarily depends on the recognition of the benefit by the group at large.

I submit that the more useful approaches to the evolution of cooperation tend to be based—implicitly if not always explicitly—on a multilevel view of selection, which assumes that a population is divided into groups and that "selection occurs within and between groups, that the balance between levels of selection can itself evolve, and that major transition occurs when election within groups is suppressed, enabling selection among groups to dominate the final vector of evolutionary change" (D. S. Wilson and E. O. Wilson 2007: 339). Previously, Redmond and I used a multilevel-selection framework to examine the evolution of the early Monte Albán state in Oaxaca, Mexico (Spencer and Redmond 2001). In the present chapter, my primary goal is to explore how the multilevel-selection perspective might shed light on the role of cooperation in the political dynamics of an ancient chiefdom in Barinas, Venezuela.

A multilevel-selection model of the evolution of cooperation has recently been presented by Traulsen and Nowak (2006). They begin by envisioning a population divided into groups, each consisting of individuals who interact with others within their group in an evolutionary "game" that determines their fitness. Cooperators collectively pay a cost, c, which allows other members of the group to receive a collective benefit, b. The total number of groups in the population is m, while n is the maximum number of individuals in any single group. Their model assumes that only individuals reproduce, although there is selection operating on both the individual and group levels. They then compute the baseline theoretical fixation probabilities for an individual cooperator in a group of $n-1$ defectors and for a cooperator group in a population of $m-1$ defector groups; the product of these probabilities, combining the effects of selection on the individual and group levels, is the baseline fixation probability of a given cooperator in the entire population (Traulsen and Nowak 2006: 10952–10953). With the additional assumption of rare group splitting (i.e., a constant number of groups), they go on to derive the following formula expressing the general conditions under which group selection would favor the evolution of cooperation:

$b/c > 1 + n/m$ (Traulsen and Nowak 2006: Equation 1)

Cooperation is favored, according to Equation 1, if the benefit/cost ratio of altruism is greater than one plus the ratio of group size to number of groups. This result, assert Traulsen and Nowak (2006: 10953), "is intuitively appealing. Smaller group sizes and larger numbers of groups favor cooperators," simply because it is easier for the benefit/cost ratio to exceed the right side of the inequality when n is small and m is large. Yet I would argue that the obverse of their statement is just as compelling: according to the Traulsen-Nowak formula, the benefits of cooperation must exceed the costs by an even greater margin in those cases that have larger group sizes relative to group number—in effect, setting the "bar" for benefits at a higher level, if cooperation is going to be favored by selection.

The application of the Traulsen-Nowak formula to empirical data presents us with a definite challenge. In most situations, the calculation of the left side of the inequality will pose more practical difficulties than the right side. This problem can be circumvented, I suggest, by focusing on the right side and employing the formula in a comparative fashion. For example, we could compute the right side of the inequality for cooperative arrangements in two or more different organizational levels of the same system. We could also carry out the computations for two points in time in a single system's trajectory, or for two different systems. A comparison of the results would suggest which of the cooperative arrangements under scrutiny is associated with a higher relative benefit/cost ratio, without requiring us to calculate that ratio directly. Moreover, this outcome can take on heuristic value if we view the calculated benefit ratio as a hypothesis, consider the nature of the benefits that multilevel selection would hypothetically provide to group members, and then proceed to evaluate the hypothesis with respect to independent lines of evidence.

Applying this approach to humans requires us to recognize the basic multilevel organization of human societies, what Crumley (1995) has called the "scalar hierarchy," whose manifestations include households composed of individuals, communities composed of households, and regional polities composed of communities (see also Spencer 1997). It is, admittedly, far from a simple undertaking to define the limits of the various organization levels in specific cultural contexts. Proximity in space and time, as well as likelihood of interaction, are useful criteria, but in some cases it may be necessary to invoke a bit of informed guesswork to identify group membership. Once the scalar hierarchy has been defined, we can use the Traulsen-Nowak method to calculate which of these organizational levels offers greater potential benefits to the corresponding cooperating social units. Like all models, the Traulsen-Nowak formula is a simplified version of real-life processes that are actually far more complicated, and so it is important to view such a calculation as a hypothesis that must be tested against the empirical record. In doing so, we should keep in mind D. S. Wilson and E. O. Wilson's (2007: 343) argument that the benefits of intragroup cooperation will positively co-vary with the intensity of intergroup selection: extremely high benefits to cooperators will be linked to very intense intergroup competition, the existence of which places a premium on the security associated with group membership, a point emphasized by Paul Roscoe in chapter 3 of this volume (see also West et al. 2006). These considerations apply to the evolution of cooperation not only in egalitarian but also in nonegalitarian societies; in the latter, both leaders and followers contribute to a negotiated cooperation and both stand to benefit from the enhanced security that results.

THE EL GAVÁN POLITY IN BARINAS, VENEZUELA

The El Gaván polity appeared around AD 550 along the banks of the Río Canaguá, in the western Venezuelan state of Barinas, where Elsa M. Redmond and I codirected a multiyear program of archaeological survey and excavation in the 1980s; these data, we have argued, indicate that the El Gaván polity was organized as a chiefdom (Redmond and Spencer 2007; Spencer 1998a; Spencer and Redmond 1992, 1998; Spencer et al. 1994). Elsewhere, I have defined the chiefdom as "a human society that has centralized political authority and institutionalized social status differentiation but lacks an internally specialized government" (Spencer 1998a: 105). The archaeological manifestations of a chiefdom include: a regional settlement hierarchy (in terms of site size) of two or three tiers, with a single first-order site at the top; an unequal distribution of public construction among the sites in this hierarchy, with the largest such buildings in the first-order center; evidence of regional political integration; and social status differentiation among basic social units such as households (Peebles and Kus 1977; Spencer 1987, 1998a; Wright 1984).

Our regional survey (Redmond and Spencer 2007) documented thirty-four habitation sites and two drained-field agricultural sites dating to the Late Gaván phase, AD 550–1000, distributed across a 290-square-kilometer zone of the high llanos (Figure 9.1). A histogram of site sizes revealed a bimodal distribution: the largest site was B12, on a contemporary *fundo* (ranch) known as El Gaván, which became the name we decided to use for the archaeological site as well as the prehistoric regional polity (Spencer and Redmond 1992). B12 covered an estimated 33 ha and was much larger than the other sites, which ranged in size from 0.5 to 9.4 ha (Redmond and Spencer 2007: Table 5.2, Figure 5.2). B12 also had the largest earthen mounds in the El Gaván region; two mounds, measuring 12 m and 10 m in height, stood on opposite sides of an avenue or elongated plaza 500 m long (Figure 9.2). At B12 we also recorded four other mounds 1–4 m in height and 130 other mounds 1 m or less high. Only the two largest mounds show signs of having been ceremonial in nature; the others probably supported residences

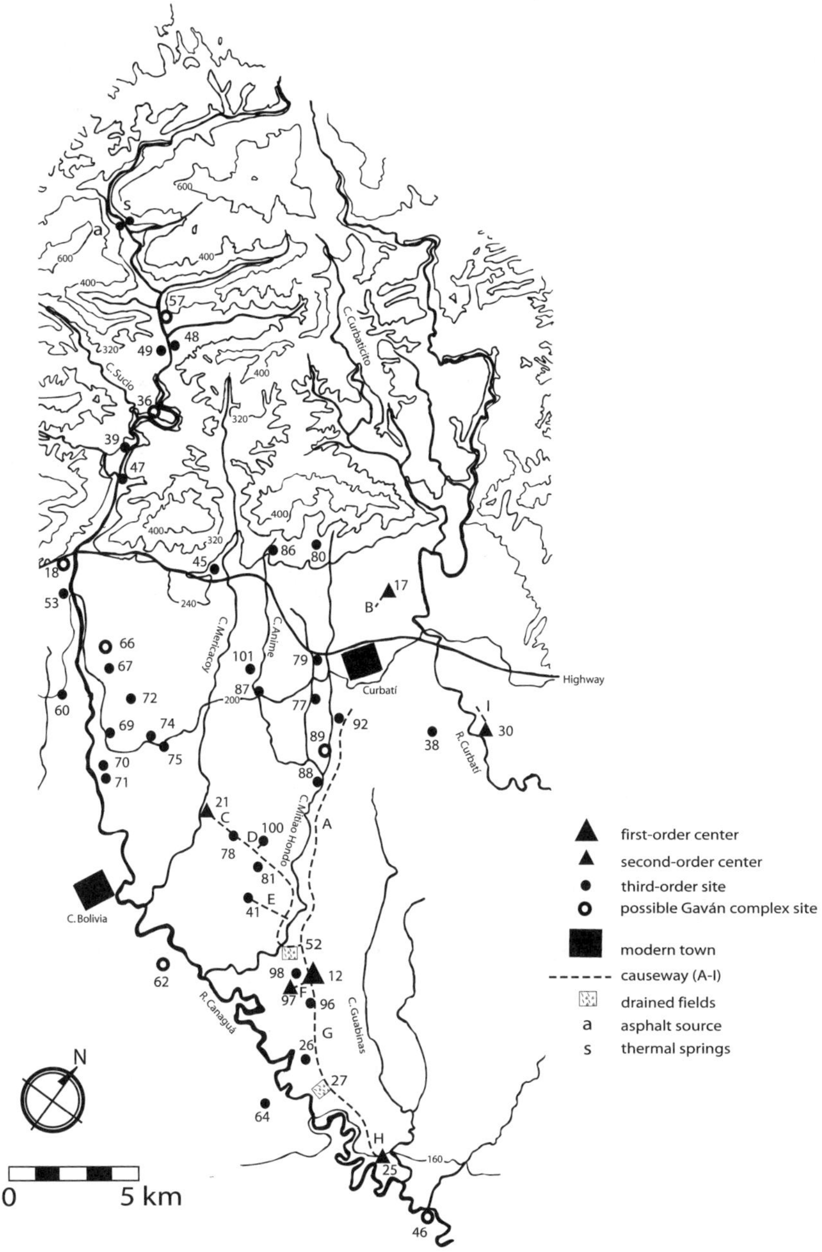

FIGURE 9.1 *Regional settlement patterns during the Late Gaván phase (AD 550–1000). Dotted lines demarcate segments of the calzada network.*

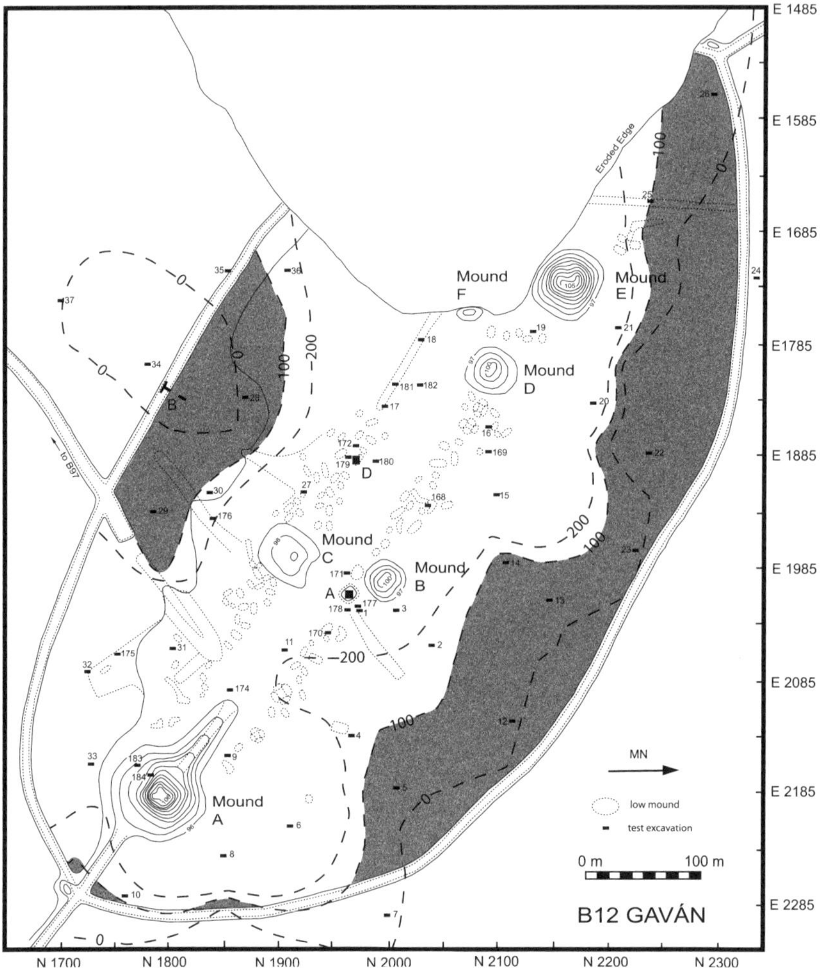

FIGURE 9.2 *Map of El Gaván (B12), showing major earthen mounds, house mounds, the encircling calzada-like earthwork, other calzada-like features within the site, and three intersite calzadas approaching the site from the northwest, southwest, and southeast. Numbered test pits are shown, along with the lettered areas of horizontal excavation. Samples recovered from the test program were used to generate a sherd-density contour map. Resulting density intervals of 0, 100, and 200 sherds per square meter are superimposed on the site map; much higher sherd densities in the center of the site were not superimposed to avoid visual clutter. The shaded area indicates a zone of low sherd density within the oval earthwork.*

(Spencer and Redmond 1998). All the mounds at B12 lie within an oval earthwork that circumscribes the site.

B12's relatively large size and impressive mounds would be consistent with a centralized regional political organization (Lathrap, Marcos, and Zeidler 1977; Peebles and Kus 1977). At the same time, the low quantity and diversity noted in B12's public (institutional/ceremonial, nonresidential) architecture (just two large mounds) would accord with chiefdom organization (Spencer 1987). By contrast, the internally specialized (i.e., bureaucratic) administration of the state is typically associated with much more diversity in public architecture at its political capital than was the case at B12 (Flannery and Marcus 1976; Spencer 1990; Spencer and Redmond 1998).

Our regional survey located five other sites that we interpret as possible second-tier settlements in the El Gaván region (B97, B21, B25, B17, and B30); these sites range in size from 4.6 to 9.4 ha, each site with two to four mounds that reach 2–6 m in height. A third settlement tier appears to be represented by the remaining 28 sites, which ranged in size from 0.5 to 4.4 ha and had no mounds detectable by surface survey (Redmond and Spencer 2007: Table 5.2). But, whether we interpret the regional settlement hierarchy as having two or three tiers, it lies within the expected range for a chiefdom. Moreover, B12 was decidedly in a class of its own in terms of site size and mounded architecture; there is little reason to doubt that B12 was the political center of the El Gaván chiefdom.

Evidence of the regional integration of the El Gaván polity can be seen in the *calzada* (earthen causeway) network that linked the first-order site (B12) to four, and perhaps all five, of the possible second-tier sites, as well as to many of the smaller sites (Figure 9.1). Three intersite calzadas approach B12 from the northwest, southeast, and southwest (Figure 9.2). We have estimated that the calzada network would have connected about two-thirds of the total population of the El Gaván regional polity (Spencer and Redmond 1998: 107). It is also notable that the two agricultural sites (B27, B52), where we recorded evidence of drained fields, lay along a calzada. The results of our survey and excavation at one of these drained-field sites (B27) revealed that maize was the primary crop (Spencer et al. 1994). We have proposed that the B27 drained fields were cultivated by collaborating farmers who lived at the nearby village of B26. Our calculations suggest that the B27 fields were capable of substantial surplus production, well beyond the needs of the local cultivators. This surplus, we suspect, was transported along the calzada network to B12, from which the regional leadership coordinated its mobilization, storage, and utilization (Spencer, Redmond, and Rinaldi 1994).

We found no evidence that the El Gaván calzada network extended to other river valleys to the east or west of the Río Canaguá. This would be consistent with our suggestion (Spencer and Redmond 1992, 1998) that the edges of the El Gaván political territory did not lie beyond 15–16 km (within a half day of travel on foot) from the regional center of B12—a political domain whose scale would be compatible with the centralized but not internally specialized (nonbureaucratic) nature of chiefdom political organization (Spencer 1987, 1990). The observed intrapolity (but not interpolity) extent of the calzada network would also imply that relations between the El Gaván polity and separate polities in other river valleys were not close enough to require calzada connections. Redmond, Gassón, and Spencer (1999) have proposed that such interpolity relations were distant and at times even hostile. In sum, based on the extent of the regional calzada network and the associated pattern of settlement, we have estimated that the El Gaván polity extended over a territory of roughly 290 square kilometers in this part of the Río Canaguá drainage (Redmond, Gassón, and Spencer 1999: 117).

Our data on interhousehold differentiation also agree with the expectations of a chiefdom model (Castaño 1987; Creamer and Haas 1985; Whalen 1983). At B12, we chose two of the 134 house mounds for horizontal excavations. We placed Area A on a house mound that reached one meter in height (Figure 9.2). Here we excavated two superimposed packed-earth house floors, Floors 1 and 2; the later floor, Floor 1, was associated with 41 postmolds that defined a roofed-over area of 27.9 square meters (Spencer and Redmond 1992: Figure 8). Most of these postmolds were carbonized, implying that the Area A house was burned upon abandonment, probably near the end of the Late Gaván phase, as suggested by the latest thermoluminescence date (AD 900 +/− 120) on pottery from the house (Spencer and Redmond 1992: Table 3). We placed Area D on a house mound that was 55 cm in height (Figure 9.2). Here we also excavated two superimposed packed-earth house floors, Floors 1 and 2; the later floor, Floor 1, was associated with 26 postmolds that demarcated a roofed-over area of 16.6 square meters, or about 59 percent of the Area A house (Spencer and Redmond 1992: Figure 9).

The difference in house size implies that the occupants of the Area A house were of higher status than those of the Area D house. This inference is supported by the excavated artifact samples. Consider the distribution of sherds from footed ceramic vessels (most of them bowls with tripod supports), which comprised 8.4 percent of the total diagnostic potsherds (all except undecorated body sherds) in our excavated samples from El Gaván polity sites (Redmond and Spencer 2007: 41). These footed vessels were elaborate serving dishes that would have been costlier to produce than most other Late Gaván vessel forms (Redmond and Spencer 2007: 34–42). So it is noteworthy that 25.0 percent of the diagnostic potsherds in our Area A excavation (Floors 1 and 2) were identified as fragments of footed vessels, whereas just 8.6 percent of the diagnostic potsherds from our Area D excavation (Floors 1 and 2) were from footed vessels. Another indicator of social differentiation is the ratio of outleaned-wall bowl rims to *olla* (necked jar) rims, which Drennan (1976: 77) has suggested is probably sensitive to status differences; the reasoning here is that, although a relatively invariable number of *ollas* is probably required for normal household activities, it is likely that higher-status households had more serving vessels like outleaned-wall bowls. In that light, it is notable that our Area A excavation (Floors 1 and 2) yielded a ratio of outleaned-wall bowl rims to olla rims of 3.25, while our Area D excavation (Floors 1 and 2) produced a value of 1.75 for the same ratio.

OTHER REGIONAL POLITIES

The El Gaván polity was not the only human occupation in this part of Venezuela during the latter half of the first millennium AD. Some 35 km south of the B12

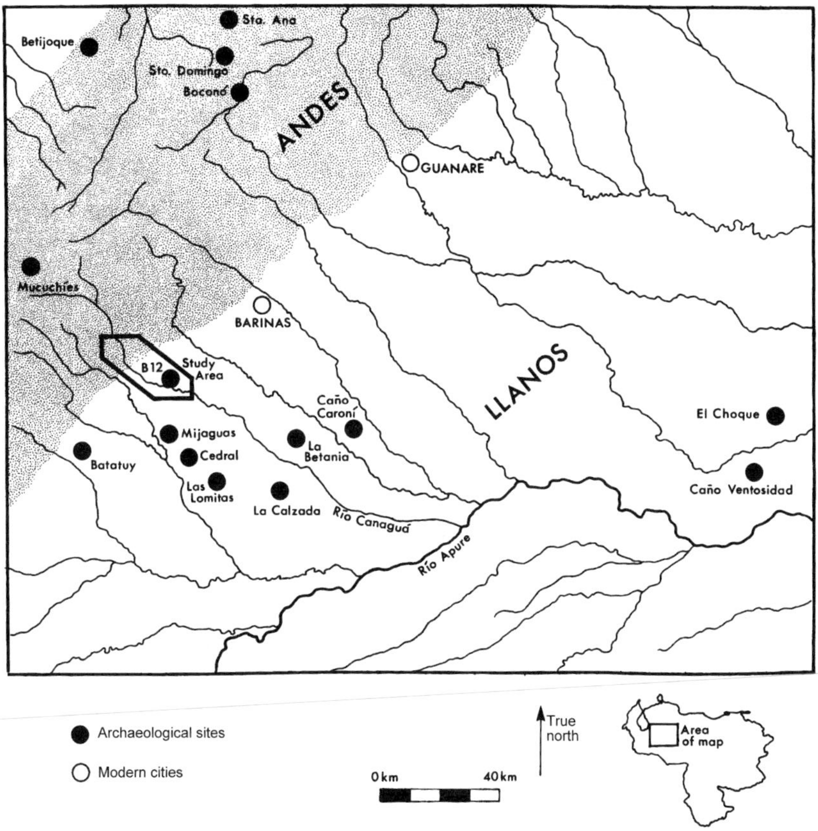

FIGURE 9.3 *Western Venezuelan llanos and adjacent Andes, with the locations of major archaeological sites and modern cities.*

site in the Río Acequia-Anaro drainage, Rafael Gassón (1998) has documented another regional chiefdom, the political center of which was the site of El Cedral (Figure 9.3). Radiocarbon dates from excavated samples at El Gaván and El Cedral reveal a substantial overlap in the occupational chronologies of the two sites, although it is possible that the initial occupation of El Gaván predated that of El Cedral (Redmond, Gassón, and Spencer 1999: Figure 7.9). There are numerous parallels between the community layouts of El Gaván and El Cedral, but it is clear that the latter site greatly surpasses the former in terms of scale. Gassón's map of El Cedral (Figure 9.4) shows that the site was ringed by an oval earthwork encompassing 150 ha, about 4.5 times the area circumscribed by the oval earthwork at El Gaván. El Cedral featured three large earthen mounds reaching 9–12 m in height, another 11–14 mounds that reach 2 m in height, and 122–125 mounds less than 1 m high. Because the site is currently occupied by a *fundo* and has been subjected to considerable earthmoving, the original number

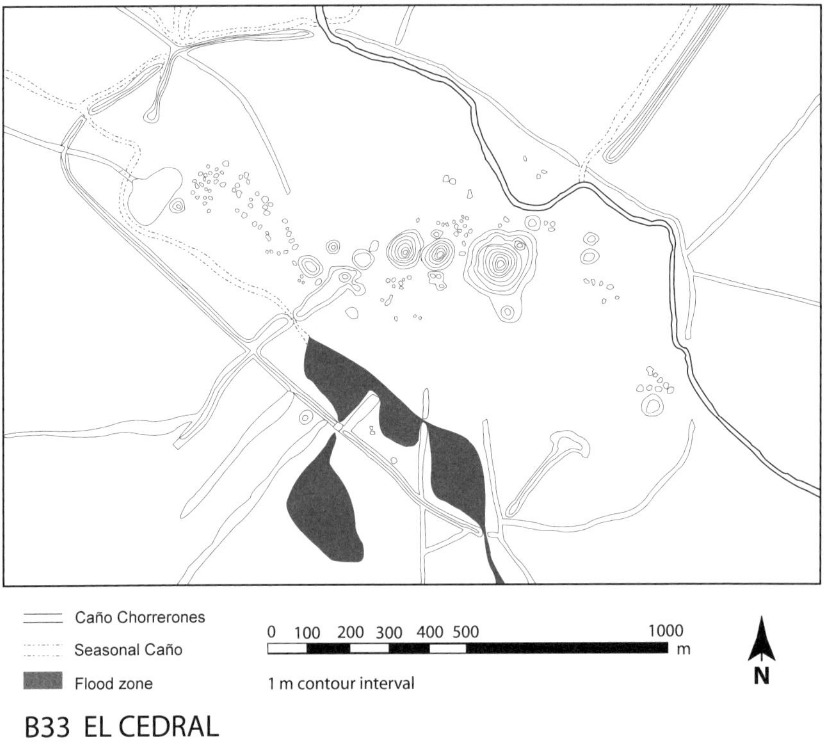

FIGURE 9.4 *Rafael Gassón's map of El Cedral (B33), showing major earthen mounds, house mounds, the encircling calzada-like earthwork, other calzada-like features within the site, and nine intersite calzadas approaching the site from the west, northwest, north, northeast, south, and southwest (redrawn from Gassón 1998: figure 2.3).*

of small mounds was probably much higher. Gassón (1998: 74–79) estimated El Cedral's population to range from 695 to 4,090 persons. He also recorded nine intersite calzadas that approached El Cedral from all sides (Figure 9.4), in contrast to the three that we found at El Gaván (Figure 9.2). In collaboration with Gassón, Juan Carlos Rey (2003) carried out a study of the regional calzada network that radiates out from El Cedral and discovered that many of them link the first-order center to smaller sites within the region. Notably, Gassón and Rey found no calzada connection between the El Cedral regional polity and the El Gaván regional polity, an observation that accords with the aforementioned results of Redmond's and my survey. A reasonable conclusion is that the calzada networks were important infrastructural features within—but not between—the independent chiefly polities of Barinas during the first millennium AD.

Gassón's survey of the El Cedral region located eight other sites 38–42 ha in size, which he suggested might represent a second tier in the regional settlement hierarchy (Gassón 1998). These eight sites are larger than the five sites that we identified as possible second-tier settlements in the El Gaván system. Gassón also conducted a systematic search for even smaller sites in a 60-square-kilometer survey zone to the northeast of El Cedral, in which he located (in addition to El Cedral itself) eight small sites, 0.5–1 ha in area. The territory occupied by the El Cedral regional polity, as defined by the distribution of the known sites and the network of calzadas, reached an estimated 448 square kilometers (Redmond, Gassón, and Spencer 1999: 122), about 1.54 times the territory that we have estimated for the El Gaván polity.

Also contemporaneous with the El Gaván and El Cedral occupations were regional polities centered on the La Calzada site (Garson 1980), about 50 km southeast of the El Gaván site, and the La Betania site (Zucchi 1967), some 45 km to the east-southeast of El Gaván (Figure 9.3). In the La Calzada region, Garson (1980) carried out a survey of 120 square kilometers and located 22 sites, of which 13 had one or more earthen mounds. He concluded that the largest site was La Calzada, whose size he did not determine, though he estimated that it was "larger than 15 hectares" (Garson 1980: 105, 294). The largest mound at La Calzada reached 13 m in height (Zucchi 1972: Figure 3), similar to the heights of the principal mounds at El Gaván and El Cedral. Two radiocarbon samples from the largest mound at La Calzada dated to the sixth century AD (Zucchi 1973). Garson recorded an extensive calzada network that linked the first-order center with many of the other sites; he noted that "the interconnection of settlements by causeways also reflects an intensity of relationship suggestive of a hierarchically organized society" (Garson 1980: 320).

In the piedmont zone immediately to the northwest of the El Gaván polity, we recorded sites of the Curbatí complex, the ceramic assemblage of which differed from that of El Gaván even though we have evidence that they were broadly contemporaneous. Our excavations at the Curbatí complex site of B8 yielded one radiocarbon sample, with a midpoint in the tenth century AD, and three thermoluminescence samples, whose dates ranged from the fourth to the eighth century AD (Spencer and Redmond 1992: Tables 2, 3). A histogram of site sizes revealed two modes; the B8 site covered 8 ha and was more than twice as large as any of the three other Curbatí complex sites that we located in our study region (Redmond and Spencer 2007: Figure 5.1). Some degree of centralization seems to be indicated, but more definitive conclusions should be deferred until further research on the Curbatí complex can be carried out.

About 40 km southwest of El Gaván, in the drainage of the Río Suripa, lies the mound site of Batatuy (Figure 9.3), which was reported by Armand (1975) to have two earthen mounds and an estimated site size of 10 ha. Two radiocarbon samples yielded dates with midpoints in the third century AD and the sixth century AD (Armand 1975:115). Although no survey has been conducted in the region surrounding Batatuy, it is likely that this area was occupied by another centralized regional polity contemporaneous with El Gaván. Finally, we have reports from local informants of other sites with impressive earthen mounds some 25–30 km to the east of El Gaván, along the banks of the Río Paguey.

Although this information has yet to be subjected to on-the-ground verification, we were not surprised to hear that this river drainage also showed signs of a complex pre-Hispanic society.

APPLYING THE MULTILEVEL-SELECTION MODEL

I suggest we can identify two major levels of cooperation in the archaeological data from first-millennium AD Barinas: (1) the community (or site) level, composed of cooperating households; and (2) the regional polity level, composed of cooperating communities (sites). We have noted that 134 house mounds were recorded at El Gaván (B12), each of which probably supported a single-family residence like the ones excavated in Area A and Area D (Spencer and Redmond 1992: Figures 8, 9). We have seen that the regional survey found a total of 34 habitation sites in the El Gaván polity (Redmond and Spencer 2007); the site-size distribution showed a two-level regional settlement hierarchy, and several of the sites were linked together by the *calzada* system that focused on B12. Finally, we have drawn upon additional data to identify four, and perhaps five, other regional polities with which the El Gaván polity could have been interacting during the mid- to late first millennium AD. Let us now consider to what extent multilevel-selection theory can help us account for patterns of cooperation on the community level and the regional polity level. In line with the earlier discussion, I will compute the right side of the Traulsen-Nowak formula separately for the two levels of cooperation proposed for the El Gaván case, and then compare the results.

Our first computation will be for the community level, composed of individual cooperating households. The total occupied area for all 34 habitation sites in El Gaván polity was about 126 ha (Redmond and Spencer 2007: Table 5.2). Drawing on the recorded density of households at B12, we can extrapolate to get an estimate of 512 households for the entire El Gaván polity. These households cooperated with others when they chose to aggregate into communities instead of dispersing across the landscape. When they did so, however, they did not aggregate into equally sized communities, but rather congregated into a regional settlement hierarchy, with the most households located at the regional center, fewer at secondary centers, and even fewer at third-order villages. Such a pattern required ordinary households to collaborate with the chiefly authority that was seeking to coordinate this system of differentially distributed households. The question here is not whether the collaborating households benefited somehow from this cooperative aggregation; it would seem that they must have for the pattern to have persisted as long as it did, even though it called for collaboration with the coordinating authority. Cooperative aggregation undoubtedly offered various benefits to the collaborating households, such as proximity to kin and a sense of community membership. But, the key question for our purposes here is to what extent we can attribute the benefits of the observed cooperation among households, and their collaboration with the chiefly authority, to the operation of intercommunity selection (which could have included competition over issues such as access to agricultural land, water sources, trade routes, or other resources). The Traulsen-Nowak approach provides us with one way to model such effects. In this application, *n* will equal the number of households in the largest community and *m* will equal the number of communities in the regional polity. The result of the computation will be an estimate of the benefit/cost ratio of cooperation (through collaboration) to individual households within a community, in order for the observed cooperation to be favored by intercommunity selection. The most complete community-level data come from our survey and excavations at El Gaván (B12). Since these excavation results are consistent with the survey-derived count of residences, we can reasonably use 134 as the value for *n* in the Traulsen-Nowak formula. The most logical value for *m* would be 34, which is the total number of habitation sites (i.e., communities) in the El Gaván regional polity. The outcome is: $b/c > 1 + 134/34$, or $b/c > 4.9$. This figure could be interpreted to indicate that there was some nontrivial benefit to cooperating (i.e., collaborating) households that derived from the operation of intercommunity selection. However, the magnitude of such a benefit is difficult to assess with this single result by itself. To gain some comparative perspective, we need to carry out a second computation of the Traulsen-Nowak formula.

Our second computation will be for the regional polity level, consisting of cooperating communities. Occurring in the context of chiefly organization, this cooperative process entailed an agreement or pact between collaborating villages and the coordinating regional authority, the material manifestations of which would have included not only the regional settlement hierarchy but also the *calzada* network that radiated out from the first-order center, connecting some two-thirds of the entire regional population in the El Gaván polity (Spencer and Redmond 1998). The Traulsen-Nowak computation requires us to estimate the number of interacting, and potentially competing, regional polities (*m*) as well as the number of collaborating subordinate communities in the largest of these chiefly polities (*n*). I have marshaled evidence indicating that the El Gaván polity was surrounded by perhaps five other polities, with which it could have been interacting and perhaps competing. The largest of these polities was probably the El Cedral polity, which, as noted earlier, covered some 448 square kilometers. Gassón (1998) pursued an informant survey over the entire territory that lay within the El Cedral polity, but his only area of intensive survey recovered 9 sites in a 60-square-kilometer zone; if we extrapolate from that density, we can estimate a total count of 67 sites for the El Cedral regional polity. Another way to estimate the total site count would be to multiply the total number of habitation sites in the El Gaván polity by 1.54, the factor by which the territory of the El Cedral polity exceed by that of El Gaván; this would yield an estimate of 51 sites in the El Cedral regional polity. To compute the Traulsen-Nowak formula for the regional polity level, let us use 5–6 as a reasonable range of values for *m*, and 51–67 as a reasonable range for *n*. The resulting computation would range from $b/c > 9.5$ through $b/c > 14.4$. It is noteworthy that these estimates of the benefits to collaborating communities that can be attributed to interpolity selection are considerably higher than the earlier estimated benefits to collaborating households that can be attributed to intercommunity selection.

To sum up, our comparative use of the Traulsen-Nowak model suggests that, although collaborating households may have benefited somewhat from intercommunity selection, collaborating communities derived even greater benefits from relatively stronger selection on the interpolity level. On both organizational levels, the outcome resulted from a cooperative process that involved the positioning of the collaborating social units within an organizational framework that was coordinated by a centralized chiefly authority. As noted earlier, it would be advisable to view this result as a hypothesis, which should then be assessed with reference to additional, independent lines of evidence. As a bridging assumption to that endeavor, let me posit that the more extreme selection hypothesized for the cooperative process on the interpolity level would likely have been expressed in the form of intense competition, probably including violence and warfare, between regional polities (West et al. 2006; D. S. Wilson and E. O. Wilson 2007). Accordingly, we should expect to see archaeological evidence of such interpolity conflict in our Barinas case.

EVIDENCE OF INTERPOLITY CONFLICT

It turns out that our five seasons of fieldwork in the region of the El Gaván polity did recover several lines of evidence that point to a condition of persistent conflict. From our very first visit to B12 in 1983, we were impressed by the oval, calzada-like earthwork that circumscribes the site (Figure 9.2); we found no such earthwork at the other sites in the El Gaván polity. Although a portion of the oval earthwork has been eroded away by a stream that skirts B12's western edge, we are reasonably confident that the entire oval (measuring 950 m by 470 m) was complete during the Late Gaván phase, circumscribing an area of some 33 ha. In our Area B, we excavated an alignment of postmolds along the centerline of this earthwork (Figure 9.5) that we interpreted as evidence of a palisade. We have argued (Spencer and Redmond 1992, 1998) that the oval earthwork at B12 served primarily as a defensive construction. Drawing on early historic documents, Morey (1975: 28) noted that the sixteenth-century societies of the Venezuelan llanos were said to have used a "palisade of tree trunks and earth" to fortify the "main village" of a regional polity. Morey also reported that a regional chief did not maintain a permanent standing army, but he was able to call up warriors from the villages of his domain to create a temporary fighting force that he deployed in offensive as well as defensive actions (1975: 96, 108, 277, 282–283, 309). In our Gaván case, we suspect such actions might have been undertaken in response to threats from the large and powerful El Cedral polity 35 km to the south of B12. The postmolds we found along the centerline of B12's oval earthwork were all carbonized and accompanied by ashy stains and areas of burned earth (Figure 9.5), indicating that the palisade was burned when B12 was abandoned toward the end of the Late Gaván phase. Redmond, Gassón, and Spencer (1999) have

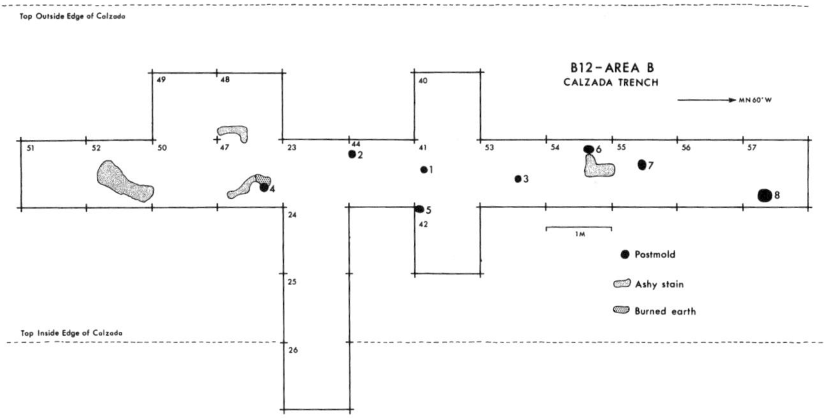

FIGURE 9.5 *Area B excavation on the circumscribing oval earthwork at El Gaván (B12); an alignment of carbonized postmolds was exposed on the centerline of the earthwork, suggesting that a palisade once stood upon it.*

argued that relations between the El Gaván and El Cedral polities included competition and intermittent conflict that lasted for much of the Late Gaván phase, but eventually El Cedral prevailed over El Gaván. B12 and all the other sites of the El Gaván polity were abandoned by AD 900–1000, although occupation apparently continued in the El Cedral polity.

Evidence of hostilities also came to light in the profile of Mound A, the largest earthen mound at B12 (Figure 9.2). We exposed the mound's southwestern profile by cleaning off the face of a bulldozer cut, which a local rancher had inflicted on the mound several years previously. A salient feature of the Mound A profile (Figure 9.6) is that three of the nine stratigraphic layers reflect episodes of burning. One of these is Layer H, which showed evidence of having been reddened by fire. The other two were Layer B and Layer D, both of which were dark brown and highly carbonaceous in appearance. If these burnings resulted from hostile actions against B12, the Mound A evidence would indicate that such actions were recurrent. It is also notable that two layers of noncarbonaceous, nonreddened brown fill lay between the reddened layer (Layer G) and the lower of the two carbonaceous layers (Layer D). Similarly, Layer D was separated from the upper carbonaceous layer (Layer B) by one layer of noncarbonaceous, nonreddened brown fill. This stratigraphic pattern implies that, after each burning episode, a new construction effort took place and the burned layer was covered with fresh fill, so that time spans of unknown length separated the three burning episodes. A pattern of recurring—but not permanent—warfare would be consistent with this stratigraphic sequence. Of course, since the Mound A profile only reflects episodes of burning at this one mound, it may well underrepresent the actual frequency of attacks that B12 experienced over the course of the Late Gaván phase. And, we can be sure that the Mound A profile does not document

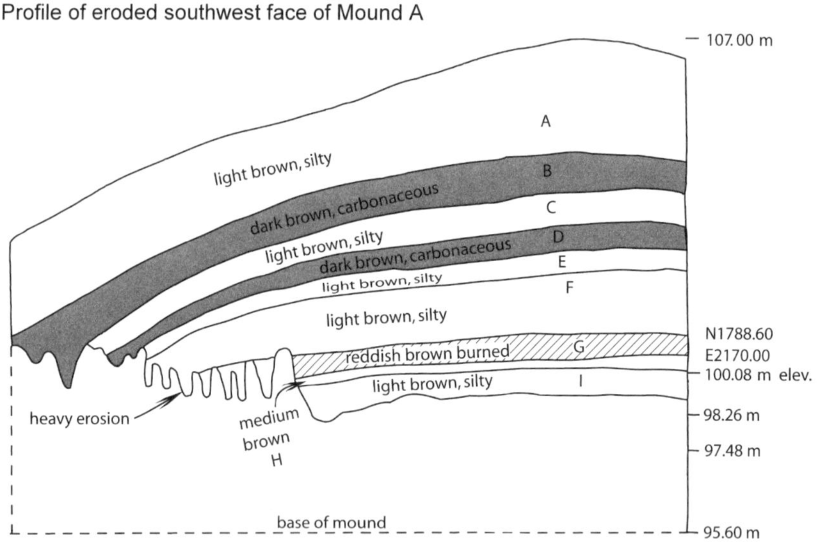

FIGURE 9.6 *Drawing of the southwest profile of Mound A at El Gaván (B12). Shaded zones represent recurring episodes of burning.*

the final years of B12's occupation, because most of the mound's top construction layer had eroded away before our fieldwork took place.

Although the inhabitants of B12 would have suffered burning and other destruction from these repeated attacks, they apparently managed to defend their community successfully each time, repelling the attackers and rebuilding the structures that had been burned. Then, after surviving three—and perhaps more—such episodes during the Late Gaván phase, the defenders experienced one final attack, which brought an end to the occupation at B12 and also an end to the El Gaván regional polity. It was this final attack that left the many carbonized postmolds we found in Floor 1 of Area A as well as those we recovered along the earthwork centerline in Area B. Moreover, this attack was undoubtedly responsible for the many fragments of burned daub that appeared in the latest occupation layer across much of B12. We recovered burned daub in the top excavation level of twenty-five test pits that lay within the oval earthwork. It is reasonable to conclude that B12 suffered catastrophic, widespread burning when it was abandoned for good near the end of the first millennium AD.

The threat of war was probably an ongoing, persistent concern to people throughout the El Gaván polity. The magnitude of this concern, I suggest, is reflected in the considerable effort they expended in the construction of the several *calzadas* that linked the first-order center to other sites. This effort required both labor and organization, including the collaboration of numerous villagers and effective coordination by the regional chiefly elite based at B12. We have argued that this regional *calzada* network would have facilitated the movement of people from smaller sites into the first-order center, not only during times of peace (perhaps to participate in large-scale ceremonies) but also—and probably more crucially—during times of war (Spencer and Redmond 1998). The inhabitants of a small village would have been especially vulnerable during an episode of interpolity warfare; one reasonable response would have been to seek shelter with others at the first-order center, a tactic that Redmond (1994: 36, 45–46) has documented for other indigenous South American societies.

Let us consider whether it would have been feasible for the community at B12 to accommodate temporary refugees from the other villages in the El Gaván polity. In addition to our horizontal excavations, Redmond and I carried out a test-pit program at B12 using a systematically stratified, random-sampling design. The test pits recovered little domestic debris beyond the extent of the mapped house mounds, suggesting there were very few, if any, residences that were not included in our original house mound count from surface indications. To depict these data, I have used the total "column density" of potsherds (i.e., all potsherds from all levels of each pit, divided by the area of the pit) as input for the Surfer 8.0 computer program (Golden Software, Inc. 2002), which generated a best-fitting contour map of column densities. The highest column densities ranged well above 1,000 sherds per square meter in the center of the site, but to avoid visual clutter I have superimposed only the three lowest column-density levels (0, 100, and 200 sherds per square meter) on the B12 site map in Figure 9.2. The shaded area indicates a zone of relatively low column density (0–100 sherds per square meter) within the oval earthwork. Because this low column-density zone also lacked house mounds, we have concluded that it was an area that had no permanent habitation. This proposed uninhabited zone covers 8.9 ha, which amounts to 31 percent of the B12 site area within the oval earthwork (28.7 ha), not counting the far western portion of the site that has suffered from erosion. Precisely because of this erosion, we do not know if an uninhabited zone also existed in the far western portion of the site. If there was no uninhabited zone there, then the total uninhabited portion of the site would constitute about 27 percent of the total estimated 33-ha site area within the oval earthwork. Thus, a reasonable estimate for the unoccupied part of the site within the oval earthwork would be 27–31 percent.

I suggest that the uninhabited 8.9-ha space within the oval earthwork at B12 was used as a place of temporary refuge by the collaborating inhabitants of subsidiary settlements within the El Gaván polity (for analogous cases, see Redmond 1994, 2007). Such an aggregation of the regional population would undoubtedly have been subject to coordination by the regional chiefly leadership based at B12. Since each temporary aggregation probably lasted only for the duration of a particular battle or campaign, it would not have required the construction of substantial residences for the refugees. Nor would we expect such temporary use to have resulted in the deposition of substantial quantities of ceramics and other artifacts, an expectation consistent with the results of our test pit program. I suspect that the uninhabited space sat empty most of the time, though always ready for a sudden influx from the smaller settlements.

But, one might ask, is it reasonable to suppose that the uninhabited space at B12 could have held, even on a temporary basis, all of the people who lived most of the time at the subsidiary settlements? If we use the aforementioned estimate of 512 households in the entire El Gaván polity, we get an estimated population for the 33 subsidiary settlements in the Gaván polity (i.e., excluding B12) of 378 households, or about 1,890 persons at 5 persons/household. Could this many people have fit on the 8.9 ha of uninhabited space at B12 for the duration of a battle or campaign? The estimated density of this temporary occupation would be 212 persons per hectare, which might seem high at first glance, but we should bear in mind that this temporary occupation would have been more of an encampment than a permanent settlement. The available space, at that density, would have allowed some 47 square meters per person or about 235 square meters for each five-person family unit, probably sufficient for a short-term occupation. By way of comparison, Flannery (1983: 133) has estimated that some 15,000 persons could have "fit comfortably" in the 4.5-ha Main Plaza of Monte Albán in Oaxaca, Mexico; this would be a density of 3,333 persons per hectare, nearly 16 times the density we have estimated for the temporary occupation of the uninhabited zone at B12. In sum, I think it would have been feasible for all the inhabitants of the subsidiary villages in the Gaván polity to take refuge during wartime within the oval earthwork of B12 for short-term periods that might have lasted several days, weeks, or possibly even months.

Such temporary aggregations would have had benefits for both the refugees and their "hosts." The former, by collaborating with the B12 leadership, would have enjoyed the security of being in a larger group, while the latter (the coordinating leadership and their close followers who lived at B12) would have benefited from the presence of additional fighting power, especially important in the absence of a permanent, standing military (no evidence of which appears in our El Gaván data). If we assume that each household could have contributed two temporary warriors (a father and an older son, perhaps), the fighting force that B12 alone could have mustered would have numbered about 268. But if a similar contribution were made by all the households from subsidiary settlements who were aggregated at B12, an additional 756 warriors would have been available, increasing the temporary force to some 1,024 warriors—a unit far more capable of mounting an effective defense than B12's inhabitants acting alone. Of course, there also would have been costs associated with these aggregations, such as the food and other supplies required by the additional occupants of B12. I should point out, however, that the surpluses generated by drained-field facilities such as B27 could have been stored at B12 and used to sustain these wartime aggregations (Spencer et al. 1994).

Although all 34 habitation sites in the El Gaván polity ceased to be occupied by AD 900–1000, it is B12 that has produced the most evidence that its abandonment was accompanied by a major conflagration. The relative frequency of burned daub in all our excavations at B12 (expressed in terms of kg of burned daub per 100 kg of all potsherds) was 6.52, while the other excavated sites in the

El Graván polity produced much lower relative frequencies: 1.1 (B97), 0.79 (B21), 1.84 (B17), and 0 (B26) (Spencer and Redmond 1998). All in all, the data we have reviewed here are consistent with the proposition that B12 was the prime target for attackers from other polities and that villagers from the entire El Graván polity came together when necessary and rallied to its defense.

CONCLUSION

I conclude that the going was often tough for the El Graván regional polity, so the inhabitants of its 34 communities chose to "think as a group," in a manner consistent with the multilevel-selection model of cooperation (Traulsen and Nowak 2006; Wilson, Timmel, and Miller 2004). This is not to say the group lacked internal differences in social status, wealth, or power. As we have seen, there is compelling evidence that the El Graván polity was organized as a chiefdom, with centralized (though nonbureaucratic) leadership and pervasive social-status differentiation. I contend that the proposed mechanism of short-term aggregations for defense was part of an ongoing process of negotiated cooperation between the political leadership at B12 and the other members of the regional polity. This cooperative process required the inhabitants of the subsidiary settlements to *collaborate* and accept the authority of the B12 leadership; what they mostly got in return was the enhanced security that followed from participation in the larger regional polity. From the perspective of the B12 leadership, the negotiated cooperation required them to *coordinate*, which included providing the venue, overall management, and provisioning for the temporary aggregations during wartime; in return, the leadership received greater access to resources and labor from the subsidiary settlements.

The position I have taken is consistent with the "competitive interaction" model that Flannery and Marcus (2000) proposed for the evolution of Mesoamerican chiefdoms. I also see compatibilities between my views and those offered by Blanton and Fargher (2008, chapter 5), who examined the role of "social negotiation," and by Turchin and Gavrilets (2009), who highlighted warfare, in the political dynamics of hierarchical societies (see also Turchin 2003, 2006). Previously, I argued that there are inherent constraints on the strategies of coordination that can feasibly be implemented by the leadership of a chiefdom (Spencer 1987, 1990, 1998b). For instance, the centralized but nonbureaucratic nature of chiefly authority tends to preclude the permanent delegation of partial authority to subordinates and thus would be incompatible with the creation of a specialized enforcement branch of government such as a permanent military or police force. Yet I have sought to show in this chapter how interpolity competition can foster a process of negotiated cooperation between chiefly leaders and followers, allowing intrapolity integration to be achieved without the need for specialized institutions of coercion. Cooperation and competition both had key roles to play in the political dynamics of the El Graván chiefdom.

ACKNOWLEDGMENTS

The archaeological fieldwork that Elsa Redmond and I carried out in Barinas was supported by the National Science Foundation (BNS–8506192) and the Wenner-Gren Foundation (Grant No. 4798). I thank the two anonymous reviewers, as well as David Carballo, Elsa Redmond, and Joyce Marcus, for their helpful commentary on an earlier draft of this chapter.

REFERENCES

Armand, Jorge. 1975. *Batatuy*. Ediciones del Museo Arqueológico, No. 1. Mérida, Venezuela: Universidad de los Andes.

Axelrod, Robert. 1984. *The Evolution of Cooperation*. New York: Basic Books.

Axelrod, Robert. 1997. *The Complexity of Cooperation: Agent-Based Models of Competition and Collaboration*. Princeton: Princeton University Press.

Blanton, Richard, and Lane Fargher. 2008. *Collective Action in the Formation of Pre-Modern States*. New York: Springer.

Castaño Uribe, Carlos. 1987. La Vivienda y el Enterramiento como Unidades de Interpretación: Anatomía de Dos Casos de Transición del Modelo de Cacicazgo. In *Chiefdoms in the Americas,* ed. Robert D. Drennan and Carlos A. Uribe, 231–249. University Press of America, Landham, MD.

Creamer, Winifred, and Jonathan Haas. 1985. "Tribe Versus Chiefdom in Lower Central America." *American Antiquity* 50 (4): 738–54. http://dx.doi.org/10.2307/280164.

Crumley, Carole L. 1995. Heterarchy and the Analysis of Complex Societies. In *Heterarchy and the Analysis of Complex Societies*, ed. Robert Ehrenreich, Carole Crumley, and J. E. Levy, 1–5. American Anthropological Association Archaeological Papers, No. 6, Washington, D.C.

Drennan, Robert D. 1976. *Fábrica San José and Middle Formative Society in the Valley of Oaxaca*. University of Michigan Museum of Anthropology Memoirs, No. 8, Ann Arbor.

Flannery, Kent V. 1983. "The Legacy of the Early Urban Period: An Ethnohistoric Approach to Monte Albán's Temples, Residences, and Royal Tombs." In *The Cloud People: Divergent Evolution of the Zapotec and Mixtec Civilizations*, ed. Kent V. and Joyce Marcus, 132–136. New York: Academic Press.

Flannery, Kent V., and Joyce Marcus. 1976. "Evolution of the Public Building in Formative Oaxaca." In *Cultural Change and Continuity: Essays in Honor of James Bennett Griffin*, ed. Charles Cleland, 205–222. New York: Academic Press.

Flannery, Kent V., and Joyce Marcus. 2000. "Formative Mexican Chiefdoms and the Myth of the 'Mother Culture.'" *Journal of Anthropological Archaeology* 19 (1): 1–37. http://dx.doi.org/10.1006/jaar.1999.0359.

Garson, Adam G. 1980. Prehistory, Settlement and Food Production in the Savanna Region of La Calzada de Páez, Venezuela. Unpublished Ph.D. dissertation, Department of Anthropology, Yale University, New Haven.

Gassón, Rafael A. 1998. Prehispanic Intensive Agriculture, Settlement Pattern and Political Economy in the Western Venezuelan Llanos. Unpublished Ph.D. dissertation, Department of Anthropology, University of Pittsburgh, Pittsburgh.

Golden Software, Inc. 2002. Surfer 8.0. Golden, Colorado.

Hamilton, William D. 1963. "The Evolution of Altruistic Behavior." *American Naturalist* 97 (896): 354–6. http://dx.doi.org/10.1086/497114.

Hamilton, William D. 1964. "The Genetical Evolution of Social Behavior, I and II." *Journal of Theoretical Biology* 7 (1): 1–52. http://dx.doi.org/10.1016/0022-5193(64)90038-4.
Lathrap, Donald A., Jorge G. Marcos, and James A. Zeidler. 1977. "Real Alto: An Ancient Ceremonial Center." *Archaeology* 30 (1): 2–13.
Morey, Nancy K. 1975. Ethnohistory of the Colombian and Venezuelan Llanos. Unpublished Ph.D. dissertation, Department of Anthropology, University of Utah, Provo.
Nowak, Martin A. 2006a. *Evolutionary Dynamics: Exploring the Equations of Life*. Cambridge, MA: Harvard University Press.
Nowak, Martin A. 2006b. "Five Rules for the Evolution of Cooperation." *Science* 314 (5805): 1560–3. http://dx.doi.org/10.1126/science.1133755.
Nowak, Martin A., and K. Sigmund. 1993. "A Strategy of Win-Stay, Lose-Shift that Outperforms Tit-for-Tat in the Prisoner's Dilemma Game." *Nature* 364 (6432): 56–8. http://dx.doi.org/10.1038/364056a0.
Nowak, Martin A., and K. Sigmund. 1998. "Evolution of Indirect Reciprocity by Image Scoring." *Nature* 393 (6685): 573–7. http://dx.doi.org/10.1038/31225.
Nowak, Martin A., and K. Sigmund. 2005. "Evolution of Indirect Reciprocity." *Nature* 437 (7063): 1291–8. http://dx.doi.org/10.1038/nature04131.
Nowak, Martin A., Corina E. Tarnita, and Edward O. Wilson. 2010. "The Evolution of Eusociality." *Nature* 466 (7310): 1057–62. http://dx.doi.org/10.1038/nature09205.
Peebles, Christopher, and Susan M. Kus. 1977. "Some Archaeological Correlates of Ranked Societies." *American Antiquity* 42 (3): 421–48. http://dx.doi.org/10.2307/279066.
Redmond, Elsa M. 1994. *Tribal and Chiefly Warfare in South America*. University of Michigan Museum of Anthropology Memoirs, No. 28, Ann Arbor.
Redmond, Elsa M. 2007. "Circum-Caribbean Chiefly Warfare." In *Latin American Indigenous Warfare and Ritual Violence*, ed. J. Richard, 91–104. Mendoza: Chacon and Rubén G.
Redmond, Elsa M., Rafael A. Gassón, and Charles S. Spencer. 1999. A Macroregional View of Cycling Chiefdoms in the Western Venezuelan Llanos. In *Complex Polities in the Ancient Tropical World*, ed. Elisabeth A. Bacus and Lisa J. Lucero, 109–129. American Anthropological Association Archaeological Papers, No. 9, Washington, D.C.
Redmond, Elsa M., and Charles S. Spencer. 2007. *Archaeological Survey in the High Llanos and Andean Piedmont of Barinas, Venezuela*. American Museum of Natural History Anthropological Papers, No. 86, New York.
Rey, Juan Carlos. 2003. Aportes al Estudio Macro-regional de las Sociedades Prehispánicas de los Llanos Occidentales Venezolanos: Configuración Regional del Area de El Cedral. Undergraduate thesis, Universidad Central de Venezuela, Facultad de Ciencias Económicas y Sociales, Escuela de Antropología, Departamento de Arqueología, Etnohistória y Ecología Cultural, Caracas.
Sober, Elliott, and David Sloan Wilson. 1998. *Unto Others: The Evolution and Psychology of Unselfish Behavior*. Cambridge: Harvard University Press.
Spencer, Charles S. 1987. "Rethinking the Chiefdom." In *Chiefdoms in the Americas*, ed. Robert D. Drennan and Carlos A. Uribe, 369–390. Landham, MD: University Press of America.
Spencer, Charles S. 1990. "On the Tempo and Mode of State Formation: Neoevolutionism Reconsidered." *Journal of Anthropological Archaeology* 9 (1): 1–30. http://dx.doi.org/10.1016/0278-4165(90)90004-W.
Spencer, Charles S. 1997. "Evolutionary Approaches in Archaeology." *Journal of Archaeological Research* 5 (3): 209–64. http://dx.doi.org/10.1007/BF02229153.

Spencer, Charles S. 1998a. "Investigating the Development of Venezuelan Chiefdoms." In *Chiefdoms and Chieftaincy in the Americas*, ed. Elsa M. Redmond, 104–137. Gainesville: University Press of Florida.
Spencer, Charles S. 1998b. "A Mathematical Model of Primary State Formation." *Cultural Dynamics* 10 (1): 5–20. http://dx.doi.org/10.1177/092137409801000101.
Spencer, Charles S., and Elsa M. Redmond. 1992. "Prehispanic Chiefdoms of the Western Venezuelan Llanos." *World Archaeology* 24 (1): 134–57. http://dx.doi.org/10.1080/00438243.1992.9980198.
Spencer, Charles S., and Elsa M. Redmond. 1998. "Prehispanic Causeways and Regional Politics in the Llanos of Barinas, Venezuela." *Latin American Antiquity* 9 (2): 95–110. http://dx.doi.org/10.2307/971989.
Spencer, Charles S., and Elsa M. Redmond. 2001. "Multilevel Selection and Political Evolution in the Valley of Oaxaca, 500–100 B.C." *Journal of Anthropological Archaeology* 20 (2): 195–229. http://dx.doi.org/10.1006/jaar.2000.0371.
Spencer, Charles S., Elsa M. Redmond, and Milagro Rinaldi. 1994. "Drained Fields at La Tigra, Venezuelan Llanos: A Regional Perspective." *Latin American Antiquity* 5 (2): 119–43. http://dx.doi.org/10.2307/971559.
Traulsen, Arne, and Martin A. Nowak. 2006. "Evolution of Cooperation by Multilevel Selection." *Proceedings of the National Academy of Sciences of the United States of America* 103 (29): 10952–5. http://dx.doi.org/10.1073/pnas.0602530103.
Turchin, Peter. 2003. *Historical Dynamics: Why States Rise and Fall.* Princeton: Princeton University Press.
Turchin, Peter. 2006. *War and Peace and War: The Life Cycles of Imperial Nations.* New York: Pi Press.
Turchin, Peter, and Sergey Gavrilets. 2009. "Evolution of Complex Hierarchical Societies." *Social Evolution & History* 8:167–98.
West, Stuart A., Andy Gardner, David M. Shuker, Tracy Reynolds, Max Burton-Chellow, Edward M. Sykes, Meghan A. Guinnee, and Ashleigh S. Griffin. 2006. "Cooperation and the Scale of Competition in Humans." *Current Biology* 16 (11): 1103–6. http://dx.doi.org/10.1016/j.cub.2006.03.069.
Whalen, Michael E. 1983. "Reconstructing Early Formative Village Organization in Oaxaca, Mexico." *American Antiquity* 48 (1): 17–43. http://dx.doi.org/10.2307/279816.
Williams, George C. 1966. *Adaptation and Natural Selection: A Critique of Some Current Evolutionary Thought.* Princeton: Princeton University Press.
Williams, George C. 1992. *Natural Selection: Domains, Levels and Challenges.* New York: Oxford University Press.
Wilson, David Sloan. 1975. "A Theory of Group Selection." *Proceedings of the National Academy of Sciences of the United States of America* 72 (1): 143–6. http://dx.doi.org/10.1073/pnas.72.1.143.
Wilson, David Sloan. 1976. "Evolution on the Level of Communities." *Science* 192 (4246): 1358–60. http://dx.doi.org/10.1126/science.1273598.
Wilson, David Sloan. 1983. "The Group Selection Controversy: History and Current Status." *Annual Review of Ecology and Systematics* 14 (1): 159–87. http://dx.doi.org/10.1146/annurev.es.14.110183.001111.
Wilson, David Sloan. 1989. "Levels of Selection: An Alternative to Individualism in Biology and the Human Sciences." *Social Networks* 11 (3): 257–72. http://dx.doi.org/10.1016/0378-8733(89)90005-1.
Wilson, David Sloan. 2007. *Evolution for Everyone: How Darwin's Theory Can Change the Way We Think About Our Lives.* New York: Random House.

Wilson, David S., John J. Timmel, and Ralph R. Miller. 2004. "Cognitive Cooperation: When the Going Gets Tough, Think as a Group." *Human Nature (Hawthorne, N.Y.)* 15 (3): 225–50. http://dx.doi.org/10.1007/s12110-004-1007-7.

Wilson, David Sloan, and Edward O. Wilson. 2007. "Rethinking the Theoretical Foundation of Sociobiology." *Quarterly Review of Biology* 82 (4): 327–48. http://dx.doi.org/10.1086/522809.

Wright, Henry T. 1984. "Prestate Political Formations." In *On the Evolution of Complex Societies: Essays in Honor of Harry Hoijer*, ed. Timothy Earle, 41–77. Mailbu, CA: Undena Press.

Wynne-Edwards, Vero C. 1962. *Animal Dispersion in Relation to Social Behavior.* Edinburgh: Oliver and Boyd.

Wynne-Edwards, Vero C. 1986. *Evolution Through Group Selection.* Oxford: Blackwell Scientific.

Zucchi, Alberta. 1967. La Betania: Un Yacimiento Arqueológico del Occidente de Venezuela. 2 vols. Unpublished Ph.D. dissertation, Universidad Central de Venezuela, Facultad de Ciencias Económicas y Sociales, Caracas.

Zucchi, Alberta. 1972. "Aboriginal Earth Structures of the Western Venezuelan Llanos." *Caribbean Journal of Science* 12:95–106.

Zucchi, Alberta. 1973. "Prehistoric Human Occupation of the Western Venezuelan Llanos." *American Antiquity* 38 (2): 182–90. http://dx.doi.org/10.2307/279364.

10

Water Control and the Emergence of Polities in the Southern Maya Lowlands

Evolutionary, Economic, and Ecological Models

Benjamin Chabot-Hanowell and Lisa J. Lucero

In the southern Maya lowlands of present-day northern Guatemala, Belize, the Yucatán and southeastern Mexico, rulers reached their apogee in the Late Classic period (c. AD 550–850) (Figure 10.1). Several factors influenced the number of supporters at any given center, the main one being the prosperity of the royal court. Powerful kings emerged in areas with noticeable seasonal variability and plentiful fertile land (Lucero 2003, 2006). Powerful Maya polities did not emerge along rivers as one finds in Egypt, Mesopotamia, China, the coastal Andes, and other regions throughout the world; nor did the Maya rely on massive irrigation systems. Instead, powerful Maya rulers emerged in areas lacking rivers or lakes, where agriculture depended primarily on rainfall. During the four- to six-month seasonal drought, people also needed drinking water in areas without access to permanent water resources. Farmers lived in such areas due to the large pockets of fertile agricultural land. Lucero has argued elsewhere (e.g., 2002, 2003, 2006) that differential access to dry-season water necessitated cooperation between political elites and thirsty farmers, laying the foundation for the emergence of complex polities. Elites, as patrons, exchanged access to potable water and hosted expensive community rituals in exchange for labor, goods, and services from client farmers. Such events acted to solidify these critical economic connections. In this scenario, patron-client exchanges originated as a compromise between the descendants of the earliest settlers, who originally controlled access to fertile land rich in *aguadas* (natural rain-fed sinkholes), and

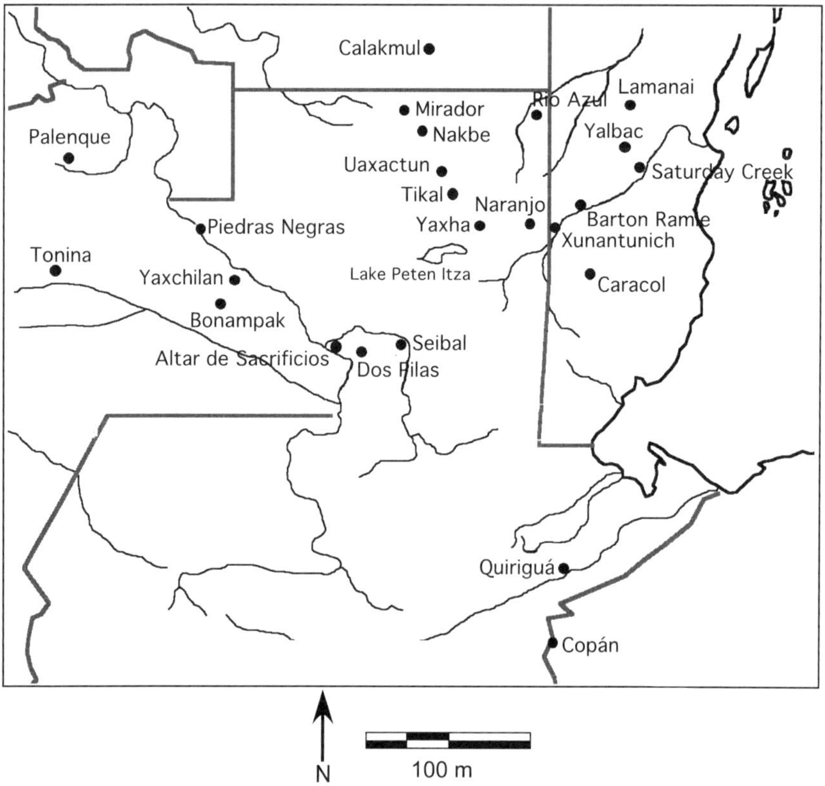

FIGURE 10.1 *Map of Maya area with sites mentioned in text (drawn by L. J. Lucero).*

latecomers, who had less access to aguadas. This compromise led to the construction of monumental buildings and artificial reservoirs. Eventually, farmers came to depend on the water reservoirs they had helped construct, while elites continued to accrue political and ritual control sufficient to demand (but not necessarily coerce) tribute from farmers.

Lucero based the above water control hypothesis on archaeological, paleoclimatic, and ethnographic evidence. Here we lend further support and theoretical rigor to the hypothesis by drawing on and extending abstract, quantitative models of territoriality and inequality developed in ecology and economics. Specifically, we review the patron-client scenario from political economics (emphasizing recent agent-based simulations) and bargaining models from economic game theory. We explore the connections between the results of these models and the social and ecological processes likely at work in the southern Maya lowlands. In doing so, we suggest a set of hypotheses that grow out of and embellish Lucero's original water control model.

PATRON-CLIENT RELATIONS: A SIMPLE SCENARIO WITH COMPLEX RESULTS

Scholars from Rousseau to Marx and beyond have discussed the role of patron-client relationships in the emergence and maintenance of institutionalized social inequality (Smith and Choi 2007). Archaeologists continue to invoke the patron-client scenario as a useful model of hierarchical social structure in several contexts (Johnson and Earle 1987), including ancient Mesoamerica (Brumfiel 1994). Most of the literature ties patron-client relationships in Mesoamerica into the broader context of craft specialization (Brumfiel 1987), while Lucero (2002, 2006) more recently considered patron-client relations as a paradigm for the political economy of Maya water control and rituals. Patron-client relationships are characterized by mutually beneficial, cooperative obligations between two social agents differentially endowed with power and resources. Patrons are wealthy and capable of defending a rich resource endowment (e.g., access to water). Clients possess a resource that is valuable to patrons (e.g., labor, political support). Patrons offer clients access to their resources in exchange for the client's resources. The putative cause of inequity in patron-client relationships is that patrons can potentially enjoy many more mutually beneficial relationships with clients than clients can with patrons. This inequity requires neither the coercion of commoner clients by elite patrons, nor constraints on socially valuable information available to clients. In this scenario, commoners are ironically complicit in their immediate or potential subjugation (Pauketat 2000). This seeming paradox arises because those who accept subjugation fare better relative to those who refuse it. As we will explain, the inequity of patron-client relationships only requires that patrons can defend exclusive access to their rich resource endowment. In addition, patron-client relationships imply no social dilemma, because relationships are mutually beneficial.

As Boone (1992) and others (Smith and Choi 2007; Summers 2005; Sutherland 1996) stress, resource defense is key to the development of patron-client relationships in particular and social hierarchy in general. Dyson-Hudson and Smith (1978) predicted that the emergence and stability of territorial strategies hinges on the predictability and density of resources. Ephemeral resource gluts are too large for a single individual to consume when encountered, which implies diminishing returns from each additional unit of resource. The presence of ephemeral resource gluts may lead to scrambles for what is available, the sharing of information about the location of temporarily abundant patches, and toleration of scrounging by other, less successful foragers (Winterhalder 1996). In contrast, resource predictability implies stable or increasing returns from each additional unit of resource, less incentive for sharing, and greater benefits to defense.[1] Where the benefits of defending a resource exceed the costs, the resource is said to be *economically defensible.*

A powerful way to understand complex systems is to distill relationships between variables through quantitative modeling and simulation. These methods allow researchers to clarify, simplify, and explicate their assumptions, then extrapolate causal relationships that follow from these assumptions. See

McElreath and Boyd (2007) and Otto and Day (2007) for concise descriptions of quantitative modeling and its benefits. Building from ecological models of territoriality, Smith and Choi (2007) developed a simple, agent-based model based on Boone's (1992) account of the patron-client scenario. The goal of the project was to show how initial variance in environmental productivity (and thus resource endowment), paired with economic defensibility of rich resources, may have allowed institutional inequality to emerge in small-scale societies, such as existed among the early Preclassic Maya. We argue that the patron-client simulation structure and results bear similarities to the Maya case. Yet in discussing these similarities, we discover the patron-client simulation's limitations. Far from discouraging such modeling exercises, the identification of model limitations supports their use because they make our verbal arguments more explicit and highlight areas for model extension.

The patron-client simulation involves agents who practice different behavioral strategies. These strategies govern their interactions with each other and with a virtual environment composed of 100 resource patches. The richness of each resource patch is randomly assigned a value on a discrete, ordinal scale from one to five. Patch (thus global) population size is constrained by resource richness, and carrying capacity is maintained via agent out-migration and (if no suitable patches are available) death. Agents asexually reproduce one offspring at a time. Time periods are abstractly defined, but could be interpreted as single years, seasons, or decades, depending on the specific context and the level of population aggregation agents are assumed to represent. During each time period, agents receive payoffs that depend on their strategy and the richness of the patch they inhabit. An agent's fertility and mortality are probabilistically defined by its previous period payoff, which increases its probability of reproduction and decreases its probability of death. Each patch is initially seeded with one agent using a strategy called "dove," which refers metaphorically to their lack of territorial behavior. Doves divide resources equally with other nonterritorial agents on their patch. With a small probability, newly born agents can mutate to one of three other strategies: solo, client, or patron (with possible back-mutation to dove). All mutational transitions occur with equal probability. Solos colonize and defend empty, rich resource patches (richness > 1). They cannot invade patches inhabited by other agents, and are ousted from inhabited patches if they have spontaneously mutated from one of the two nonterritorial strategies (dove or client). Clients behave just as doves, but form economic relationships with patrons. Patrons provide their clients with limited access to their rich resources in exchange for labor, and are otherwise identical to solos. Clients may have only one patron, but patrons can have many clients. Smith and Choi iterated the simulation for 2,000 time periods, and ran it 200 times under each of the specified parameter settings.[2] Figure 10.2 provides a graphical depiction of the simulation's demographic processes. Table 10.1 is a table of strategy-dependent payoffs.

Our approach makes two general assumptions: (1) cultural evolution shares several characteristics with genetic evolution (selection, drift, and mutation); and

TABLE 10.1 Per-period payoff structure for Smith and Choi's (2007) patron-client simulation: P is the richness of an agent's current patch of residence; n is the number of agents currently residing on an agent's current patch of residence; δ is the cost of defending sole use of a patch; q is a binary indicator for whether patrons are in the current population; π is the share of patron's returns from a single client that is allocated to the client; τ is the patron's returns from a single client; λ is the cost of labor to a client; and m is the number of clients a patron currently has (which is a function of the number of patrons and clients currently in the population). Roman characters indicate variables dependent on current population structure. Greek characters are parameters exogenously set by the simulator.

Strategy	*Per-period payoff*
dove	P/n
client	$P/n + q(\pi\,\tau - \lambda)$
solo	$P - \delta$
patron	$P - \delta + m(1 - \pi)\,\tau$

(2) patron-client relationships are not maladaptions brought on by rapid environmental or social change. Drift is an evolutionary process that occurs because traits (genetic or otherwise) are randomly sampled from the population (e.g., through reproduction or imitation). Drift inevitably leads to the fixation of some traits and the extinction of others from the population, even in the absence of selection pressures (Wright 1988).

We now summarize the results of Smith and Choi's simulations. The largest positive effect on the emergence of patron-client regimes (measured as the proportion of runs in which greater than 90 percent of agents played patron or client by the last 10 periods) is the value of surplus generated by patron-client interactions. Here, the surplus is definied as:

$$(10.1)\ Z = \{[P/n + (\pi\tau - \lambda)] - P/n\} + \{[(P - \delta) + (1 - \pi)\tau] - [P - \delta]\} = (\pi\tau - \lambda) + (1 - \pi)\tau$$

The left-hand side of the equation above is the sum of two terms (each in curly braces), which define the payoff advantage of clients over doves and patrons over solos, respectively. This sum simplifies on the right-hand side to the sum of the benefits clients and patrons receive, respectively, from their mutualistic relationship. The surplus is positive as long as the client's share of the patron's increased productivity due to client labor is greater than the cost of labor ($\pi\tau > \lambda$), the patron retains a positive share of increased productivity from interactions with clients ($\pi < 1$), and patrons and clients coexist (i.e., q and m from Table 10.1 are greater than 0).

This result is not surprising given that the selective advantage of patrons and clients is a simple function of the surplus they generate, which solos and doves do not. A more complete (but untested) explanation is that the evolutionary force of drift weakens relative to the force of selection as surplus benefits increase the payoffs of patrons and clients relative to solos and doves. Consequently, patrons and clients (hence, patron-client regimes) increase and dominate with higher

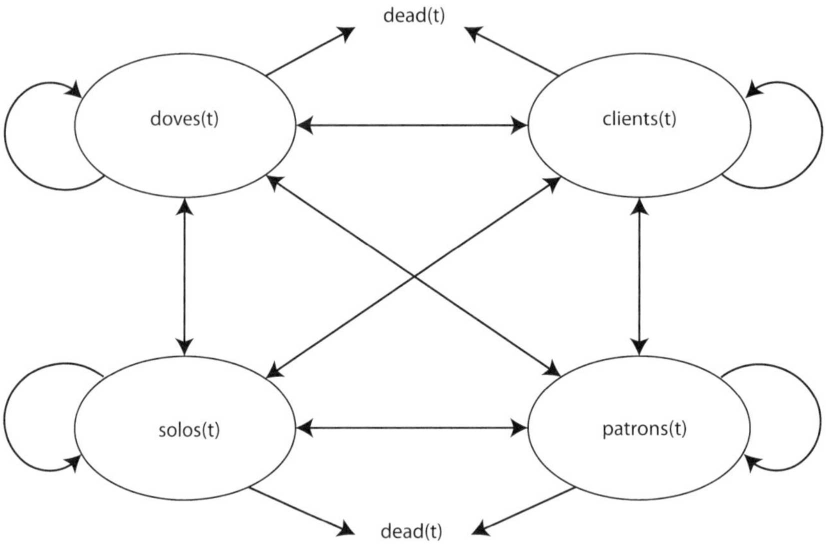

FIGURE 10.2 *Flow chart of Smith and Choi patron-client simulation dynamics: Solid arrows indicate direct flows of agent types (indicated as bubbles) in the simulation. Flows are due to mutation (arrows from one strategy into another, with mutation rates between two different strategies indicated by μ), reproduction (reflexive arrows, with strategy-specific reproduction probability indicated by β) and death (arrows pointing to skull and crossbones, with strategy-specific mortality indicated by Ω).*

probability. This result is consistent with a drift-selection balance predicted to emerge in stochastic evolutionary processes (Nowak 2006; Otto and Day 2007) such as the one occurring in the simulation.

Another result is that labor cost to clients is inversely related to the emergence of patron-client regimes because they lessen the surplus benefits of patron-client interactions. In this case, the (again untested) intuition is that drift is strong within and among patches inhabited by nonterritorial strategies. If drift keeps the client strategy at low frequency (recall that it is initially rare because the environment is seeded with doves), patrons on average will interact with fewer clients. If patrons have fewer clients, their payoff is more similar to the solo payoff. Among rich patches colonized by territorial agents, the power of drift is strong relative to the selective advantage to patrons over solos.

As the surplus and labor cost results suggest, drift is important to the simulation outcome and more generally to stochastic evolutionary processes. In a deterministic evolutionary process, patrons and clients would be favored over doves and solos as long as patron-client relationships are profitable enough to both parties. Yet Smith and Choi's simulation took place on a relatively small grid of 100 patches, and no patch could support more than five agents. In most simulations, patch richness was assigned randomly, following a uniform distribution, making three agents the average patch population capacity. As was likely the case for many real prehistoric societies (e.g., the Preclassic Maya), small local and global population size enhances the force of drift and limits the probability that patron-client regimes will emerge and become stable even if they are favored by the selection process. These results may have important implications for the distribution of polity emergence in the southern lowlands, as we discuss below. What about the effects of resource distribution across the landscape?

In the simulation, the effects of spatial resource distribution on the emergence of patron-client regimes are especially informative. Specifically, environmental conditions must favor the coexistence of territorial and nonterritorial strategies. A necessary but insufficient condition for this coexistence is that resources must be economically defensible ($P > \delta$). The more costly resource defense ($\delta \rightarrow P$), the more likely territorial agents are to drift out of the population, and the less likely patrons and clients are to interact. In addition, if resources are uniformly abundant (patch richness uniformly set to five) in the simulated environment, then reproductive rates are uniformly high. Deceased nonterritorial agents are replaced too rapidly by their progeny (and mutation rates are too low) for territorial strategies to colonize empty patches. If resources are uniformly scarce (patch richness uniformly set to two), then nonterritorial agents reproduce too slowly to prevent territorial agents from colonizing. Once established, territorial agents will reproduce rapidly via a monopoly of rich resources. Under either condition, patrons and clients are obviously unable to interact, and patron-client regimes would not emerge. Patrons may be favored over solos if mutation in strategies is rapid enough (due to the occasional presence of a client, and vice versa), but patrons and clients will rarely coexist long enough for patron-client regimes to dominate.

We will see in a subsequent section that these conditions, necessary for the emergence of patron-client regimes in Smith and Choi's simulation, match those that appear to have constrained the emergence of similar institutions in the southern lowlands. Before this discussion, we briefly review issues of surplus distribution in the patron-client model.

SURPLUS DISTRIBUTION IN THE PATRON-CLIENT SCENARIO AND SIMULATION

For simplicity, Smith and Choi (2007) forced patrons and clients to split the surplus benefits of cooperation equally. More realistic scenarios might dictate that the surplus is unequally distributed. Here, we provide two logical reasons why inequity in surplus distribution is critical to patron-client dynamics. First, recall that the patron-client scenario assumes clients can only have one patron at a time. If clients can have more than one patron, we would expect more of the income inequality between patrons and clients to be accounted for by heterogeneity in resource endowment rather than by inequality in the number of social partnerships. The income inequality that emerges would be less distinguishable from what would exist in a population of doves and solos, who do not interact. Unless the resource patches that patrons in the southern lowlands inhabited were many times more productive than those of clients, patron lineages might have been less capable of accumulating the capital that would one day fund the public works characteristic of Maya polities. What could stimulate greater capital accumulation for one party or the other? Unequal surplus distribution becomes an attractive alternative explanation. But who would be favored in such distribution processes: patrons or clients? While the intuitive answer might be the patrons, the patron-client model described above cannot confirm this assumption, nor can it predict how unequal the distribution should be.

Second, Smith and Choi's agents cannot choose strategically from among alternative social partners. In evolutionary models, the presence of alternative social partners is referred to as a "biological market." Economists and evolutionary biologists have convincingly argued that competition among potential social partners within biological markets can strongly influence social interactions (Noë and Hammerstein 1994, 1995), including mutualistic interactions between different classes of social agent (e.g., between members of biological species: Bowles and Hammerstein 2003). Boone (1992) suggested the possibility of competition between patrons over access to clients, and between clients over access to patrons. To attract more clients, patrons may compete by offering higher wages. Yet offering higher wages decreases the force of selection favoring patrons over other territorial agents who offer lower wages or who do not take clients. The same argument applies to clients who would accept lower wages or who do not interact with patrons. In the patron-client scenario, the success of one strategy that demands a larger proportion of the surplus depends on the success of another strategy that will accept the smaller proportion. If one strategy drifts out of the population, the other will follow. Only if strategy mutation rates are rapid enough to support frequent interactions between patrons and clients will selection favor one mutualistic strategy over its "antisocial" counterpart. The same logic would likely apply for a conditional strategy that acts as a patron if rich resource patches are available and as a client if not.

Therefore, one result may be that patron-client regimes with unequal surplus division will be unlikely to dominate unless the surplus, the population, or preferably both is large enough. If these conditions are met, what will influence the distribution of the surplus? As stated above, the patron-client scenario is unable to make such predictions. However, we can draw on economic game theory to understand the outcomes of such interactions, which may be conceptualized as part of a bargaining process. Before doing so, we turn to the empirical evidence to demonstrate the applicability of Smith and Choi's simulation to the water control hypothesis and to further motivate the analysis of surplus distribution in the context of Maya patron-client relationships.

WATER CONTROL AND THE PATRON-CLIENT SCENARIO IN THE SOUTHERN MAYA LOWLANDS

The contents of this section derive from earlier work by Lucero (2006: Chapters 2–6), where they are presented in greater detail. After 800 BC, Maya families began leaving densely populated coastal, riverine, and lacustrine areas. They migrated to the interior, which was rich in fertile land but lacked year-round surface water. The earliest settlers laid claim to the richest resource patches. In those areas with seasonal drought and no perennial water, land adjacent to aguadas would have been particularly valuable. By the Late Preclassic (c. 300 BC) some families acquired sufficient capital to build larger homes, sponsor the construction of small temples, and acquire exotic prestige items. This incipient inequality resulted from increasingly saturated interior resources as more migrants arrived. Habitat saturation encouraged interdependence between elites and nonelites. In exchange for services rendered, early elites sponsored traditional ceremonies and feasts for laborers and their families. These events provided respite from work, the opportunity to socialize, and a venue for elites to competitively signal status. Only in some regions did patron-client relations develop into complex polities by c. AD 200, if not earlier.

By c. AD 300 in areas without lakes or rivers, the Maya relied on massive artificial reservoirs during the annual four- to six-month drought; kings performed water rites and organized the maintenance of reservoirs to safeguard water quality (Scarborough 2003). People at centers near rivers also relied on royal capital to repair subsistence systems and to tide people over for losses suffered when heavy rain or flooding damaged crops. Regional rulers at river centers such as Copán, Seibal, and Palenque, and nonriver centers such as Tikal, Calakmul, and Caracol, also monopolized nearby resources and prestige-goods exchange, and integrated commoners through large-scale ceremonies (Lucero 2003). Kings at secondary centers such as Yalbac, Altar de Sacrificios, Quiriguá, and Bonampak acquired wealth by dominating prestige-goods exchange and nearby agricultural land. Secondary-center rulers had access to fewer laborers than those in regional centers because they were unable to control widely dispersed pockets of agricultural land and small-scale subsistence systems, not to mention access farmers who were sparsely scattered across the landscape. Elites at minor centers such as Saturday Creek and Barton Ramie in the Belize Valley had little, if any, political hold over the populace because agricultural land was extensive, water was plentiful year-round, and farmers were relatively dispersed. Farmers did not rely on elites, rulers, or subsistence systems for prosperity, but on the seasonal ebb and flow of the river (recession agriculture).

In combination with the patron-client simulation results, the water control model poses two alternative explanations for why patron-client relationships emerged and gave rise to Maya polities in particular regions of the southern lowlands. First, patron-client relationships (regardless of how the surplus was distributed) could have been more likely to emerge and become stable where water resources were patchily distributed and predictable. The expected waiting time until an event is inversely related to the probability that the event occurs. Thus we should expect polities to have emerged earlier in areas where the force of selection (which likely favored mutualistic patron-client relationship) was relatively more powerful than the force of drift. If polities emerged earlier in the southern lowlands, their ruling lineages would have had more time to accumulate capital, and the polity more time to grow. Second, patron-client relationships with unequal surplus distribution could have been more likely to emerge and become stable in the southern lowlands, where resources were patchily distributed and predictable. More unequal surplus division would have led to more rapid accumulation of social and material capital by elite lineages in the southern lowlands, which would have fostered more rapid polity growth. These explanations are not mutually exclusive. Yet one (the second) may be more plausible than the other, as we argue below. Smith and Choi's patron-client model provides some basis for why exchange between elites and farmers was more likely to emerge in areas with seasonal drought and patchily distributed water resources: patron-client regimes were unlikely to emerge in simulations where resources were uniformly scarce or uniformly dense. Inequality in initial resource endowment is required for territorial and nonterritorial agents to coexist. In areas of the southern lowlands with abundant and perennial water resources, farmers could remain productive and autonomous. If the demographic results of the patron-client simulation held in the southern lowlands, then areas with uniformly distributed but scarce water resources were also less likely to yield the necessary mixture of more and less territorial strategies. By contrast, patrons and clients could coexist in environments characterized by seasonally predictable, highly concentrated, patchily distributed, and limited water resources.

In the simulation, patron-client regimes depend on the value of the surplus generated by patron-client exchange. In regions where farmers did not depend as much on elites for potable water access, the surplus benefits of interaction would have been smaller. Accordingly, the average surplus value of patron-client relationships may have been minuscule. In centers of the Preclassic southern lowlands, most of which were small, minor economic advantages to a rare patron-client pair would entail weak selection on such social norms relative to the power of drift. In a simple stochastic birth-death process, a rare behavior with as much as a 10 percent selective advantage has only a 9 percent chance of taking over in a population of 100, and would have to emerge 7 times to have even a 50 percent chance of taking over the population (Nowak 2006). Patrons and clients must arise simultaneously in order to begin making exchanges. Thus, the probability of taking over the population was even lower for patron-client regimes. On the other hand, in areas where farmers depended more on elites for access to water at critical times, selection would have been stronger. Consequently, a rare patron-client relationship would be more likely to "catch on." If patron-client regimes could have emerged earlier in water-scarce regions, then patron lineages would have had more time to accumulate social and material capital. With this capital, they could attract larger numbers of clients to support the construction of more magnificent monumental architecture and reservoirs. But what if patron-client regimes did not emerge earlier in areas with seasonal droughts and patchily distributed water sources?

Indeed, patron-client regimes were not unique to areas lacking surface water. Elites sponsored ceremonies and rituals across the southern lowlands in exchange for labor, goods, and services from commoners. Critical to the water control hypothesis is that elites in areas without permanent surface water were capable of demanding greater tribute than elites in other regions with plentiful water sources. In the previous section, we suggested that the emergence and stability of patron-client regimes marked by unequal surplus distribution would be unlikely if the benefits of patron-client relationships are small. Extrapolating to the Maya case, regions where potable water resources were seasonally scarce and patchily distributed would entail greater surplus benefit to cooperation between elite and nonelite families. Elites would have more to gain from mounting competitive campaigns against other water resource holders, and farmers in the hinterlands would have more to gain with less water for drinking and agriculture. Thus the surplus benefits of patron-client relationships were larger, and the inequality in surplus distribution less constrained. Unequal surplus distribution would support more rapid capital accumulation by elites, culminating in the historical dynasties of the Classic period. The task now is to develop a more rigorous theoretical basis for why unequal surplus distribution should have favored incipient Maya elites. Such a framework has existed in economic game theory for six decades, and the time is ripe for extending it to address archaeological problems.

ECONOMIC BARGAINING MODELS

Moving from the basic patron-client model to a one of unequal resource distribution brings us into the realm of economic bargaining theory (Binmore 1987; Muthoo 1999). Bargaining theory developed to answer just the sort of question that now stands: How will social partners distribute a cooperative surplus? John Nash (1950) used the techniques of cooperative game theory (Myerson 1991; Von Neumann and Morgenstern 1953) to derive a unique solution to a simple, two-person bargaining game. He asserted a set of properties that a bargaining outcome should reasonably have and developed a function that satisfies these requirements. In this section, we review this model, showing that, in a bargaining game with committed players, resource distribution should be skewed toward individuals with stronger bargaining power and better opportunities at points of impasse. Next, we will review Rubinstein's (1982) noncooperative bargaining model. This model does not assume players are committed to the bargaining agreement, and predicts that bargaining power is heavily influenced by the importance agents place on future payoffs, the opportunities players have if agreement is never reached, and opportunities outside the social partnership. Together these discussions will aid our systematization of the water control hypothesis.

The Nash bargaining solution is defined by two vectors. First, $p = (p_i, p_j)$ refers to the distribution of productivity to players i and j. The vector $b = (b_i, b_j)$ refers to the impasse point, or the payoff profile that makes players indifferent between the social agreement and their outside opportunities. For simplicity and without loss of generality, we will assume $p_i + p_j = 1$ and $b_i + b_j \leq 1$ so that we can interpret our analysis in terms of the proportion of productivity acquired by each player. To derive b (hereafter described as the threshold share), we define three additional variables. The variables in vector $L = (M_i, M_j)$ refer to the value of an impasse payoff for players i and j, respectively. The variable S refers to the total productivity of a social partnership. We will assume for simplicity that payoffs are measured in units of fitness. The fitness payoff to player i if a bargaining agreement is reached is Sp_i. If the players reach an impasse (e.g., the players can never reach an agreement), player i receives fitness payoff M_i, and player j receives M_j. Player i is thus indifferent between social partnership and its impasse payoff if $Sp_i = M_i$, and the condition $Sp_j = M_j$ applies for player j. Setting p = b and solving for b, we obtain $b_i = M_i/S$, $b_j = M_j/S$. This result reveals that larger social productivity and smaller impasse payoffs lead to smaller threshold shares and a larger surplus over which the players bargain. Recalling that $b_i + b_j \leq 1$ and substituting in our solutions for b, we also find that partnerships are stable only if $S - (M_i + M_j) \geq 0$ (if there is mutual benefit to cooperation). Now define $a = (a_i, a_j)$, which refers to the bargaining power of players i and j, respectively. Bargaining power is defined verbally as the degree to which a player is advantaged or disadvantaged by the structure of the bargaining process (Binmore 2007). Since only relative bargaining power is important, assume for simplicity and without loss of generality that $a_i + a_j = 1$.

With these conditions for group stability established, we now present the Nash bargaining solution. We find this solution using what is commonly called the *asymmetric Nash product* (Myerson 1991), defined as:

(10.2) $NP(p, b, a) = (p - b)^a$

To find the Nash bargaining solution, maximize NP(p, b, a) with respect to p, recalling that $p_j = 1 - p_i$ in a two-person game. The proportion of group productivity S enjoyed by i (ego) is thus:

(10.3) $b_i + a_i(1 - [b_i + b_j])$

The first term above shows that players are awarded at least their threshold share before bargaining over the remaining surplus ensues. Then, players receive a share of the remaining surplus that is proportional to their bargaining power.

An important issue is how the value of group productivity (S) relative to the impasse point (M) influences the relative importance of the impasse point and bargaining power. To begin answering this question, recall that the surplus is a function relating group productivity to the impasse point (Specifically, $S - [M_i + M_j]$). The smaller each players' impasse payoffs are relative to group productivity, the larger the surplus they will split ($S - [M_i + M_j] \rightarrow S$ as $S >> M_i + M_j$). As a result, a player's payoff in a partnership far more valuable than either player's impasse payoff will depend more on its bargaining power than its threshold share ($b_i + a_i(1 - [b_i + b_j]) \rightarrow a_i$ as $b = (M_i/S, M_j/S) \rightarrow 0$). Conversely, the larger players' impasse payoffs are relative to group productivity, the smaller the surplus ($S - [M_i + M_j] \rightarrow 0$ as $M_i + M_j \rightarrow S$). So, a players' threshold share becomes more important as the surplus becomes smaller ($b_i + a_i(1 - [b_i + b_j]) \rightarrow b_i$ as $b_i + b_j \rightarrow 1$). An example of a surplus function with these properties is Equation 10.1, where we calculated the surplus generated by a single patron-client relationship (Z).

The Nash bargaining solution is problematic because it assumes players can signal, maintain, and enforce commitment to the bargaining agreement. The model specifies neither the context nor the structure of the bargaining process, which if known could elicit the factors that determine relative bargaining power (Binmore 1987, 2007; Binmore, Rubinstein, and Wolinsky 1986), and the reasons why players should adhere to an agreement. To address these issues, Nash called upon researchers to develop noncooperative bargaining models with explicit game structures that reasonably captured the reality of bargaining processes, and then determine the conditions under which they converge on the Nash bargaining solution. Game theorists call this paradigm the *Nash program* (Binmore 1987, 2007; McNamara, Binmore, and Houston 2006). We argue that the Nash program is just as useful to archaeologists studying the emergence of complex societies as it has been to economists studying behavior in real estate and auctioning markets.

The venues for the bargaining process in the southern Maya lowlands likely included interpersonal interactions between patrons and clients. Yet public displays and rituals—evidenced by the ball courts, plazas, palaces, temples, and monuments—could also have acted as the theater for the production and reproduction of social norms of surplus division (see also Stanish, chapter 4). In these rituals, Maya rulers signaled their utility to commoners over competing elites. As Roscoe (2009, chapter 3) recently suggested for contact-era New Guinea groups, community rituals provided Maya social agents with reliable information about their own status, and that of their social partners. In this way, Maya rituals may be analogized as part of a bargaining game that played out over a period of centuries. No doubt, rituals reminded the Maya, especially those without reliable water access, of all at stake in this game. How did ecological conditions influence the bargaining outcome, in particular through the factors implicated by the Nash bargaining solution?

Following the Nash program, Binmore, Rubinstein, and Wolinsky (1986) used Rubinstein's (1982) noncooperative, sequential bargaining model to show how time preferences influence bargaining power. In this model, players make sequential offers to one another, and the other player accepts or rejects. If the other player rejects, the roles switch. This process continues until the players reach an agreement. Binmore and colleagues found that as time between subsequent offers decreases to zero (a reasonable assumption to make if waiting is costly and players prefer to end negotiations quickly), the Rubinstein solution converges to the Nash bargaining solution where bargaining power is inversely proportional to the rate at which players discount future payoffs (Binmore 1987,

2007). Players with higher discount rates will be prone to accept earlier and likely lower offers. While this result provides insight into the determinants of bargaining power, it begs another question: what factors determine a player's discount rate? Behavioral ecologists and ecological anthropologists have identified two factors that influence this rate: the probability that future benefits will be realized, and the rate at which investments are compounded (Alvard 1998). For present purposes, we focus narrowly on how the probability of realized future benefits affects the discount rate. The lower the probability of future benefit, the higher the discount rate. As will become clear, differential discount rates may have been important to social outcomes in the Maya area.

In the Maya area, access to potable water would influence time preferences. Early elites had access to both fertile land and potable water. Because of their limited access to water, farmers would stand greater risk of crop failure and mortality the longer the bargaining process was drawn out. Thus, the descendants of settlers who controlled access to aguadas would have lower discount rates and stronger bargaining power. In those regions where the surplus benefit of patron-client relationships was large, differential time preferences could lead to unequal surplus distribution favoring patrons even if elites possessed threshold shares less than or equal to those of their clients. Thus, patrons in regions with seasonal drought and patchily distributed water resources could more rapidly accumulate capital from tribute than their counterparts in other regions. Feedback between rapid capital accumulation and increasing control over water through the construction of artificial reservoirs would have led to earlier and more pronounced polity emergence in these areas.

Another important determinant of bargaining outcomes is how the bargaining process can break down, which influences the impasse point (Muthoo 1999). One way that bargaining can break down is if individuals temporarily disagree on how to distribute the surplus. Inside options are the payoffs that players receive during the intervening period before bargaining begins again. During the temporary breakdown, the game is akin to a war of attrition. Muthoo (1999) showed that, in the time-limiting case described above, a player's threshold share will be larger with better inside options, and smaller with a larger discount rate. Note the additional effect of discount rates aside from its influence on bargaining power. The intuition being that temporary breakdown is more costly to individuals with higher discount rates. In the Maya case, the income accrued in the periods leading up to agreements on the construction of public works and exchange rates between water access and farm productivity would influence impasse points. Prior to labor specialization, commoners and elites incurred income from their farms. Commoners would discount this income more heavily due to lack of potable water. Even if elites and commoners had equal inside options before time discounting (likely the case early on in the southern lowlands), differential discount rates would favor elites through larger impasse payoffs. Where potable water was patchily distributed, this advantage was greater than where commoners had more reliable access to water.

Players may also have opportunities outside the social partnership. Tying these possibilities into economic bargaining theory invokes the outside option principle (Muthoo 1999). This principle asserts that players must make credible threats to affect bargaining outcomes. The threat to take an outside option is only credible if its value is greater than the payoff obtained in the agreement defined by a Rubinstein bargaining game without outside options. If neither player could do better than the bargaining process, no one can make a credible threat. If both players can make credible threats, both players will leave for their outside options. Therefore, only one player's outside options can be active at a time in a two-player bargaining game, although multiple players can collude to threaten others in a game with three or more players (Cant and Johnstone 2009). If one player's outside option is active, its share of the surplus will be just greater than the value of the outside option. For Maya elites, outside opportunities could have included the expulsion of commoners through indiscriminate territory defense (as solos do in the Smith-Choi simulation, described above). For commoners, the outside option could have been to disperse to previously unoccupied or sparsely occupied hinterlands.

Under what conditions were either, both, or neither of these strategy-shifting threats credible? Elites likely had too much to gain from commoners to prefer indiscriminate defense of territory, and territory defense against commoners is costly. Commoner threats were credible when they had reliable access to water and could live autonomously. Such was not the case in parts of the southern lowlands. The threat could have been credible in other regions, leading to more equal distributions of cooperative surplus and slower rates of capital accumulation by elites.

Another outside option is to shift allegiance to another patron, or hire another client. The value of this outside option depends on the availability of alternative social partners and the share of productivity individuals would expect in another group. When the supply of elites is large relative to demand for water access, elites will compete with one another by offering greater compensation to clients. Interelite competition would increase the expected value of clients' allegiance-shifting outside option. When the supply of commoners is large relative to labor demand, commoners will compete with one another by demanding lower wages (or paying greater tribute). Intercommoner competition would increase the expected value of patrons' allegiance-shifting outside option (Boone 1992; Noë and Hammerstein 1995). Since each additional client provides additional revenue and political support to the elite, we would expect elites to continue engaging in costly displays of their quality as patrons to maintain the largest and most effective retinue possible.

To summarize, during the Preclassic (before AD 250) the surplus value of cooperation in areas of the Maya lowlands with patchily distributed water resources was likely greater than in riverine and lacustrine regions. Fertile soils were plentiful in these regions, making the value of inside options of early commoners and elites approximately equal. Yet these inside options were mediated by time preferences. Because commoners had high discount rates due to great risk of crop failure and starvation during drought months, they stood to gain much from conscription over autonomy. In sum, the principle variable that governed the rate at which Maya kings accrued power and the spatial distribution of complex polity emergence in the Maya region was the degree of inequality in access to water, the variation in time preferences that it generated, and the resulting differentials of bargaining power over surpluses generated by cooperation between elites and commoners.

The relative importance of bargaining power and the impasse point in determining bargaining outcomes depends on the surplus value of cooperation. If the surplus is smaller, the impasse point becomes more important. If the surplus is larger, bargaining power becomes more important. Under the water control hypothesis, population was growing in the Maya region into the Classic period (AD 250–850). As population grew, the construction of public works became more efficient, labor specialization diversified, patrons and clients became increasingly interdependent, and the productivity of polities increased relative to impasse opportunities. As such, the surplus value of cooperation between elites and commoners increased. Therefore, bargaining power—mediated by discount rates that arose from concerns over water access and crop failure—likely grew in importance over time over the entire Maya region.

FROM THE EMERGENCE OF POLITIES TO THEIR DECLINE

The relationship between resource defensibility and bargaining outcomes may help explain the collapse or failure of patron-client regimes as well as their emergence and embellishment. The Maya prospered under a patron-client system in a semitropical setting for 800 years. The more Maya kings depended on large-scale reservoirs to attract subjects and their labor, the more vulnerable they were to changing rainfall patterns. As cases demonstrate worldwide, inflexible subsistence systems are less able to adapt to changing conditions (e.g., Crumley 1994, 1995; Fagan 2004, 2008). In the Maya case, while political systems were somewhat fluid and flexible, one aspect of the subsistence system was not—their dependency on rainfall and water systems. By the end of the Classic period, royal ceremonies and class-structured production no longer worked due to long-term droughts. Perhaps responding to the increase in the outside opportunities of commoners, kings sponsored ever-more-ornate ceremonies, consuming resources that should have been used to shore up production against impending water shortages. To farmers, kings failed to reach the gods. People blamed them for the lack of water. Affiliation with elites no longer generated a surplus, so they abandoned rulers and centers. Monumental architecture fell into disrepair and interior trade economies recessed in the face of bankrupt symbolic capital. Commoners aggregated into smaller communities in the hinterlands or dispersed in all directions in search of wetter prospects (e.g., to the northern lowlands where centers experienced a florescence).

LIMITATIONS OF THE MODELS

We recognize that the models we present possess several theoretical weaknesses. First, the models assume dyadic interactions. In economic bargaining models, games with greater than two players quickly yield continuous solution sets, or even dominance cycles that render solutions infeasible (Binmore 2007; Torstensson 2009). Yet if bargaining processes took place on a dyadic basis between elite and commoner individuals and their families, then our analysis retains much of its validity. Second, each model assumes players have complete information about other's preferences, available social partners, and dispersal openings. Evolutionary bargaining models support the Rubinstein bargaining solution and its convergence to the Nash bargaining solution when players are not perfectly rational or omniscient (Young 1993). An extension of evolutionary bargaining models supports the effect of time preferences on relative bargaining power (Robles 2008). Third, the models are abstract. Future models should extend the general insights produced here using currencies, and units (social and spatial) relevant to the Maya case. Fourth, the bargaining models we have presented assume players can commit to a long-term agreement and that social interactions are limited to a single instance of bargaining. Since bargaining behavior among the ancient Maya was likely influenced by preexisting social norms that were reinforced through ritual experience and shared beliefs, the assumption that agents play as if adhering to a long-term social contract may be appropriate.

CONCLUDING REMARKS AND SYNTHESIS

A combination of ecological, economic, and evolutionary models lend additional support to Lucero's water control hypothesis for the emergence and collapse of Maya polities. Mutualistic relationships between early Maya elites and commoners were more likely to emerge where the surplus benefits to cooperation were large. Unequal distributions of patron-client surplus would place greater upward constraints on the surplus size. Variance in time preferences, which depended on water access and the size of cooperative surpluses, are likely the most important variables influencing geographic variation in Maya polity growth. These processes appear self-evident. But our contribution is the implication of ultimate mechanisms linking the distribution of cooperative benefits to the rate at which elites acquire (and lose) power in the absence of coercion.

This chapter ties to several of this volume's main themes (see Carballo, chapter 1). First, we show how cooperation and competition are intertwined even when there are no social dilemmas that threaten the production of social goods. In our case, competition arises through bargaining over the distribution of cooperative benefits. In this sense, we agree with Blanton and Fargher (chapter 5) that the emergence of complex, hierarchical societies is likely the product of manipulation and negotiation as much as the purportedly prosocial tendencies of humans. We take our cue from Roscoe (chapter 3) that archaeologists must separate conceptually the vertical and horizontal components of polities. Our discussion also suggests a way to integrate the two. The mechanism that assures cooperation between patrons and clients (in our case mutual benefit) interacts with but is not necessarily the same as the mechanism that determines surplus distribution (in our case, sequential bargaining). Second, we show how polity growth in the Maya region could have emerged from a self-organizing, bottom-up process involving strategic interactions between elites and commoners. Third, the bargaining models we describe touch on the fourth "R" (reciprocity, retribution, reputation, *reward*), which has been neglected in the evolutionary game theory literature in favor of the other three. Finally, we suggest that elite-sponsored rituals could have acted as an institution to negotiate and solidify norms of surplus distribution.

NOTES

1. We should note that if resources are especially dense and predictable to the point where they are nonlimiting (i.e., there is no competition for access to them), there is no need for individuals to claim territory. Conversely, if resources are sparse, but still dense and predictable, then individuals benefit from claiming them.

2. See Smith and Choi (2007) for a comprehensive description of methods, tables of results, lists of parameter settings, and example images of the graphic user interface (and for another model of social inequality they call "managerial mutualism." Managerial mutualism does not lead to stable inequality as often as the patron-client model, which is predictable because "managers" act to resolve difficult public-goods problems).

REFERENCES

Alvard, Michael S. 1998. "Evolutionary Ecology and Resource Conservation." *Evolutionary Anthropology* 7 (2): 62–74. http://dx.doi.org/10.1002/(SICI)1520-6505(1998)7:2<62::AID-EVAN3>3.0.CO;2-I.

Binmore, Ken. 1987. Nash Bargaining Theory I, II, &III. In *The Economics of Bargaining*, ed. K. Binmore and P. Dasgupta, 27–46, 61–76, and 239–256. Blackwell, Malden, MA.

Binmore, Ken. 2007. *Playing for Real: A Text on Game Theory*. Oxford: Oxford University Press. http://dx.doi.org/10.1093/acprof:oso/9780195300574.001.0001

Binmore, Ken, Ariel Rubinstein, and Asher Wolinsky. 1986. "The Nash Bargaining Solution in Economic Modelling." *Rand Journal of Economics* 17 (2): 176–88. http://dx.doi.org/10.2307/2555382.

Boone, James L. 1992. "Competition, Cooperation and the Development of Social Hierarchies." In *Evolutionary Ecology and Human Behavior*, ed. E. A. Smith and B. Winterhalder, 301–338. New York: Aldine de Gruyter.

Bowles, Samuel, and Peter Hammerstein. 2003. "Does Market Theory Apply to Biology?" In *Genetic and Cultural Evolution of Cooperation*, ed. P. Hammerstein, 167–184. Cambridge, MA: MIT Press.

Brumfiel, Elizabeth M. 1987. "Elite and Utilitarian Crafts in the Aztec State." In *Specialization, Exchange, and Complex Societies*, ed. E. M. Brumfiel and T. K. Earle, 102–118. Cambridge: Cambridge University Press.

Brumfiel, Elizabeth M. 1994. "Factional Competition and Political Development in the New World: An Introduction." In *Factional Competition and Political Development in the*

New World, ed. E. M. Brumfiel and J. W. Fox, 3–14. Cambridge: Cambridge University Press. http://dx.doi.org/10.1017/CBO9780511598401.002

Cant, Michael A., and Rufus A. Johnstone. 2009. "How Threats Influence the Evolutionary Resolution of Within-Group Conflict." *American Naturalist* 173 (6): 759–71. http://dx.doi.org/10.1086/598489.

Crumley, Carole L. 1994. "The Ecology of Conquest: Contrasting Agropastoral and Agricultural Societies' Adaptation to Climate Change." In *Historical Ecology: Cultural Knowledge and Changing Landscapes*, ed. C. L. Crumley, 183–201. Santa Fe, NM: School of American Research Press.

Crumley, Carole L. 1995. "Building an Historical Ecology of Gaulish polities." In *Celtic Chiefdom: Celtic State: The Evolution of Complex Social Systems in Prehistoric Europe*, ed. B. Arnold and D. B. Gibson, 26–33. Cambridge: Cambridge University Press.

Dyson-Hudson, Rada, and Eric Alden Smith. 1978. "Human Territoriality: An Ecological Reassessment." *American Anthropologist* 80 (1): 21–41. http://dx.doi.org/10.1525/aa.1978.80.1.02a00020.

Fagan, Brian. 2004. *The Long Summer: How Climate Changed Civilization*. New York: Basic Books.

Fagan, Brian. 2008. *The Great Warming: Climate Change and the Rise and Fall of Civilization*. New York: Bloomsbury Press.

Johnson, Allen W., and Timothy Earle. 1987. *The Evolution of Human Societies: From Foraging Group to Agrarian State*. Stanford, CA: Stanford University Press.

Lucero, Lisa J. 2002. "The Collapse of the Classic Maya: A Case for the Role of Water Control." *American Anthropologist* 104 (3): 814–26. http://dx.doi.org/10.1525/aa.2002.104.3.814.

Lucero, Lisa J. 2003. "The Politics of Ritual: The Emergence of Classic Maya Rulers." *Current Anthropology* 44 (4): 523–58. http://dx.doi.org/10.1086/375870.

Lucero, Lisa J. 2006. *Water and Ritual: The Rise and Fall of Classic Maya Rulers*. Austin: University of Texas Press.

McElreath, Richard, and Robert Boyd. 2007. *Mathematical Models of Social Evolution: A Guide for the Perplexed*. Chicago: University of Chicago Press.

McNamara, John M., Ken Binmore, and Alasdair I. Houston. 2006. "Cooperation Should Not be Assumed." *Trends in Ecology & Evolution* 21 (9): 476–8. http://dx.doi.org/10.1016/j.tree.2006.07.005.

Muthoo, Abinay. 1999. *Bargaining Theory with Applications*. Cambridge: Cambridge University Press.

Myerson, Roger B. 1991. *Game Theory: Analysis of Conflict*. Cambridge, MA: Harvard University Press.

Nash, John F., Jr. 1950. "The Bargaining Problem." *Econometrica: Journal of the Econometric Society* 18 (2): 155–62. http://dx.doi.org/10.2307/1907266.

Noë, Ronald, and Peter Hammerstein. 1994. "Biological Markets: Supply and Demand Determine the Effect of Partner Choice in Cooperation, Mutualism and Mating." *Behavioral Ecology and Sociobiology* 35 (1): 1–11. http://dx.doi.org/10.1007/BF00167053.

Noë, Ronald, and Peter Hammerstein. 1995. "Biological Markets." *Trends in Ecology & Evolution* 10 (8): 336–9. http://dx.doi.org/10.1016/S0169-5347(00)89123-5.

Nowak, Martin A. 2006. *Evolutionary Dynamics: Exploring the Equations of Life*. Cambridge, MA: Harvard University Press.

Otto, Sarah P., and Troy Day. 2007. *A Biologist's Guide to Mathematical Modeling in Ecology and Evolution*. Princeton: Princeton University Press.

Pauketat, Timothy R. 2000. "The Tragedy of the Commoners." In *Agency in Archaeology*, ed. M. A. Dobres and J. Robb, 113–129. London: Routledge.
Robles, Jack. 2008. "Evolution, Bargaining, and Time Preferences." *Economic Theory* 35 (1): 19–36. http://dx.doi.org/10.1007/s00199-007-0227-6.
Roscoe, Paul. 2009. "Social Signaling and the Organization of Small-Scale Society: The Case of Contact-Era New Guinea." *Journal of Archaeological Method and Theory* 16 (2): 69–116. http://dx.doi.org/10.1007/s10816-009-9062-3.
Rubinstein, Ariel. 1982. "Perfect Equilibrium in a Bargaining Model." *Econometrica: Journal of the Econometric Society* 50 (1): 97–109. http://dx.doi.org/10.2307/1912531.
Scarborough, Vernon L. 2003. *The Flow of Power: Ancient Water Systems and Landscapes*. Santa Fe: School of American Research Press.
Smith, Eric Alden, and Jung Kyoo Choi. 2007. "The Emergence of Inequality in Small-Scale Societies: Simple Scenarios and Agent-Based Simulation." In *The Model-Based Archaeology of Socionatural Systems*, ed. T. A. Kohler and S. E. van der Leeuw, 105–120. Santa Fe: School for Advanced Research Press.
Summers, Kyle. 2005. "The Evolutionary Ecology of Despotism." *Evolution and Human Behavior* 26 (1): 106–35. http://dx.doi.org/10.1016/j.evolhumbehav.2004.09.001.
Sutherland, William J. 1996. *From Individual Behaviour to Population Ecology*. Oxford: Oxford University Press.
Torstensson, Pär. 2009. "An n-Person Rubinstein Bargaining Game." *International Game Theory Review* 11 (1): 111–5. http://dx.doi.org/10.1142/S0219198909002170.
Von Neumann, John, and Oskar Morgenstern. 1953. *Theory of Games and Economic Behavior.* Princeton: Princeton University Press.
Welensky, Uri. 1999. NetLogo. http://ccl.northwestern.edu/netlogo/. Center for Connected Learning and Computer-Based Modeling, Northwestern University. Evanston, IL.
Winterhalder, Bruce. 1996. "Social Foraging and the Behavioral Ecology of Intragroup Resource Transfers." *Evolutionary Anthropology* 5 (2): 46–57. http://dx.doi.org/10.1002/(SICI)1520-6505(1996)5:2<46::AID-EVAN4>3.0.CO;2-U.
Wright, Sewall. 1988. "Surfaces of Selective Value Revisited." *American Naturalist* 131 (1): 115–23. http://dx.doi.org/10.1086/284777.
Young, H. Peyton. 1993. "An Evolutionary Model of Bargaining." *Journal of Economic Theory* 59 (1): 145–68. http://dx.doi.org/10.1006/jeth.1993.1009.

11

Labor Collectives and Group Cooperation in Pre-Hispanic Central Mexico

David M. Carballo

> Cooperation is a powerful symbol used in the definition and social construction of community . . . Thus, cooperation is not simply a story told about village life. Rather, it is a set of practices through which Santa Ana as a social entity is created. Cooperation and reciprocity are the arenas through which socially significant actions take place and through which the practices and commitments of Santañeros are measured . . . [C]ooperative structures are found in most rural, indigenous communities in Mexico and Mesoamerica. (Cohen 1999: 9)

Individuals cooperate as parts of communities all over the world, but the particular manner in which they do forms a central, determinative component of community structure and identity. As Cohen notes in the epigraph, individuals living in more rural parts of Mesoamerica reckon their particular nested scales of community based largely on cooperative relations, and the variable nature of these relations captures a significant portion of the heterogeneity between communities. Mesoamerican communities also share much in how they organize and sustain cooperative undertakings, attributable to entangled culture histories stretching back through pre-Hispanic times. Among these similarities, cooperation through group labor collectives is well documented within contact-period Mesoamerica, and continues, albeit modified, as an important organizing principle among contemporary rural communities.

The potential implications of self-organization through cooperative relations have been insufficiently theorized in studies of pre-Hispanic Mesoamerican social evolution—topics such as sedentarization, the development of economic specialization and exchange, the creation of social institutions, and urbanization. Greater attention has thus far been given to the role of political competition and ecological factors in these broad transformations. While critical, such foci are often on different extremes of the continuum of scales of analysis; they tend to grant agency to only a few aggrandizing individuals or they emphasize the expansive scales of environments and social systems. More recent consideration of group behavior within the Mesoamerican past assists in explaining how the aggregate effects of individual goal-seeking behaviors may lead to major social change. Examples include studies of factional competition, intermediate elites, and commoners (e.g., Brumfiel and Fox 1994; Elson and Covey 2006; Gonlin and Lohse 2007; Joyce, Bustamante, and Levine 2001). They also include differentiating more corporate (group-oriented) from more exclusionary (elite-oriented) political strategies between and among communities (e.g., Blanton and Fargher 2008; Blanton et al. 1996; Fargher, Blanton, and Heredia Espinoza 2010; Feinman 2001; Peregrine 2012).

In this chapter I examine how cooperation in collective labor defined and constituted communities in pre-Hispanic central Mexico (Figure 11.1). I combine ethnographic and ethnohistoric accounts with interdisciplinary models of cooperation and collective action in order to interpret the archaeological record from earlier periods, for which historical sources do not exist. In particular, I evaluate the fit between Nahua (I reserve the term *Aztec* for the Postclassic period) practices of group labor and earlier urban societies of the first millennium AD, and the relationship between group labor and ritual within more recent rural highland Mexican communities and the archaeological record of pre-/protourban communities of the first millennium BC. I propose that the development and maintenance of collective labor obligations (known as *tequitl* among the Nahuas) were critical to, and causal in, the evolution of complex polities in Formative and Classic period central Mexico, and that individuals defined and organized such obligations through public rituals that emphasized corporate ideologies yet permitted the emergence of heightened status differentiation. I suggest that during the process of increasing urbanization and hierarchy that characterized the later Formative through Classic periods, more voluntary institutions of collective labor were manipulated by social and political elites, becoming transformed into the labor tax systems backed by coercion that underlay urban political economies.

LABOR COLLECTIVES AND GROUP COOPERATION IN THE ETHNOGRAPHIC AND ETHNOHISTORIC RECORDS

Reciprocal labor exchanges are known from rural societies across the globe (Dietler and Herbich 2001; Erasmus 1956), and one could create an A-to-Z list spanning Amish barn raisings to the festive work parties of eastern Zambia

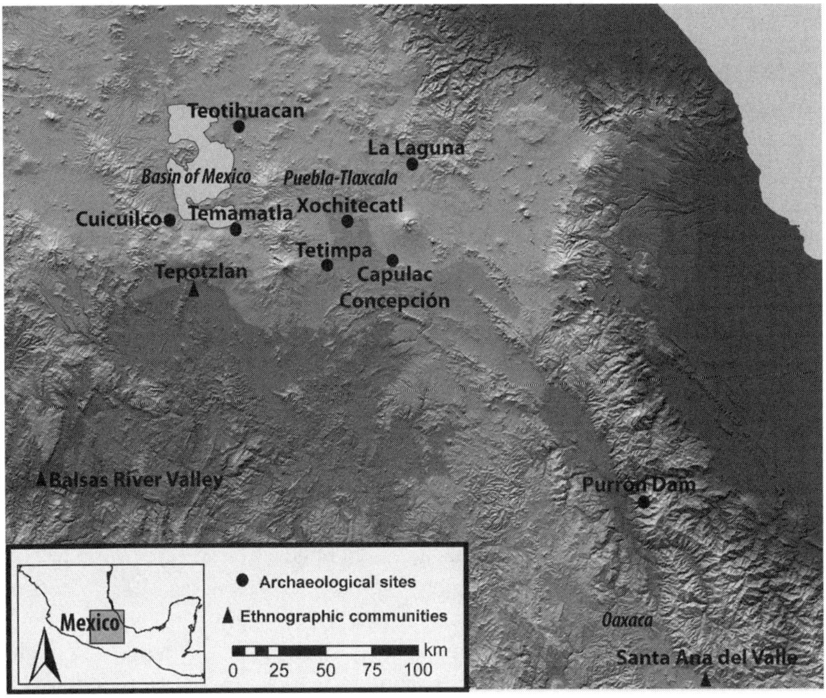

FIGURE 11.1 *Central Mexico with sites and regions discussed in text (ASTER GDEM is a product of METI and NASA [https://wist.echo.nasa.gov/]).*

(Hoon 2007; Marglin 2007: 1–19). These are essentially self-regulating systems, as poor reciprocators are easily shunned by their bad reputation and face the threat of ostracism from their communities—examples of reputation and retribution, following the mechanisms for sustaining cooperation outlined in chapter 1 (see also Gintis, Smith, and Bowles 2001; Henrich et al. 2006). Building on earlier terminology presented by Erasmus (1956) and Moore (1975), Dietler and Herbich (2001) outline a useful continuum of collective labor practices. In smaller-scale *work exchanges* labor itself is reciprocated, usually between households; in larger-scale *work feasts* labor is reciprocated by a sponsoring individual or institution through ritualized consumption, either during or following the labor event. Work feasts are further parsed as being more voluntary or more obligate, with *corvée* labor an example of the latter (Dietler and Herbich 2001: 244).

Ethnographers discuss cooperative labor within highland Mesoamerica in several guises, and by using various Spanish or indigenous terms. Within Spanish-speaking mestizo communities, *vuelta mano* (turning hand) labor involves reciprocity on the part of households or corporate-kin groups in tasks such as planting and harvesting (e.g., Nutini 1968: 177). While common in most societies, the scale and organization of agricultural work exchanges within central Mexico are impacted by systems of corporate land management, a well-documented component of modern and pre-Hispanic societies (Caso 1958; Hicks 1986: 48–50; Lewis 1951–1963: 114–118; Lockhart 1992: 142–149, Table 5.3; Sarukhán and Larson 2001). The prevalence of land worked jointly by numerous households means that agricultural fields were often a common-pool resource, and were therefore not as excludable as fully private land, as outlined in chapter 1 (see also Bayman and Sullivan 2008; Eerkens 1999; Kohler 1992; Ostrom 1990; Ostrom, Gardner, and Walker 1994). Accordingly, agricultural work exchanges are larger in scale and more open to free-riding in highland Mexico than in many other parts of the world.

Cooperation at scales larger than kin-group work exchange is illustrated in mestizo towns and villages by a form of labor often called *comunidad* (community). For instance, in his study of drained-field agriculture during the 1960s, Wilken (1968: 228) notes: "Towns in Tlaxcala adhere to the custom of comunidad in which one day each week, usually Monday, is devoted to community projects. All adult male townsmen otherwise unoccupied are expected to participate." Wilken records one event in which over 100 individuals worked collectively to fell trees and widen a canal—accompanied, not surprisingly, by decorous speeches on the part of government officials. He concludes, "The job was accomplished in two days, an impressive example of the power of massed cooperative effort" (Wilken 1968: 228).

Comunidad is the mestizo equivalent of a longstanding indigenous highland Mexican tradition of community labor obligations more commonly called by its Nahuatl name, *tequitl* (or its modified version, *tequio*). Among the definitions of tequitl are "task," "work," "tribute," "duty," "assignment," and "work deed" (Lockhart 2001: 234; Molina 2008 [1571]: 105). The institution is well documented in the works of Carrasco (1978), Rojas Rabiela (1977, 1986), and others. I use the Nahuatl term, but other Mesoamerican societies possessed related institutions of work as part of a collectivity. Among Maya communities, the Spanish term *faena* (from Latin *facienda*, "things to do," also spelled *fagina*) is often used (Redfield and Villa Rojas 1967: 77–80; Wells 2007). A comparable Mixtec term is *saa sa'a* (Monaghan 1996). Within Santa Ana del Valle, the Zapotec community studied by Cohen (1999), individuals differentiate a range of cooperative relations including tequio, *compadrazgo* (godparenthood), *cargos* (community service positions), and *guelaguetza* (gift exchanges following delayed reciprocity). Good Eshelman (2005) notes that the Nahua communities along the Balsas River define themselves on the basis of tequitl relations, which cover a similar range of practices as those described by Cohen and are grounded in notions of mutual respect and shared resources. Mesoamerican archaeologists are well aware of these types of institutions, as we often participate in the cooperative networks of the communities with whom we conduct fieldwork.

Indigenous cooperative institutions are resilient, but ethnographic studies demonstrate that they take a great deal of work to maintain and they evolve over time along with broad-scale social transformations, such as how increased market and monetary exposure allows individuals to hire substitutes for their tequio duties (Cohen 1999: 114; see also Dehouve 1987; Erasmus 1956; Monaghan 1996; Moore 1975). This last point is particularly well illustrated for highland Mexico by Foster's (1967) study of Tzintzuntzan, a mestizo and Tarascan community in the state of Michoacán, where, he argued, individualism and a conceptualization of limited good permeated the community through increasing economic marginality (see also Brandes 1988). Such cases demonstrate how cooperative institutions might break down.

Within contact-period central Mexico, labor obligations at the largest intra- and intercommunity scales were usually designated *coatequitl* (literally "snake/twin work" but glossed as public works or community works [Lockhart 1992: 345; Molina 2008 (1571): 23]). Colonial-period documents provide suggestions for how pre-Hispanic tequitl and coatequitl functioned within the Aztec empire and other portions of the Mexican highlands. These forms of labor could be mobilized through the *calpolli* (literally, "big house"), which represented a fundamental, corporate social organization combining notions of shared land, labor, and kinship. They could also be mobilized through the *tecalli* (literally, "lord's house"), which represented the noble estates that controlled most land and provided usufruct rights in return for tax or tribute. Tequitl and coatequitl labor involved duties such as digging irrigation canals or making crafts within palaces or temple precincts (e.g., Díaz del Castillo 1956: 211–212; Gibson 1964: 220–256; Hicks 1986; Katz 1966; Rojas Rabiela 1977, 1986; Zorita 1963: 202–207).

In her detailed consideration of coatequitl, Rojas Rabiela (1977) argued that work gangs (*cuadrillas*) organized on a base-20 numeric system were the pre-Hispanic norm, but that temporal and regional heterogeneity existed in the organization and activities of labor groups within Nahua communities. Several terms for institutionalized overseers of coatequitl obligations are recorded, such as the *calpixque* and the *tequitlato* (Lockhart 1992: 43–44; Rojas Rabiela 1977: 48–53; Smith n.d.; Zorita 1963: 114). These individuals possessed the coercive ability to punish transgression (Motolinía 1555–1971: 25–26), consistent with hierarchical state systems elsewhere (see Stanish [2010] for examples from the Andes). Contact-period Nahua nobles were exempt from labor tribute, but paid taxes in goods to superiors at rates reflecting their greater means (Lockhart 1992: 96). A strong norm, and even an enjoyment, of work is apparent in the Nahua informants' commenting on virtues and vices to the Spanish chronicler Bernardino de Sahagún, and in the writings of another one of their early Spanish defenders, Alonso de Zorita:

> The good man [is] a worker, a sage, a willing worker—one who works willingly. He works energetically; he is resolute; he is a steadfast worker. The bad mature man [is] uncoöperative, irresponsible; he is impetuous; he acts without consideration. (Sahagún 1961: 12)

> In the old days they performed their communal labor in their own towns . . . They did their work together and with much merriment, for they are people who do little work alone, but together they accomplish something . . . The building of the temples and the houses of the lords and public works was

always a common undertaking, and many people worked together with much merriment. (Zorita 1963: 203)

Stated ideals and human practice can often vary widely, particularly in hierarchical societies in which dominant ideologies may serve to reify inequalities of power. Contact-era Nahua society was highly stratified and, through coatequitl, cooperative norms could be used as an obligate labor tax that supported the interests of states and upper classes. Hicks (1996) discusses the persuasive power of elite ideology emphasizing class consciousness and loyalty to class structures, and of interclass ideology stressing state consciousness and patriotic loyalty to political structures. These ideologies incorporated the existential roles of humans and gods, including the idea that irrespective of class, individuals must merit their fortune in life (*macehua*) through the tequitl they perform (Hicks 1996: 263–265). Blanton and Fargher (2008; Fargher, Blanton, and Heredia Espinoza 2010) and Hicks (1999) provide insightful overviews of the relative power of commoners within Postclassic Nahua (or "Aztec") communities.

Coatequitl persisted as a recognized practice in the rural village of Tepotzlan (rendered *cuatequitl*) as recently as fifty years ago, where Lewis (1951–1963) recorded it on three scales. At the smallest scale of neighborhoods, coatequitl was relatively common for tasks such as road repair and water works; at the barrio scale coatequitl was often directed as maintaining churches; and at the village level, the various wards of the community could assemble as many as 600 men for clearing brush to define boundaries during a dispute with a neighboring village (Lewis 1951–1963: 108–111). Lewis observed two interesting dimensions in the practice of coatequitl in Tepotzlan: it could be used coercively by people in positions of authority, which he viewed as a recent development within this relatively egalitarian community (Lewis 1951–1963: 110; see also Cohen 1999: 8); and free alcoholic beverages characterized all events, which he viewed as a needed stimulus, and not a recent development (Lewis 1951–1963: 111).

Ritual networks, often involving food and drink, are critical components for organizing cooperation within Mesoamerican communities (Good Eshelman 2004; Monaghan 1990, 1996; Wells 2007). Participation in such networks comes at great expense for contemporary Nahuas, yet they provide an important means of reaffirming community identity within modern industrialized-labor contexts, including work in factories (Lazcano and Barrientos 1999). Feasts and other forms of ritual consumption have been a productive venue of archaeological theorizing, but such research more commonly emphasizes their role in generating and intensifying inequality (e.g., Clark and Blake 1994; Hayden 1995). Not all feasts are about empowerment (Dietler and Hayden 2001; Kohler, VanBuskirk, and Ruscavage-Barz 2004; M. L. Smith 2010: 91–95). An equally important dimension of ritual consumption is that individuals participate fully cognizant of the fact that strings are attached to gifts and other forms of largesse, and they act with the goal of strategically advancing their position within social networks and hierarchies (e.g., Kelly 2001; Pauketat et al. 2002). Various forms of feasting were practiced across the socioeconomic spectrum of Aztec society, and while many served to construct and reify politico-religious hierarchies, others functioned less hierarchically within rural provinces of the empire or among commoner households (Smith, Wharton, and Jan Marie Olson 2003). The terms *fiesta finance* (Monaghan 1996) or *work feasts* (Dietler and Herbich 2001) encapsulate the organization of labor in central Mexican ritual networks, but the distinction drawn by Dietler and Herbich between voluntary and obligate works feasts is slippery for tequitl. Unlike the institutionalized labor tax of central Mexican states, the sense of duty that accompanies tequitl in smaller-scale settings makes it virtually obligate, but it is backed by less coercive, more collective community sanctions along the lines of what cooperation theorists refer to as *strong reciprocity* (Bowles and Gintis 2002, 2004; Boyd, Gintis, and Bowles 2010; Dubreuil 2008; Ostrom 2003).

Some of the burden of collective labor within the hierarchical polities of central Mexico, such as the Aztec *altepemeh* ("city-states"), was apparently ameliorated by the food and festivities associated with these undertakings (e.g., Katz 1969: 222–223; Lockhart 1992: 344–345). It likely accounts for much of the "merriment" with which Nahuas performed collective labor, according to Zorita. Reyes (1977: 67–68) emphasized reciprocal food and drink in organizing coatequitl, and suggested it was used as a stimulus in earlier central Mexican village settings, as basic sustenance in the labor tax system of the Aztecs, and was not provided by the Spanish rulers who later co-opted the indigenous system, thereby altering the system of reciprocity and rewards that characterized pre-Hispanic forms of cooperation (see also Zorita 1963: 115). In smaller-scale settings then, reciprocity was immediate in its temporality and direct in its relation to the work event, while city-state labor tax systems featured low levels of these types of reciprocity as politico-religious institutions were supposed to reciprocate in other ways (such as coordination in public works, defense, ritual) that were often delayed and indirect in relation to the work event.

In sum, institutions of collective labor are pervasive among recent and contact-period central Mexican societies. They form the building blocks of community and social complexity. A review of the ethnographic and ethnohistoric literature highlights three dimensions of collective labor practices relevant to the archaeology of central Mexico and to the application of cooperation theory through resource problems, institutions and strategies, and material culture traits, as reviewed in chapter 1.

(1) Resource Problems. Collective labor is commonly directed at undertakings such as planting and harvesting on communally managed lands, irrigation projects, landscape reclamation and maintenance, craft production and exchange networks, construction projects such as roads and civic or ceremonial buildings, and the mobilization of consumable or durable resources to fulfill religious obligations.

(2) Institutions and Strategies. Within rural Mexican villages, the institution of tequitl labor can be organized at various scales, from reciprocity among extended families to community-wide projects that involve group cooperation by non-kin. Strategies for defining reciprocal obligations, monitoring participation, and reputation or low-level retribution are critical to this scale of organization. At the much larger scale of hierarchical Aztec city-states, these more conventional forms of cooperation were also operational but could additionally be manipulated as coatequitl labor tax, forming a cornerstone of the urban political economy that was backed by institutionalized punishment.

(3) Material Symbols. At all scales, rituals or other festivities involving food and drink played important roles in motivating collective labor, but the directness of reciprocity likely varied based on the degree of institutionalized inequality; it was more direct and immediate in smaller-scale settings, such as villages, and more indirect and delayed in larger-scale ones, such as urban or palatial labor taxes. Rituals and ritual consumption events were materialized in their culturally appropriate spaces and accoutrements.

Although institutions of collective labor and their associated norms are deep seated in central Mexico, how may they have functioned in practice within communities for which we lack documentary records? Is it possible to archaeologically infer their emergence, transformations, and strategic usages by elite and nonelite segments of earlier pre-Hispanic societies?

LABOR COLLECTIVES AND GROUP COOPERATION IN THE LATE FORMATIVE TO CLASSIC PERIODS

It is very likely that cooperative institutions similar to those documented ethnographically and ethnohistorically characterized pre-Hispanic central Mexican societies prior to the Aztecs, but lacking the detailed texts of the last 500 years their organization and associated practices remain decidedly more speculative. The centrality of such institutions within more recent societies makes this a significant void in our understanding of pre-Hispanic ones, however, and a worthwhile endeavor to attempt to deduce through the archaeological record. Likewise, the corpus of comparative cases that the social sciences draw from in interpreting human behavior is enriched through archaeological perspectives that reconstruct cooperative networks within prehistoric societies, or by their emphasis on the material dimensions of how these networks are created, sustained, and abandoned in both historic and prehistoric societies.

The Late Formative through Classic periods (c. 600 BC–AD 600) span over a millennium of significant social transformations in central Mexico, including increased social inequality and exchange, initial urbanization, and the rise and collapse of Teotihuacan—a city that was the most populous in the Americas during its height and that served as the capital for the most influential state polity in Mesoamerica (Cowgill 2000; Manzanilla 1999; Plunket and Uruñuela 2012). Only pictographic writing exists from this period, which does not offer details regarding collective labor, group cooperation, or any of the other topics discussed above;

these must instead be reconstructed archaeologically. As an additional obstacle, scholars currently lack consensus regarding the dominant ethnolinguistic group(s) in the region during this time. Pre-Aztec Nahuas or Otomis are the most likely, but multiple ethnicities are documented and Teotihuacan was surely a polyglot city (compare Cowgill 1992; Kaufman and Justeson 2007; King and Gómez Chávez 2004; Macri 2005; Taube 2000). For these reasons I abandon the linguistically specific terms discussed above and refer more generally to cooperative relations and institutions that may have operated analogously in different times and places, and how these may be reflected in, and causal to, large-scale social change. My intention is to outline testable models for pre-Hispanic group behaviors, rather than propose any direct-historical linkages. By combining regionally specific data from central Mexico with contemporary theory on cooperation and collective action, I consider the resource problems, institutions and strategies, and material symbols of pre-Hispanic group cooperation through collective labor, and their fit with the archaeological record of the Late Formative through Classic periods.

Resource Problems

Central Mexico's semiarid climate, seasonally concentrated precipitation, and shallow lake systems combine to make water management an important class of resource problem. Irrigation canals are constructed as early as the Middle Formative (c. 900–600 BC) in certain parts of the Basin of Mexico (Nichols 1982). They were present in the earliest urban societies, such as Cuicuilco and Teotihuacan, and continued through the pre-Hispanic sequence (Doolittle 1990, 2006; Nichols and Frederick 1993; Nichols, Spence, and Borland 1991; Palerm 1973; Scarborough 2003: 115–124). In the arid Tehuacan Valley a large reservoir was created through the construction of the Purrón Dam, beginning as early as the Middle Formative and in continued use until the Classic period (Spencer 2000; Woodbury and Neely 1972). Spencer (2000: 175–176) notes that, while impressive in scale, the dam appears to have been built and maintained through small, decentralized labor groups. The case provides a compelling scenario for how formal inequality and leadership might be tolerated in a more egalitarian society due to a group's desire for institutionalized retribution to effectively manage a common-pool resource (Spencer 1993). A third type of water management was drained-field agriculture, which is documented in the Basin of Mexico and for the wetter portions of southern Puebla-Tlaxcala, including near the important Formative regional center of Xochitecatl (Serra Puche, Jesús, and Manuel de la Torre 2004b; Serra Puche and Palavicini Beltrán 1996). This portion of Tlaxcala is the same general area as Wilken's (1968) study involving comunidad labor discussed above. It is therefore reasonable to suspect that similar practices were in effect over two millennia earlier at Xochitecatl.

Collective action in managing such water management systems has been explored in detail by Ostrom (1990, 1992; Ostrom and Gardner 1993). Based on her analysis of cross-cultural cases with diachronic depth, Ostrom stresses the importance of mutual monitoring in upholding a system deemed to be fair by its participants. She proposes that even large watersheds used by 15,000 individuals—a scale similar to more densely settled regions of Formative central Mexico—can evolve cooperatively if: (1) the watershed is divided into various tributaries; (2) one subgroup of farmers along a tributary can organize and agree on norms of work, monitoring, and sanctioning; (3) others see the advantages of this organization and copy it; (4) subgroups begin working collectively across tributaries; and (5) these associations aggregate to eventually cover the whole watershed (Ostrom 2003: 60). Such criteria are applicable to the spring-fed farmland of the lower Teotihuacan Valley, and could be used as a model of more bottom-up, self-organized initial urbanization, similar to the one proposed by Angulo (1993, 2007; see also Millon, Hall, and Díaz 1962; Webster 1996). The model does not preclude high levels of competition between early settlements in the valley, but offers a less centralized perspective on the city's initial stages, which is more consistent with the heterogeneity in material culture documented archaeologically (Angulo 2007; Sanders et al. 1975a, 1975b; West 1965). Successful systems based on monitoring, reputation, and retribution may have been copied (conformist transmission), with the cumulative, initially unintended result of a form of landscape capital (Brookfield 2001) that may, or may not, have been centrally managed during the Classic period. Testing this model requires excavations of late first millennium BC deposits near Teotihuacan's springs, which unfortunately are buried under the water table and 2–3 m of alluvial sediments and later cultural layers. Yet at the rate that the valley's water is being siphoned away by Mexico City, excavations may be possible in the future.

Forest management would have been another common-pool resource issue in central Mexico, and illuminating comparative cases from the Alps and Japan were discussed by Ostrom (1990). Based on a study of over 2,600 charcoal samples from the Teotihuacan Valley, Adriano Morán and McClung de Tapia (2008) argue for the persistence of at least patches of wood resources throughout the city's apogee. The analysis of paleosols and phytoliths from the valley by Solleiro-Rebolledo et al. (2011) indicate periods of deforestation, however, and that any management of such resources may have eventually run its course after some six centuries of dense occupation. Our recent excavations at the Formative period regional center of La Laguna provide an earlier, smaller-scale case of forest management. In an analysis of over 3,300 charcoal samples the dominant wood is pine (75 percent), seen today only on the highest hilltops, while the secondary growth of species like the juniper that characterizes the modern landscape is scarce archaeologically (*Pinus* = 75 percent, *Quercus* = 17 percent, *Prunus* = 5 percent, other hardwood = 1.5 percent; *Juniperus* and other softwood = 1.5 percent; 300 samples were unidentifiable; see Carballo et al. 2011; Ortmann and Carballo 2010, 2011). Yet stratigraphic evidence suggesting that members of the community practiced a destructive swidden system (Borejsza et al. 2008) provides a possible axis along which agricultural and forest resources were mediated, either more or less successfully.

Formative-period conflict within the Teotihuacan Valley and other protourban settings constitute another potential form of resource problem involving warfare, where collective benefits would take the form of gains through either offensive or defensive measures (e.g., Roscoe 2009, chapter 3; Spencer 1993, chapter 9). While architectural fortifications are rare in Formative central Mexico (Armillas 1951), elevated ceremonial centers may have served both ritual and defensive functions (e.g., García Cook and Rodríguez 1975; Serra Puche and Palavicini Beltrán 1996). Evidence for village-level raiding is suggested by osteological remains (e.g., Pijoan Aguadé and Mansilla Lory 1997) and the periodic burning of residences and temples (e.g., Carballo 2009; Uruñuela and Plunket 2007).

Drennan and Haller (2007) identify corporate land management and economies of scale as two other collective enterprises that would have encouraged and sustained increasingly larger-scale social interaction in Formative Mexico. Land tenure is one of the most elusive social institutions to reconstruct prehistorically, but the prevalence of forms of corporate land management in historic periods make it reasonable to infer archaeologically, if corporate groups like those analogous to historical systems of land management can be identified. In contrast, archaeological evidence for cooperative economic activities is abundant, providing one of our best opportunities to explore the institutions and strategies underlying group cooperation through collective labor. I therefore discuss the data and the institutions and strategies they implicate jointly in the next section.

Institutions and Strategies

Theoretical research on cooperation and collective action underscores the importance of repeat interactions to maintaining the necessary bonds of trust and reciprocity between actors (Baumard 2010; Boyd and Richerson 2008; Gächter and Herrmann 2009; Gurven and Winking 2008; Henrich and Henrich 2007; Ostrom, Gardner, and Walker 2003; Tarnita et al. 2009). Repeat interactions are the basis for the sorts of "thick relationships" (Hardin 2003) that cooperative social institutions require for success, and within which individuals negotiate particular strategies. Within the pre- or protourban communities of much of Formative central Mexico, and the rural communities outside of Teotihuacan and other Classic period cities, repeat interactions were common at the settlement level. As is the case in rural Mexico today, individuals in communities of hundreds to low thousands had high likelihood of repeat interaction, particularly at events such as community rituals, which would make people cognizant of other community members. In large cities such as Teotihuacan, however, repeat interactions with a significant portion of the population were impossible, and these were instead concentrated at the neighborhood level. Indeed, following Michael Smith (2010), neighborhoods may be defined as settlement aggregations where face-to-face interactions are common and essential to bottom-up integration, which have characteristic social and/or material attributes, such as ethnic or other corporate symbols, and distinctive or delimiting architecture.

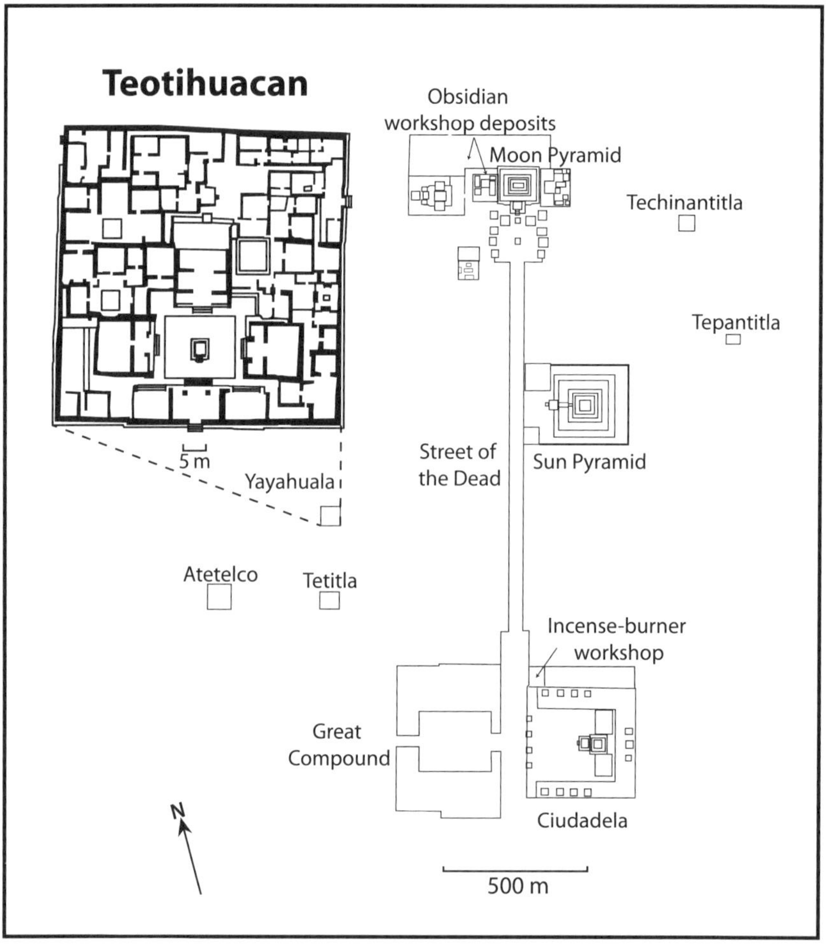

FIGURE 11.2 *Simplified map of central Teotihuacan (based on Millon 1973), depicting two documented temple production contexts and inset detail of an apartment compound (redrawn and modified from Manzanilla 1996: fig. 7).*

As one of the more populous residences of the ancient world, the over 2,300 apartment compounds of Teotihuacan would have incorporated groups of dozens to over 100 cooperating individuals, bound together by variable kinship ties, and interacting with masses of non-kin within structured neighborhoods (Cowgill 2007, 2008; Manzanilla 1996, 2009; Millon 1976, 1981). These rectilinear compounds were enclosed by bounding walls, providing a degree of privacy, and were oriented along a standardized grid plan sprawling over 25 km^2 (Figure 11.2). The multiple families living within apartment compounds were integrated through cooperative household and neighborhood craft production and ritual (Manzanilla 2002, 2009). Teotihuacan apartment compounds certainly match the archaeological criteria used to define corporate groups elsewhere (e.g., Schachner 2010), and the relatively flexible organization of central Mexican corporate groups from later periods—together with some level of collective land management—provides a reasonable model for how they were organized at Teotihuacan. Nevertheless, the compounds are uniquely orderly, leading Millon (1981: 209) to suggest their construction in the middle of the city's history represented a strategy on the part of state leaders to administer corporate groups for the purposes of taxation and labor recruitment, citing the fact that the residence type does not survive the collapse of the state. Building on such arguments, Kurtz (1996; Kurtz and Nunley 1993) argued that state leaders inculcated of an ideology of work at Teotihuacan through hegemonic, but not coercive, processes, such as the deemphasis on individual achievements and subordination of individuals to gods, natural forces, and social roles in art (see also Blanton et al. 1996).

In order for individuals to relinquish facets of their production activities and divide labor to permit economies of scale they must either (1) trust in those individuals with whom they are cooperating; (2) trust that cheaters or free-riders will be punished or excluded; or (3) have a system imposed on them by representatives of a hegemonic institution. Whereas the first two (bottom-up) scenarios are more characteristic of domestic or market economies, the third (top-down) is characteristic of political economies. All three economic systems operated simultaneously and recursively in Late Formative through Classic central Mexico. State-organized labor at Teotihuacan would have been directed primarily at the massive construction projects throughout the city, military recruitment, and the production of goods that generated and extended state power. The latter is documented in deposits associated with two major temple complexes. Next to the Moon Pyramid we have recovered evidence of intensive but apparently episodic production of obsidian dart points and martial-themed ceremonial eccentrics (Carballo 2007b, 2011). The production of ceramic censers with mold-made symbols associated with war, fertility, and ancestors was documented earlier at the Ciudadela by Carlos Múnera Bermúdez (1985). In these two examples, evidence of high-intensity production is documented next to temple complexes, but the same types of products were also manufactured in apartment compounds and the shared plazas of neighborhood centers. Both cases are therefore consistent with the operation of coatequitl-like labor organization at Teotihuacan, while the organization of production activities in domestic contexts is equally suggestive of collective labor akin to tequitl duties. In other words, a more organic form of household or suprahousehold cooperation may have been mobilized as an obligate labor tax as part of the city's political economy.

Crafting next to temples predates Teotihuacan at Xochitecatl, where Blanco (1998) documented dense concentrations of obsidian on the hilltop ceremonial center, including production debris and finished blades that were likely used in other craft activities (Figure 11.3). Several other production activities have been documented in domestic areas away from the ceremonial center, including lapidary production involving imported jade (Hirth et al. 2009). Taken together the

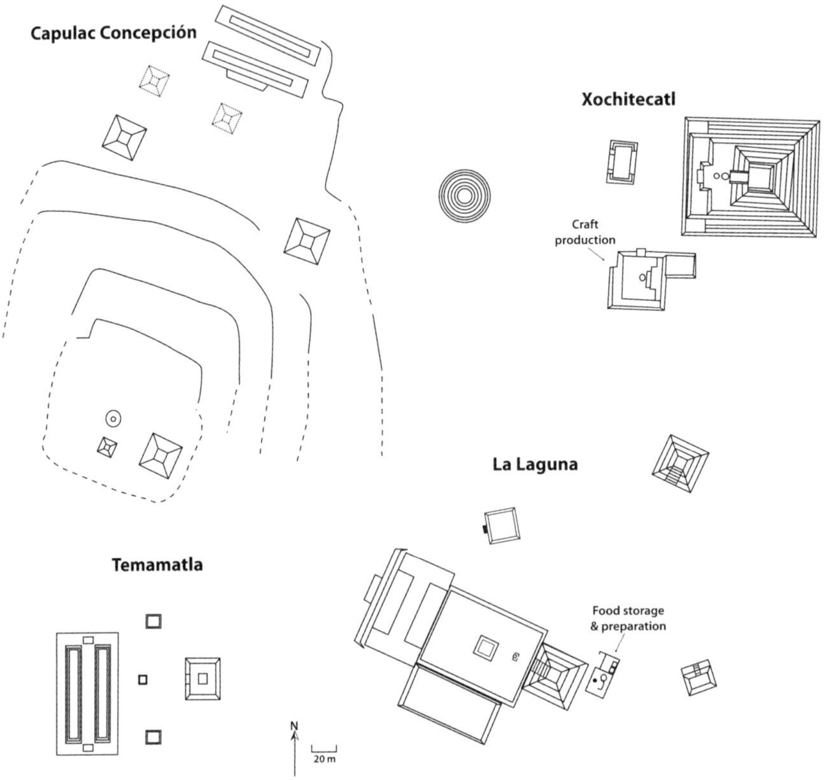

FIGURE 11.3 *Ceremonial centers of three Late Formative sites, showing temple craft production at Xochitecatl, structure associated with food storage and preparation at La Laguna, and certain shared conventions in all four (Capulac Concepción redrawn and modified from García Cook 1983: fig. 3; Xochitecatl redrawn and modified from Serra Puche, Lazcano Arce, and Mendoza 2004a: fig. 4; Temamatla based on Serra Puche 1996: fig. 2).*

two examples are suggestive of the operation of both suprahousehold, independent production and exchange, and politically organized facets of Xochitecatl's economy, implicating different forms of collective labor analogous to what is observed among later societies.

Variability in food storage provides another index of cooperative intuitions and practices of collective labor involved in land management and/or planting and harvesting, like vuelta mano (Eerkens, chapter 7; Flannery 2002; Pluckhahn, chapter 8). As noted by Grove and Gillespie (2002: 11): "If contemporary village life in central Mexico can be used as an interpretive guide, household membership was not based solely on the biology of kinship. It was continually enacted in practices confirming joint investment in a single shared food supply and by ritual references to common origins." More or less conspicuous storage practices also relate to a community's ethos of sharing, as visible storage facilities generate common knowledge regarding the resources of a household or corporate group (Hendon 2000). At the exceptionally well preserved (due to a volcanic eruption) Terminal Formative site of Tetimpa, Uruñuela and Plunket (n.d.) document a shift from subterranean storage to large external granaries in front of houses, and a correspondence between larger storage capacity and the status of the families that inhabited them. They relate this shift to changes in the ethos of resource display and the formation of patron-client–type relations between households. This transformation in social relations may have parallels to what Pluckhahn (chapter 8) documents during the Late Woodland period at Kolomoki, but in the case of Tetimpa the sequence is truncated by the volcanic eruption that resulted in the abandonment of the site and the nucleation of population into Cholula and Teotihuacan. At Teotihuacan apartment compounds, domestic storage is again hidden behind large exterior walls, reflecting the more corporate nature of inter-household social relations (Manzanilla 1996).

The decline of corporate relations at Teotihuacan and its destabilization of the state polity, as discussed by Manzanilla (2007, 2009), provides a potential, and potentially cataclysmic, example of the breakdown of cooperative institutions. Manzanilla's detailed investigations at the neighborhood center of Teopancazco suggest that noble houses grew in power later in the city's history at the expense of centralized government, creating a less stable system involving greater factional competition and elite networking. Based on multiple GIS analyses of surface deposits and architecture, Robertson (2005) also suggests later social instability at Teotihuacan due to a trend of increased spatial segregation by status, which may have created class tensions that led to political destabilization. Both scenarios are not mutually exclusive, and in either case the material symbols and ritual practices of the state religious system apparently were not enough to maintain the scale of political and urban centralization of the city during its apogee. These ritual dimensions are explored in the next section.

Material Symbols and Ritual

Theoretical work and historic and ethnographic cases demonstrate how symbolic communication and formalized ritual are critical to fostering cooperation (e.g., Bowles and Gintis 2002; Boyd and Richerson 2008; Gintis, Smith, and Bowles 2001; Henrich and Henrich 2007; Richardson and McBride 2009; Ruffle and Sosis 2007; Smith and Bliege Bird 2005; Sosis and Alcorta 2003). The idea is not new to archaeology, but several insights derived from this work are useful for better understanding symbols and ritual remains in the archaeological record. Three that are important for pre-Hispanic central Mexico, and that I focus on here, are (1) mutual monitoring and the generation of common knowledge as part of collective rituals held in open spaces such as plazas; (2) reciprocity for participation in collective labor through ritual consumption; and the (3) codification of symbols representing widely shared conceptualizations of sacred entities and the concomitant obligations they entail.

Ritual spaces have various scales in Mesoamerica—including households, neighborhoods, and community plazas—and archaeologists reason that the rules and key elements of ritual practices are those that are mostly likely to be visible in the patterning within such spaces (Marcus 2007). Increasing exclusivity in ritual is characteristic of social evolutionary sequences documented archaeologically, as egalitarian bands with ad hoc rituals precede more formal but collective village rituals and the hierarchical rituals involving institutionalized specialists characteristic of larger and more complex societies (Marcus and Flannery 2004). Within this general trend, important cultural and temporal heterogeneity exists in the relative degrees of inclusivity of ritual (Blanton and Fargher 2008; Blanton et al. 1996; Fargher et al. 2010), particularly apparent in societies where differentiation is intermediate in the continuum between egalitarianism and hierarchy (Adler and Wilshusen 1990; Feinman, chapter 2).

Significant variability is observable in the ritual architecture of Formative central Mexico, suggesting that community leaders were experimenting with what features and layouts made compelling venues for collective experiences and politico-religious performance (Plunket and Uruñuela 2012). Nevertheless, during the Late Formative period several communities developed a related architectural grammar (*sensu* Lewis and Stout 1998) of public ritual space that persisted in large part for the remainder of the pre-Hispanic period. The most essential component was a primary temple to the east, facing west over a plaza. This feature is shared by central Mexico's greatest cities, including Cuicuilco, Teotihuacan, Cholula, Tula, and Tenochtitlan. A second common component is that the second largest temple is to the north, facing south. Mastache and Cobean (2006) discuss the evolution of this partitioning, which included the greatest separation between the two temples at Teotihuacan (c. AD 250), their arrangement around a plaza at Tula (c. AD 1000), and their conflation into a double temple at Tenochtitlan and other Aztec cities (c. AD 1500). A final common component of the architectural grammar of central Mexican ritual precincts is an I-shaped ballcourt on the western side of the plaza, across from the main temple. Smith (2008a, 2008b) has designated the temple-plaza-ballcourt configuration the "Tula plaza plan," noting how the Aztecs emulated the plan of the former Toltec capital.

Elements of the Tula plaza plan have deeper roots than the Toltec, and I build on Smith (2008b: Table 1) in arguing that many of the components of a distinctively central Mexican architectural grammar are discernable in the Late Formative, to various degrees. As a heuristic tool, considering the layouts of ritual architecture as arrangements of signs following grammar-like conventions focuses attention on the fact that ancient central Mexicans both drew on their predecessors and, like with language, reworked these structures consciously and unconsciously as part of dynamically evolving systems. Early urban or protourban centers such as Cuicuilco and Xochitecatl feature massive eastern temples, followed in the Classic period by Teotihuacan and Cholula. Our recent investigations at La Laguna have documented the full temple-plaza-ballcourt configuration (Barba et al. 2009), exhibiting similarities and differences with other towns of the period such as Temamatla (Serra Puche 1996; Serra Puche, personal communication 2010) and Capulac Concepción (García Cook 1983) (Figure 11.3). What factors may account for the successful spread and resilience of these grammar-like conventions in ritual architecture? An observation by Bowles and Gintis is relevant:

> [T]he unravelling of co-operation that often afflicts communities can be averted if opportunities for mutual monitoring and punishment of non-co-operators are built into the structure of social interactions. Policies to increase the visibility of the actions of peers in communities, along with policies to enhance the effectiveness of forms of multilateral sanctioning of shirkers may thus contribute to co-operative solutions to problems, even if a majority of members are self-interested. (Bowles and Gintis 2002: 430)

Much greater attention has thus far been given to central Mexico's awe-inspiring pyramids, and their clear connections to powerful political and religious institutions, than to the much humbler plazas they abut. Yet plazas are the loci where relations of rule and resistance are negotiated (Beezley, Martin, and French 1994; Low 2000), including mutual monitoring of participation in, and commitment to, collective actions. Earlier philosophical treatises on monitoring and social structure (e.g., Bentham 1791; Foucault 1979) have been elaborated upon by experimental work showing that individuals are more likely to be generous in partitioning resources with others if only a set of artificial eyes appear to be watching them (Haley and Fessler 2005). Many public ritual precincts in the archaeological record suggest that mutual visibility was purposefully built to facilitate common knowledge: that community members know the affiliations and collective involvements of one another, and the knowledge of this knowledge is shared or transparent (Chwe 2001; Graves and Van Keuren 2011). Stanish has applied such perspectives to the coevolution of public architecture and economies of scale in the Formative period Titicaca Basin of Peru and Bolivia, incorporating game theoretic perspectives of punishment, coordination through mutual monitoring, and notions of fairness (Stanish 2004, chapter 4; Stanish and Haley 2005). Following Stanish's model, the temple platforms and open plazas of Formative central Mexican centers would have been venues for public rituals and the open negotiation of labor duties, identities, and emergent hierarchies.

Increasing exclusivity and ritual violence characterize the layouts of Classic through Postclassic ceremonial spaces—including rituals in the enclosed central courtyards of palaces, and a dramatic escalation in human sacrifice—but the essential grammar developed in the villages and towns of the Late Formative continued (see also Uruñuela and Plunket 2007). Kurtz and Nunley (1993) argue that even human sacrifice, that most overtly coercive ritual of central Mexico, may have related to hegemonic discourse of collective work in that the rarer sacrifice of citizens involved honoring them as productive, responsible, even godly, while the more common sacrifice of noncitizen war captives and culturally debased slaves set them apart as nonproductive members of society.

Consumption events would have been integral parts of community ritual connected to collective labor, but, based on ethnohistoric and ethnographic precedents from central Mexico, likely involved more indirect, delayed reciprocity in large cities like Teotihuacan and more direct, immediate reciprocity in rural settings. Work feasts in highland Mexico more often serve as the proverbial carrot; not dangled by aggrandizing leaders, but rather demanded and repeatedly referenced by participants through all stages of the undertaking (e.g., Good Eshelman 2004; Monaghan 1990, 1996). In other words, the terms of social contracts of labor and reciprocity are continually renegotiated. Though examples of exclusionary, power-seeking ritual consumption certainly exist from pre-Hispanic central Mexico (Pohl 2003; Smith, Wharton, and Olson 2003), their role in reciprocity for collective labor was of equal or greater importance.

An example of suprahousehold ritual consumption comes from La Laguna, where a structure within the three major temples at the site contained abundant evidence of food storage and production, including a likely granary, grinding tools, and semicomplete storage and serving vessels (Carballo 2012; Carballo and Barba Pingarrón 2011). The structure differs from residences excavated at the site in that the facilities related to food exceed the space that could have potentially been used for dwelling and, though large, the simple architecture is unlike elite residences at the site, which were built on tall masonry platforms (Carballo 2009). Artifact assemblages, floor chemistry signatures, and the architecture of this structure are consistent with ritual consumption at the level of the community, or at least a significant portion.

One of the semicomplete vessels recovered from the food storage and production structure at La Laguna was an effigy vessel depicting an early version of the central Mexican Storm God, which had been ritually cached with a *mano* (maize pestle), suggestive of a symbolic meaning relating to fertility or agricultural abundance. Ceramic effigies with depictions of the Storm God and Old God of Fire became formalized during the later Formative period, indicative of increased interaction and integration in central Mexico, and the emergence of widely shared conceptualizations of divine entities (Carballo 2007a). Effigy vessels depicting these two deities were part of ritual practice at cities such as Cuicuilco, towns such as La Laguna, and in more rural villages of the period. The domestic associations of the Old God of Fire and more public contexts of the Storm God are known from Teotihuacan—where the former is the most prominent household effigy vessel and the latter the most prominent effigy vessel in temple offerings—and these distinctions are apparent in depositional contexts of the later Formative. Whereas the Old God was a domestic god of the hearth and home, the Storm God was a very public god associated with rain and fertility. The common goods of water and regeneration were appealed for in Formative public rituals, sometimes involving large basins or effigy vessels depicting the Storm God, but those individuals leading the appeals were increasingly differentiated from the rest of the community based on these, and other, abilities (see Barba de Piña Chán 2002; Manzanilla 2000).

The form and contexts of later Formative effigy vessels depicting the two deities are consistent with uses in rituals of exchange and communion (e.g., Bell 1997: 108–114). In such rituals, attentive care given to the effigy is seen to enforce reciprocal obligations between practitioners and the deity. They are therefore symbolic parts of the larger moral system defining reciprocal obligations between people and deities, and among people, and are at the root of collective labor organization in central Mexico from historic to contemporary times (e.g., Lazcano and Barrientos 1999; Rojas Rabiela 1977).

CONCLUSION

Collective labor is the most conspicuous form of group cooperation in the ethnographic and ethnohistoric records of central Mexico. Labor obligations such as tequitl are not merely how individuals get things done: they construct and continually redefine communities; they are interwoven with systems of ritualized consumption and reciprocity that set the standards for evaluating social roles and responsibilities; and they are the building blocks of more complex sociopolitical systems, which mobilized cooperative labor following earlier, more organic templates expanded to obligate labor tax systems operating at the scale of tens or hundreds of thousands of individuals. For the towns and villages of the Formative period, and more rural regions of later times, the norms, practices, and institutions known ethnographically from rural Mexico likely provide appropriate models to evaluate in light of the archaeological record. At a scale much larger than is seen ethnographically, the political-economic mobilization of collective labor in pre-Hispanic urban capitals may in certain cases be analogous to what is documented ethnohistorically from Aztec cities, suggested by the temple craft production documented near Teotihuacan's Moon Pyramid and Ciudadela.

Like other parts of the world, important resource problems to consider for central Mexico include land, water, forests, warfare/defense, and various forms of economic production and exchange. Due to the prevalence of corporate land management, the scale of corporate-group cooperation in agricultural work exchanges was on the larger end of the spectrum, compared with this nearly universal form of group labor elsewhere. Though land tenure can be difficult to reconstruct archaeologically, focusing on the relative scales of corporate groups may allow archaeologists to indirectly infer these relations. Land would be a more valuable common-pool resource through water management projects including irrigation canals, dams, and drained fields. Such projects are still organized by tequitl in Mexico today, and their proliferation during the Middle to Late Formative may have provided opportunities for individuals who successfully sustained such undertakings to breach egalitarian norms through differentiated use of material goods and divisive ideologies, as has been suggested by Spencer (1993, 2000) for the Purrón Dam and contexts of intervillage raiding. In an example of how cooperation and competition are entwined, leaders who succeed in coordinating collective endeavors may have been permitted by others to exhibit certain noncooperative behaviors that became the foundations of the more elaborate sociopolitical inequalities of the Classic period (e.g., Boyd and Richerson 2008; Kaplan, Hooper, and Gurven 2009). Such bottom-up processes may have stimulated population nucleation and initial urbanization in more complex settlement systems like Teotihuacan.

The repeat interactions and thick relationships essential to sustaining cooperative institutions were built into Late Formative towns in the open plaza plans that allowed mutual monitoring and the generation of common knowledge (e.g., Bowles and Gintis 2002; Chwe 2001; Ostrom 1990; Stanish and Haley 2005). Public ritual space became increasingly formalized during the later Formative, resulting in the emergence of a common architectural grammar with conventions that persisted for nearly 2,000 years, up until the Spanish conquest. Symbolic representations of central deities, and the reciprocal obligations they implicated, became simultaneously formalized. Ritual consumption was likely part of this reciprocity, serving as a motivating stimulus and in the negotiation of social roles and responsibilities.

Repeat interactions were built into exceptionally large residences—the apartment compounds—and structured neighborhoods in Classic period Teotihuacan, each featuring their own spaces for economic production and ritual practices. Teotihuacano leaders may have promoted strong norms of work, like those noted by Sahagún and Zorita for the Nahua, through hegemonic means involving centralized urban planning and political rituals (e.g., Kurtz and Nunley 1993). Concerted ethnic migrations to Teotihuacan and the intense involvement of ethnic migrants in production and exchange activities (Spence et al. 2005) suggest that the potential for collective gains was high for a critical mass of the city's inhabitants. Expected reciprocity for this productive commercial economy probably took the form of coatequitl-like labor tax, mobilized for construction and production projects as part of the city's political economy.

This case demonstrates the archaeological applicability of contemporary models for the cultural and evolutionary dynamics of cooperation, and the relevance of archaeological data to refining abstracted models by comparing how people construct community through cooperation, drawing on evolutionary sequences with centuries-long depth. Multiactor modeling combines evolutionary concerns with strategic decision making on the part of individuals, illuminating the conditions that may encourage or discourage cooperation. Differentiation of resource problems appropriate to particular cultural settings is useful in defining types of cooperation issues, and in prioritizing the mechanisms that foster cooperative strategies and institutions in different settings. In central Mexico, institutions of communal labor known from historical periods also appear to have stimulated important cultural change in earlier ones, with potential origins relating to cooperation dilemmas, the formalization of deities and public ritual spaces, and the simultaneous integration and differentiation of the region's communities.

ACKNOWLEDGMENTS

Investigations of La Laguna's ritual precinct have been possible through grants from the National Science Foundation (BCS 0941278), National Geographic Society (CRE 8634–09), and UCMEXUS-CONACYT (awarded jointly to the author and Luis Barba), and by the support of the Consejo Nacional de Antropología, INAH, México. I thank these institutions and the individuals who have provided helpful feedback on facets of this chapter, including Luis Barba, Jennifer Carballo, Kenneth Hirth, Carlos Lazcano, Aurelio López, Agustín Ortíz, Mari Carmen Serra, David Walton, and two anonymous reviewers. I also thank the De Haro González family for their hospitality at La Laguna, and the inhabitants of San José Laguna for inspiring my thinking on collective labor and ritual consumption.

REFERENCES

Adler, Michael A., and Richard H. Wilshusen. 1990. "Large-Scale Integrative Facilities in Tribal Societies: Cross-Cultural and Southwestern US Examples." *World Archaeology* 22 (2): 133–46. http://dx.doi.org/10.1080/00438243.1990.9980136.

Adriano-Morán, Carmen Cristina, and Emily McClung de Tapia. 2008. "Trees and Shrubs: The Use of Wood in Prehispanic Teotihuacan." *Journal of Archaeological Science* 35 (11): 2927–36. http://dx.doi.org/10.1016/j.jas.2008.06.001.

Angulo, Jorge. 1993. Water Control and Communal Labor during the Formative and Classic periods in Central Mexico. In *Economic Aspects of Water Management in the Prehispanic New World*, ed. Vernon L. Scarborough and Barry L. Isaac, 151–220. Research in Economic Anthropology Supplement 7. JAI Press, Greenwich.

Angulo, Jorge. 2007. "Early Teotihuacan and Its Government." In *The Political Economy of Ancient Mesoamerica: Transformations during the Formative and Classic Periods*, ed. Vernon L. Scarborough and John E. Clark, 83–99. Albuquerque: University of New Mexico Press.

Armillas, Pedro. 1951. "Mesoamerican Fortifications." *Antiquity* 96: 77–86.

Barba, Luís, Jorge Blancas, Agustín Ortíz, and David Carballo. 2009. "Geophysical Prospection and Aerial Photography in La Laguna, Tlaxcala, Mexico." *Revue d'Archéométrie* 33: 17–20.

Barba de Piña Chán, Beatriz. 2002. "Tlapacoya, probable centro de peregrinaciones a las deidades del agua." In *Pasado, presente y futuro de la arqueología en el Estado de México: Homenaje a Román Piña Chán*, ed. Argelia Montes and Beatriz Zúñiga, 27–40. México, D.F.: Instituto Nacional de Antropología e Historia.

Baumard, Nicolas. 2010. "Has Punishment Played a Role in the Evolution of Cooperation? A Critical Review." *Mind and Society* 9 (2): 171–92. http://dx.doi.org/10.1007/s11299-010-0079-9.

Bayman, James M., and Alan P. Sullivan, III. 2008. "Property, Identity, and Macroeconomy in the Prehispanic Southwest." *American Anthropologist* 110 (1): 6–20. http://dx.doi.org/10.1111/j.1548-1433.2008.00004.x.

Beezley, William H., Cheryl English Martin, and William E. French, eds. 1994. *Rituals of Rule, Rituals of Resistance: Public Celebrations and Popular Culture in Mexico.* Wilmington, DE: SR Books.

Bell, Catherine. 1997. *Ritual: Perspectives and Dimensions.* Oxford: Oxford University Press.

Bentham, Jeremy. 1791. *Panopticon: Or, the Inspection-House. Containing the idea of a new principle of construction applicable to any sort of establishment, in which persons of any Description are to be kept under Inspection. and in particular to penitentiary-houses, prisons, Houses of Industry, Work-Houses, Poor-Houses, Manufactories, Mad-Houses, Hospitals, and Schools. With a plan of Management adapted to the Principle. In a series of letters, written in the Year 1787, From Crecheff in White Russia, to a Friend in England. By Jeremy Bentham, of Lincoln's Inn, Esq.* Dublin: Gale, Eighteenth Century Collections Online.

Blanco, Mónica. 1998. La obsidiana de Xochitécatl. In *Anthropología e Historia del Occidente de México.* XXIV Mesa Redonda de la Sociedad Mexicana de Antropología, 1719–32. Sociedad Mexicana de Antropología and Universidad Nacional Autónoma de México, México, D.F.

Blanton, Richard, and Lane Fargher. 2008. *Collective Action in the Formation of Pre-Modern States.* New York: Springer.

Blanton, Richard E., Gary M. Feinman, Stephen A. Kowalewski, and Peter N. Peregrine. 1996. "A Dual-Processual Theory for the Evolution of Mesoamerican Civilization." *Current Anthropology* 37 (1): 1–14. http://dx.doi.org/10.1086/204471.

Borejsza, Aleksander, Isabel Rodríguez López, Charles D. Frederick, and Mark D. Bateman. 2008. "Agricultural Slope Management and Soil Erosion at La Laguna, Tlaxcala, Mexico." *Journal of Archaeological Science* 35 (7): 1854–66. http://dx.doi.org/10.1016/j.jas.2007.11.024.

Bowles, Samuel, and Herbert Gintis. 2002. "Social Capital and Community Governance." *Economic Journal* 112 (483): 419–36. http://dx.doi.org/10.1111/1468-0297.00077.

Bowles, Samuel, and Herbert Gintis. 2004. "The Evolution of Strong Reciprocity: Cooperation in Heterogeneous Populations." *Theoretical Population Biology* 65 (1): 17–28. http://dx.doi.org/10.1016/j.tpb.2003.07.001.

Boyd, Robert, Herbert Gintis, and Samuel Bowles. 2010. "Coordinated Punishment of Defectors Sustains Cooperation and Can Proliferate When Rare." *Science* 328 (5978): 617–20. http://dx.doi.org/10.1126/science.1183665.

Boyd, Robert, and Peter J. Richerson. 2008. "Gene-Culture Coevolution and the Evolution of Social Institutions." In *Better than Conscious? Decision Making, the Human Mind, and Implications for Institutions*, ed. Christoph Engel and Wolf Singer, 305–323. Cambridge: MIT Press.

Brandes, Stanley. 1988. *Power and Persuasion: Fiestas and Social Control in Rural Mexico.* Philadelphia: University of Pennsylvania Press.

Brookfield, Harold. 2001. *Exploring Agrodiversity.* New York: Columbia University Press.

Brumfiel, Elizabeth M., and John W. Fox, eds. 1994. *Factional Competition and Political Development in the New World.* Cambridge: Cambridge University Press.

Carballo, David M. 2007a. "Effigy Vessels, Religious Integration, and the Origins of the Central Mexican Pantheon." *Ancient Mesoamerica* 18 (1): 53–67. http://dx.doi.org/10.1017/S0956536107000028.

Carballo, David M. 2007b. "Implements of State Power: Weaponry and Martially Themed Obsidian Production near the Moon Pyramid, Teotihuacan." *Ancient Mesoamerica* 18 (1): 173–90. http://dx.doi.org/10.1017/S0956536107000144.

Carballo, David M. 2009. "Household and Status in Formative Central Mexico: Domestic Structures, Assemblages, and Practices at La Laguna, Tlaxcala." *Latin American Antiquity* 20 (3): 473–501.

Carballo, David M. 2011. *Obsidian and the Teotihuacan State: Weaponry and Ritual Production at the Moon Pyramid. La obsidiana y el Estado teotihuacano: La producción militar y ritual en la Pirámide de la Luna. University of Pittsburgh Memoirs in Latin American Archaeology No.*

21. Center for Comparative Archaeology, Department of Anthropology, University of Pittsburgh, and Instituto de Investigaciones Antropológicas, Universidad Nacional Autónoma de México. Pittsburgh, México: D.F.

Carballo, David M. 2012. "Public Ritual and Urbanization in Central Mexico: Plaza and Temple Offerings from La Laguna, Tlaxcala." *Cambridge Archaeological Journal* 22(3): 329–352.

Carballo, David M., and Luis Barba Pingarrón, eds. 2011. "Proyecto Arqueológico La Laguna, Tlaxcala: Temporada 2010." Technical report on file at the Instituto Nacional de Antropología e Historia, México, D.F.

Carballo, David, Luis Barba, Agustín Ortíz, Jorge Blancas, Jorge H. Toledo Barrera, and Nicole Cingolani. 2011. "La Laguna, Tlaxcala: Ritual y Urbanización en el Formativo." *Revista Teccalli: Estudios Puebla-Tlaxcala* 1(2): 1–11. http://vir.inah.gob.mx/images/stories/Publicaciones/2012/Teccalli_no_1_vol_2.pdf#page=4.

Carrasco, Pedro. 1978. "La economía del México prehispánico." In *Economía política e ideología en el México prehispánico*, ed. Pedro Carrasco and Johana Broda, 13–74. Editorial Nueva Imagen, México, D.F.

Caso, Alfonso. 1958. *Indigenismo.* D.F., México: Instituto Nacional Indigenista.

Chwe, Michael Suk-Young. 2001. *Rational Ritual: Culture, Coordination, and Common Knowledge.* Princeton, NJ: Princeton University Press.

Clark, John E., and Michael Blake. 1994. "The Power of Prestige: Competitive Generosity and the Emergence of Rank Societies in Lowland Mesoamerica." In *Factional Competition and Political Development in the New World*, ed. Elizabeth Brumfiel and John Fox, 17–30. Cambridge: Cambridge University Press. http://dx.doi.org/10.1017/CBO9780511598401.003.

Cohen, Jeffrey H. 1999. *Cooperation and Community: Economy and Society in Oaxaca.* Austin: University of Texas Press.

Cowgill, George L. 1992. "Teotihuacan Glyphs and Imagery in the Light of Some Early Colonial Texts." In *Art, Ideology, and the City of Teotihuacan*, ed. Janet Catherine Berlo, 231–246. Washington, D.C.: Dumbarton Oaks Research Library and Collection.

Cowgill, George L. 2000. The Central Mexican Highlands from the Rise of Teotihuacan to the Decline of Tula. In *The Cambridge History of the Native Peoples of the Americas, Vol. II: Mesaomerica, Part 1*, ed. Richard E.W. Adams and Murdo J. MacLeod, 250–317. Cambridge University Press, Cambridge.

Cowgill, George L. 2007. The Urban Organization of Teotihuacan, Mexico. In *Settlement and Society: Essays Dedicated to Robert McCormick Adams*, ed. Elizabeth L. Stone, 261–295. Cotsen Institute of Archaeology, University of California, Los Angeles, and Oriental Institute, University of Chicago, Chicago.

Cowgill, George L. 2008. "An Update on Teotihuacan." *Antiquity* 82:962–75.

Dehouve, Danièle. 1987. "La communauté indienne, une survivance? Une municipalité mexicaine du XIXe siècle à nos jours." *Social Sciences Information. Information Sur les Sciences Sociales* 26 (2): 345–68. http://dx.doi.org/10.1177/053901887026002005.

Díaz del Castillo, Bernal. 1956. *The Discovery and Conquest of Mexico, 1517–1521.* Translated by Alfred P. Maudslay. Farrar, Straus and Cudahy, New York.

Dietler, Michael, and Brian Hayden, eds. 2001. *Feasts: Archaeological and Ethnographic Perspectives on Food, Politics, and Power.* Washington, D.C.: Smithsonian Institution Press.

Dietler, Michael, and Ingrid Herbich. 2001. "Feasts and Labor Mobilization: Dissecting a Fundamental Economic Practice." In *Feasts: Archaeological and Ethnographic Perspectives on Food, Politics, and Power*, ed. Michael Dietler and Brian Hayden, 240–264. Washington, D.C.: Smithsonian Institution Press.

Doolittle, William E. 1990. *Canal Irrigation in Prehistoric Mexico: The Sequence of Technological Change.* Austin: University of Texas Press.

Doolittle, William E. 2006. "An Epilogue and Bibliographic Supplement to Canal Irrigation in Prehistoric Mexico: The Sequence of Technological Change." *Mono y Conejo: Journal of the Mesoamerican Archaeological Research Lab, University of Texas at Austin* 4:3–15.

Drennan, Robert D., and Mikael J. Haller. 2007. "The Local Village Community and the Larger Political Economy: Formative and Classic Interaction Patterns in the Tehuacán Valley Compared to the Valley of Oaxaca and the Basin of Mexico." In *The Political Economy of Ancient Mesoamerica: Transformations during the Formative and Classic Periods*, ed. Vernon L. Scarborough and John E. Clark, 65–81. Albuquerque: University of New Mexico Press.

Dubreuil, Benoît. 2008. "Strong Reciprocity and the Emergence of Large-Scale Societies." *Philosophy of the Social Sciences* 38 (2): 192–210. http://dx.doi.org/10.1177/0048393108315509.

Eerkens, Jelmer W. 1999. "Common Pool Resources, Buffer Zones, and Jointly Owned Territories: Hunter-Gatherer Land and Resource Tenure in Fort Irwin, Southeastern California." *Human Ecology* 27 (2): 297–318. http://dx.doi.org/10.1023/A:1018777311943.

Elson, Christina M., and R. Alan Covey, eds. 2006. *Intermediate Elites in Pre-Columbian States and Empires.* Tucson: University of Arizona Press.

Erasmus, J. Charles. 1956. "Culture Structure and Culture Process: The Occurrence and Disappearance of Reciprocal Farm Labor." *Southwestern Journal of Anthropology* 12 (4): 444–69.

Fargher, Lane F., Richard E. Blanton, and Verenice Y. Heredia Espinoza. 2010. "Egalitarian Ideology and Political Power in Prehispanic Central Mexico: The Case of Tlaxcallan." *Latin American Antiquity* 21 (3): 227–51.

Feinman, Gary M. 2001. "Mesoamerican Political Complexity: The Corporate-Network Dimension." In *From Leaders to Rulers*, ed. Jonathan Hass, 151–75. New York: Kluwer Academic/Plenum Publishers. http://dx.doi.org/10.1007/978-1-4615-1297-4_8.

Flannery, Kent V. 2002. "The Origins of the Village Revisited: From Nuclear to Extended Households." *American Antiquity* 67 (3): 417–33. http://dx.doi.org/10.2307/1593820.

Foster, George M. 1967. *Tzintzuntzan: Mexican Peasants in a Changing World.* Boston: Little, Brown and Company.

Foucault, Michel. 1979. *Discipline and Punish: The Birth of the Prison.* Trans. Alan Sheridan. New York: Vintage Press.

Gächter, Simon, and Benedikt Herrmann. 2009. "Reciprocity, Culture, and Human Cooperation: Previous Insights and a New Cross-Cultural Experiment." *Philosophical Transactions of the Royal Society B —Biological Sciences* 364 (1518): 791–806. http://dx.doi.org/10.1098/rstb.2008.0275.

García Cook, Ángel. 1983. "Capulac-Concepción (P–211): un juego de pelota temprano en el Altiplano Central de México." *Jahrbuch für Geschichte Lateinamerikas* 20:1–16.

García Cook, Ángel, and Felipe Rodríguez. 1975. "Excavaciones arqueológicas en "Guadalupe Las Dalias", Puebla. *Comunicaciones.*" *Fundación Alemana para la Investigación Científica* 12:1–8.

Gibson, Charles. 1964. *The Aztecs under Spanish Rule: A History of the Indians of the Valley of Mexico, 1519–1810.* Stanford, CA: Stanford University Press.

Gintis, Herbert, Eric Alden Smith, and Samuel Bowles. 2001. "Costly Signaling and Cooperation." *Journal of Theoretical Biology* 213 (1): 103–19. http://dx.doi.org/10.1006/jtbi.2001.2406.

Gonlin, Nancy, and John C. Lohse, eds. 2007. *Commoner Ritual and Ideology in Ancient Mesoamerica.* Boulder: University Press of Colorado.

Good Eshelman, Catharine. 2004. Trabajando juntos: Los vivos, los muertos, la tierra y el maíz. In *Historia y vida ceremonial en las comunidades mesoamericanas: los ritos agrícolas*, ed. Johana Broda and Catharine Good Eshelman, 151–174. CONACULTA-INAH and UNAM, México, D.F.

Good Eshelman, Catharine. 2005. "Ejes conceptuales entre los Nahuas de Guerrero: expresión de un modelo fenomenológico mesoamericano." *Estudios de Cultura Nahuatl* 36: 87–113.

Graves, Willaim M., and Scott Van Keuren. 2011. "Ancestral Pueblo Villages and the Panoptic Gaze of the Commune." *Cambridge Archaeological Journal* 21 (2): 263–82. http://dx.doi.org/10.1017/S0959774311000278.

Grove, David C. Susan D. and Gillespie. 2002. Middle Formative Domestic ritual at Chalcatzingo, Morelos. In *Domestic Ritual in Ancient Mesoamerica*, ed. Patricia Plunket, 11–19. Cotsen Institute of Archaeology, University of California, Los Angeles.

Gurven, Michael, and Jeffrey Winking. 2008. "Collective Action in Action: Pro-Social Behavior In and Out of the Laboratory." *American Anthropologist* 110 (2): 179–90. http://dx.doi.org/10.1111/j.1548-1433.2008.00024.x.

Hardin, Russell. 2003. "Gaming Trust." In *Trust and Reciprocity: Interdisciplinary Lessons from Experimental Research*, ed. Elinor Ostrom and James Walker, 80–101. New York: Russell Sage Foundation.

Haley, Kevin A., and Daniel M.T. Fessler. 2005. "Nobody's Watching? Subtle Cues Affect Generosity in an Anonymous Economic Game." *Evolution and Human Behavior* 26 (3): 245–56. http://dx.doi.org/10.1016/j.evolhumbehav.2005.01.002.

Hayden, Brian. 1995. Pathways to Power: Principles for Creating Socioeconomic Inequalities. In *Foundations of Social Inequality*, ed. T. Douglas Price and Gary M. Feinman, 15–86. Plenum Press, New York.

Hendon, Julia A. 2000. "Having and Holding: Storage, Memory, Knowledge, and Social Relations." *American Anthropologist* 102 (1): 42–53. http://dx.doi.org/10.1525/aa.2000.102.1.42.

Henrich, Natalie, and Joseph Henrich. 2007. *Why Humans Cooperate: A Cultural and Evolutionary Explanation.* Oxford: Oxford University.

Henrich, Joseph, Richard McElreath, Abigail Barr, Jean Ensminger, Clark Barrett, Alexander Bolyanatz, Juan Camilo Cardenas, Michael Gurven, Edwins Gwako, Natalie Henrich, et al. 2006. "Costly Punishment Across Human Societies." *Science* 312 (5781): 1767–70. http://dx.doi.org/10.1126/science.1127333.

Hicks, Frederic. 1986. Prehispanic Background of Colonial Political and Economic Organization in Central Mexico. In *Supplement to the Handbook of Middle American Indians, Vol. 4, Ethnohistory*, ed. Victoria Reifler Bricker, and Ronald Spores, 35–54. University of Texas Press, Austin.

Hicks, Frederic. 1996. "Class and State in Aztec Official Ideology." In *Ideology and the Formation of Early States*, ed. Henri J.M. Claessen and Jarich G. Oosten, 256–277. Leiden: E.J. Brill.

Hicks, Frederic. 1999. "The Middle Class in Ancient Central Mexico." *Journal of Anthropological Research* 55 (3): 409–27.

Hirth, Kenneth G., Mari Carmen Serra Puche, Jesús Carlos Lazcano Arce, and Jason De León. 2009. Intermitent Domestic Lapidary Production during the Late Formative Period at Natavitas, Tlaxcala, Mexico. In *Housework: Specialization, Risk, and Domestic*

Craft Production in Mesoamerica, ed. Kenneth G. Hirth, 157–173. Archeological Papers No. 19, American Anthropological Association, Washington, DC.

Hoon, Parakh. 2007. "Working is Celebrating: The Syncretic Politics of Labor Transformation in Rural Zambia." In *Reconfiguring Institutions across Time and Space*, ed. Denis Galvan and Rudra Sil, 188–204. New York: Palgrave Macmillan.

Joyce, Arthur A., Laura Arnaud Bustamante, and Marc N. Levine. 2001. "Commoner Power: A Case Study from the Classic Period Collapse on the Oaxaca Coast." *Journal of Archaeological Method and Theory* 8 (4): 343–85. http://dx.doi.org/10.1023/A:1013786700137.

Kaplan, Hillard S., Paul L. Hooper, and Michael Gurven. 2009. "The Evolutionary and Ecological Roots of Human Social Organization." *Philosophical Transactions of the Royal Society of London. Series B, Biological Sciences* 364 (1533): 3289–99. http://dx.doi.org/10.1098/rstb.2009.0115.

Kaufman, Terrence, and John Justeson. 2007. "The History of the Word for Cacao in Ancient Mesoamerica." *Ancient Mesoamerica* 18 (2): 193–237. http://dx.doi.org/10.1017/S0956536107000211.

Katz, Friedrich. 1966. *Situación Social Y Económica De Los Aztecas Durante Los Siglos XV y XVI*. D.F., México: Universidad Nacional Autónoma de México.

Katz, Friedrich. 1969. *The Ancient American Civilisations*. London: Weidenfeld and Nicolson.

Kelly, Lucretia S. 2001. "A Case of Ritual Feasting at the Cahokia Site." In *Feasts: Archaeological and Ethnographic Perspectives on Food, Politics, and Power*, ed. Michael Dietler and Brian Hayden, 334–367. Washington, D.C.: Smithsonian Institution Press.

King, Timothy, and Sergio Gómez Chávez. 2004. "Avances en el desciframiento de la escritura jeroglífica de Teotihuacan." In *La costa del Golfo en tiempos teotihuacanos: propuestas y perspectivas. Memoria de la Segunda Mesa Redonda de Teotihuacan*, ed. Elena Ruiz Gallut María and Arturo Pascual Soto, 201–244. D.F., México: Instituto Nacional de Antropología e Historia.

Kohler, Timothy A. 1992. "Field Houses, Villages, and the Tragedy of the Commons in the Early Northern Anasazi Southwest." *American Antiquity* 57 (4): 617–35. http://dx.doi.org/10.2307/280826.

Kohler, Timothy A., Stephanie VanBuskirk, and Samantha Ruscavage-Barz. 2004. "Vessels and Villages: Evidence for Conformist Transmission in Early Village Aggregations on the Pajarito Plateau, New Mexico." *Journal of Anthropological Archaeology* 23 (1): 100–18. http://dx.doi.org/10.1016/j.jaa.2003.12.003.

Kurtz, Donald V. 1996. "Hegemonic Culturation and Work in State Formations." In *Ideology and the Formation of Early States*, ed. Henri J.M. Claessen and Jarich G. Oosten, 278–297. Leiden: E.J. Brill.

Kurtz, Donald V., and Mary Christopher Nunley. 1993. "Ideology and Work at Teotihuacan: A Hermeneutic Interpretation." *Man* 28 (4): 761–78. http://dx.doi.org/10.2307/2803996.

Lazcano, Olga, and Gustavo Barrientos. 1999. "Ritual and Community Networks among Laborer Groups in Mexico." *Annals of the American Academy of Political and Social Science* 565 (1): 207–17. http://dx.doi.org/10.1177/0002716299565001014.

Lewis, R. Barry, and Charles Stout (editors). 1998. *Mississippian Towns and Sacred Spaces: Searching for an Architectural Grammar*. University of Alabama Press, Tuscaloosa.

Lewis, Oscar. 1951–1963. *Life in a Mexican Village: Tepoztlán Restudied*. Urbana: University of Illinois Press.

Lockhart, James. 1992. *The Nahuas After the Conquest: A Social and Cultural History of the Indians of Central Mexico, Sixteenth Through Eighteenth Centuries*. Stanford, CA: Stanford University.

Lockhart, James. 2001. *Nahuatl as Written: Lessons in Older Written Nahuatl, with Copious Examples and Texts.* Stanford University Press, Stanford, and UCLA Latin American Center Publications, Los Angeles.

Low, Setha M. 2000. *On the Plaza: The Politics of Public Space and Culture.* Austin: University of Texas Press.

Macri, Martha J. 2005. "Nahua Loan Words from the Early Classic Period: Words for Cacao Preparation on a Río Azul Ceramic Vessel." *Ancient Mesoamerica* 16 (02): 321–6. http://dx.doi.org/10.1017/S0956536105050200.

Manzanilla, Linda. 1996. "Corporate Groups and Domestic Activities at Teotihuacan." *Latin American Antiquity* 7 (3): 228–46. http://dx.doi.org/10.2307/971576.

Manzanilla, Linda. 1999. "The Emergence of Complex Urban Societies in Central Mexico: the Case of Teotihuacan." In *Archaeology in Latin America*, ed. Gustavo G. Politis and Benjamin Alberti, 93–129. New York: Routledge.

Manzanilla, Linda. 2000. "The Construction of the Underworld in Central Mexico." In *Mesoamerica's Classic Heritage: From Teotihuacan to the Aztecs*, ed. David Carrasco, Lindsay Jones, and Scott Sessions, 87–116. Boulder: University Press of Colorado.

Manzanilla, Linda. 2002. Living with the Ancestors and Offering to the Gods: Domestic Ritual at Teotihuacan. In *Domestic Ritual in Ancient Mesoamerica*, ed. Patricia Plunket, 43–52. Cotsen Insitute of Archaeology Monograph 46. University of California, Los Angeles.

Manzanilla, Linda. 2007. Las 'casas' nobles de los barrios de Teotihuacan: estructuras exclusionistas en un entorno corporativo. *La Memoria 2007*, 453–470. Mexico: El Colegio Nacional.

Manzanilla, Linda. 2009 Corporate Life in Apartment and Barrio compounds at Teotihuacan, Central Mexico: Craft Specialization, Hierarchy, and Ethnicity. In *Domestic Life in Prehispanic Capitals: A Study of Specialization, Hierarchy, and Ethnicity*, ed. Linda R. Manzanilla, L., and Claude Chapdelaine, 21–42. Museum of Anthropology, University of Michigan, Ann Arbor.

Marcus, Joyce. 2007. "Rethinking Ritual." In *The Archaeology of Ritual*, ed. Evangelos Kyriakidis, 43–76. Los Angeles: Cotsen Institute of Archaeology, University of California.

Marcus, Joyce, and Kent V. Flannery. 2004. "The Coevolution of Ritual and Society: New 14C Dates from Ancient Mexico." *Proceedings of the National Academy of Sciences of the United States of America* 101 (52): 18257–61. http://dx.doi.org/10.1073/pnas.0408551102.

Marglin, Stephen A. 2007. *The Dismal Science: How Thinking like and Economist Undermines Community.* Cambridge, MA: Harvard University Press.

Mastache, Alba Guadalupe, and Robert H. Cobean. 2006. "El recinto sagrado de Tula." In *Arqueología e historia del Centro de México: Homenaje a Eduardo Matos Moctezuma*, ed. Leonardo López Luján, Davíd Carrasco, and Lourdes Cué, 203–216. D.F., México: Instituto Nacional de Antropología e Historia.

Millon, René. 1973. *Urbanization at Teotihucán, Mexico, Vol. I, Part 1: The Teotihuacan Map, Text.* University of Texas Press, Austin.

Millon, René. 1976. "Social Relations in Ancient Teotihuacan." In *The Valley of Mexico: Studies in Pre-Hispanic Ecology and Society*, ed. Eric R. Wolf, 205–248. Albuquerque: University of New Mexico Press.

Millon, René. 1981. Teotihuacan: City, State, and Civilization. In *Supplement to the Handbook of Middle American Indians, Vol. 1, Archaeology*, ed. Jeremy Sabloff, 198–243. University of Texas Press, Austin.

Millon, René, Clara Hall, and May Díaz. 1962. "Conflict in the Modern Teotihuacan Irrigation System." *Comparative Studies in Society and History* 4 (4): 494–524. http://dx.doi.org/10.1017/S0010417500001432.

Molina, Fray Alonso. 2008 [1571]. *Vocabulario en Lengua Castellana/Mexicana*. México, D.F.: Editorial Porrúa.

Monaghan, John. 1990. "Reciprocity, Redistribution, and the Transaction of Value in the Mesoamerican Fiesta." *American Ethnologist* 17 (4): 758–74. http://dx.doi.org/10.1525/ae.1990.17.4.02a00090.

Monaghan, John. 1996. "Fiesta Finance in Mesoamerica and the Origins of Gift Exchange Systems." *Journal of the Royal Anthropological Institute* 2 (3): 499–516. http://dx.doi.org/10.2307/3034900.

Moore, M. P. 1975. "Co-operative Labour in Peasant Agriculture." *Journal of Peasant Studies* 2 (3): 270–91. http://dx.doi.org/10.1080/03066157508437936.

Motolinía, Fray Toribo de Benavente. 1555–1971. *Memoriales: Libro de las cosas de la Nueva España y de los naturales de ella*. Universidad Nacional Autónoma de México, México, D.F.

Múnera Bermúdez, Luis Carlos. 1985. Un taller de cerámica ritual en la Ciudadela, Teotihuacan. Professional Thesis, Escuela Nacional de Antropología e Historia, México.

Nichols, Deborah L. 1982. "A Middle Formative Irrigation System near Santa Clara Coatitlan in the Basin of Mexico." *American Antiquity* 47 (1): 133–44. http://dx.doi.org/10.2307/280059.

Nichols, Deborah L., and Charles D. Frederick. 1993. Irrigation Canals and Chinampas: Recent Research in the Northern Basin of Mexico. In *Economic Aspects of Water Management in the Prehispanic New World*, ed. Vernon L. Scarborough and Barry L. Isaac, 123–150. Research in Economic Anthropology Supplement 7. JAI Press, Greenwich.

Nichols, Deborah L., Michael W. Spence, and Mark D. Borland. 1991. "Watering the Fields of Teotihuacan: Early Irrigation at the Ancient City." *Ancient Mesoamerica* 2 (01): 119–29. http://dx.doi.org/10.1017/S0956536100000432.

Nutini, Hugo G. 1968. *San Bernardino Contla: Marriage and Family Structure in a Tlaxcalan Municipio*. Pittsburgh: University of Pittsburgh Press.

Ortmann, Nicole, and David M. Carballo. 2010. Análisis arqueobotánico. In Proyecto Arqueológico La Laguna, Tlaxcala: Temporada 2009, ed. David M. Carballo and Luis Barba Pingarrón, 159–171. Technical report on file at the Instituto Nacional de Antropología e Historia, México, D.F.

Ortmann, Nicole, and David M. Carballo. 2011. Análisis arqueobotánico. In Proyecto Arqueológico La Laguna, Tlaxcala: Temporada 2010, ed. David M. Carballo and Luis Barba Pingarrón, Chapter 7. Technical report on file at the Instituto Nacional de Antropología e Historia, México, D.F.

Ostrom, Elinor. 1990. *Governing the Commons: The Evolution of Institutions for Collective Action*. Cambridge: Cambridge University Press.

Ostrom, Elinor. 1992. *Crafting Institutions for Self-Governing Irrigation Systems*. San Francisco: Institute for Contemporary Studies Press.

Ostrom, Elinor. 2003. "Toward a Behavioral Theory Linking Trust, Reciprocity, and Reputation." In *Trust and Reciprocity: Interdisciplinary Lessons from Experimental Research*, ed. Elinor Ostrom and James Walker, 19–79. New York: Russell Sage Foundation.

Ostrom, Elinor, and Roy Gardner. 1993. "Coping with Asymmetries in the Commons: Self-Governing Irrigation Systems Can Work." *Journal of Economic Perspectives* 7:93–112.

Ostrom, Elinor, Roy Gardner, and James M. Walker. 1994. *Rules, Games, and Common-Pool Resources.* Ann Arbor: University of Michigan Press.
Ostrom, Elinor, Roy Gardner, and James M. Walker, eds. 2003. *Trust and Reciprocity: Interdisciplinary Lessons from Experimental Research.* New York: Russell Sage Foundation.
Palerm, Angel. 1973. *Obras hidráulicas prehispánicas en el sistema lacustre del Valle de México.* D.F., México: Instituto Nacional de Antropología e Historia.
Pauketat, Timothy R., Lucretia S. Kelly, Gayle J. Fritz, Neal H. Lopinot, Scott Elias, and Eve Hargrave. 2002. "The Residues of Feasting and Public Ritual at Early Cahokia." *American Antiquity* 67 (2): 257–79. http://dx.doi.org/10.2307/2694566.
Peregrine, Peter. 2012. "Power and Legitimation: Political Strategies, Typology, and Cultural Evolution." In *The Comparative Archaeology of Complex Societies*, ed. Michael E. Smith, 165–191. Cambridge: Cambridge University Press.
Pijoan Aguadé, Carmen Ma, and Josefina Mansilla Lory. 1997. Evidence for Human Sacrifice, Bone Modification and Cannibalism in Ancient Mexico. In *Troubled Times: Violence and Warfare in the Past*, ed. Debra L. Martin and David W. Frayer, 217–40. Overseas Publishers Association, Amsterdam.
Plunket, Patricia, and Gabriela Uruñuela. 2011. "Where East Meets West: The Formative in Mexico's Central Highlands." *Journal of Archaeological Research* 20: 1–51.
Pohl, John M. 2003. "Royal Marriage and Confederacy Building among the Eastern Nahuas, Mixtecs, and Zapotecs." In *The Postclassic Mesoamerican World*, ed. Michael E. Smith and Francis F. Berdan, 243–48. Salt Lake City: University of Utah Press.
Redfield, Robert, and Alfonso Villa Rojas. 1967. *Chan Kom: A Maya Village.* Abridged Edition, Third Impression. Phoenix Books. Chicago: University of Chicago Press.
Reyes, G. Luis. 1977. Comentarios. In *El trabajo y los trabajadores en la historia de México/Labor and Laborers through Mexican History*, ed. Elsa Cecilia Frost, Michael C. Meyer, Josefina Zoraida Vázquez, and Lilia Díaz, 66–69. El Colegio de México, D.F., and University of Arizona Press, Tucson.
Richardson, Gary and Michael McBride. 2009. "Religion, Longevity, and Cooperation: The Case of the Craft Guild." *Journal of Economic Behavior & Organization* 71 (2): 172–86. http://dx.doi.org/10.1016/j.jebo.2009.03.012.
Robertson, Ian G. 2005. "Patrones diacrónicos en la constitución de los vecindarios teotihuacanos." In *Arquitectura y urbanismo: pasado y presente de los espacios en Teotihuacan*, ed. Elena Ruiz Gallut María and Jesús Torres Peralta, 277–294. D.F., México: Instituto Nacional de Antropología e Historia.
Rojas Rabiela, Teresa. 1977. La organización del trabajo para las obras públicas: el coatequitl y las cuadrillas de trabajadores. In *El trabajo y los trabajadores en la historia de México/Labor and Laborers through Mexican History*, ed. Elsa Cecilia Frost, Michael C. Meyer, Josefina Zoraida Vázquez, and Lilia Díaz, 41–66. El Colegio de México, D.F., and University of Arizona Press, Tucson.
Rojas Rabiela, Teresa. 1986 El sistema de organización en cuadrillas. In *Origen y Formación el Estado en Mesoamérica*, ed. Andrés Medina, Alfredo López Austín, and Mari Carmen Serra, 135–150. UNAM, México, D.F.
Roscoe, Paul. 2009. "Social Signaling and the Organization of Small-Scale Society: The Case of Contact-Era New Guinea." *Journal of Archaeological Method and Theory* 16 (2): 69–116. http://dx.doi.org/10.1007/s10816-009-9062-3.
Ruffle, Bradley J., and Richard Sosis. 2007. "Does It Pay To Pray? Costly Ritual and Cooperation." *B.E. Journal of Economic Analysis & Policy* 7 (1): 1–35. http://dx.doi.org/10.2202/1935-1682.1629.

Sahagún, Bernardino de. 1961. *General History of the Things of New Spain. Book 10: The People.* Translated and ed. Charles E. Dibble and Arthur J.O. Anderson, No. 14, Part XI. School of American Research and Museum of New Mexico, Santa Fe.

Sanders, William T., Michael West, Charles Fletcher, and Joseph Marino. 1975a. *The Teotihuacan Valley Project Final Report, Volume 2: The Formative Period Occupation of the Valley, Part I, Texts and Tables.* Occasional Papers in Anthropology 10, Department of Anthropology, The Pennsylvania State University, University Park.

Sanders, William T., Michael West, Charles Fletcher, and Joseph Marino. 1975b. *The Teotihuacan Valley Project Final Report, Volume 2: The Formative Period Occupation of the Valley, Part II, Plates and Figures.* Occasional Papers in Anthropology 10, Department of Anthropology, The Pennsylvania State University, University Park.

Sarukhán, José, and Jorge Larson. 2001. "When the Commons Become Less Tragic: Land Tenure, Social Organization, and Fair Trade in Mexico." In *Protecting the Commons: A Framework for Resource Management in the Americas*, ed. Johanna Burger, Elinor Ostrom, Richard B. Norgaard, David Policansky, and Bernard D. Goldstein, 45–70. Washington, D.C.: Island Press.

Scarborough, Vernon L. 2003. *The Flow of Power: Ancient Water Systems and Landscapes.* Santa Fe, NM: School of American Research.

Schachner, Gregson. 2010. "Corporate Group Formation and Differentiation in Early Puebloan Villages of the American Southwest." *American Antiquity* 75 (3): 473–96.

Serra Puche, Mari Carmen. 1996. "El sur de la Cuenca de México; el lago, la ribera y la sierra." In *Homenaje a William T. Sanders.* vol. I. ed. Alba Guadalupe Mastache, Jeffery Parsons, Robert Santley, and Carmen Serra Puche Mari, 161–172. D.F., México: Instituto Nacional de Antropolgía e Historia and Arqueología Mexicana.

Serra Puche, Mari Carmen, Jesús Carlos Lazcano Arce, and Manuel de la Torre Mendoza. 2004a. *Cerámica de Xochitécatl.* D.F., México: Universidad Nacional Autónoma de México.

Serra Puche, Mari Carmen, Jesús Carlos Lazcano Arce, and Manuel de la Torre Mendoza. 2004b. Explotación prehispánica de recursos en el sur del valle de Tlaxcala: una perspectiva de género. In *Género, ritual y desarrollo sostenido en comunidades rurales de Tlaxcala*, ed. Pilar Alberti Manzanares, 199–226. Plaza y Valdés, México, D.F.

Serra Puche, Mari Carmen, and Beatriz Palavicini Beltrán. 1996. "Xochitécatl, Tlaxcala, en el periodo Formativo (800 a.C. —100 d.C.)." *Arqueología* 16:43–57.

Smith, Eric A., and Rebecca Bliege Bird. 2005. "Costly Signaling and Cooperative Behavior." In *Moral Sentiments and Material Interests: The Foundations of Cooperation in Economic Life*, ed. Herbert Gintis, Samuel Bowles, Robert Boyd, and Ernst Fehr, 115–148. Cambridge: MIT Press.

Smith, Michael E. 2008a. *Aztec City-State Capitals.* Gainesville: University of Florida Press.

Smith, Michael E. 2008b. "City Planning: Aztec City Planning." In *Encyclopedia of the History of Non-Western Science, Technology, and Medicine*, revised Internet edition (second edition), ed. Helaine Selin, 577–587. Dordrecht: Springer. http://dx.doi.org/10.1007/978-1-4020-4425-0_8512

Smith, Michael E. 2010. "The Archaeological Studies of Neighborhoods and District in Ancient Cities." *Journal of Anthropological Archaeology* 29 (2): 137–54. http://dx.doi.org/10.1016/j.jaa.2010.01.001.

Smith, Michael E. n.d. Aztec Taxation at the City-State and Imperial Levels. Paper for the conference, "Fiscal Regimes and the Political Economy of Early States," May, 2010, Stanford University, in press.

Smith, Michael E., Jennifer B. Wharton, and Jan Marie Olson. 2003. "Aztec Feasts, Rituals, and Markets: Political Uses of Ceramic Vessels in a Commercial Economy." In *The Archaeology and Politics of Food and Feasting in Early States and Empires*, ed. Tamara L. Bray, 235–268. New York: Kluwer Academic/Plenum Publishers. http://dx.doi.org/10.1007/978-0-306-48246-5_9.

Smith, Monica L. 2010. *A Prehistory of Ordinary People*. Tucson: University of Arizona Press.

Solleiro-Rebolledo, E., S. Sycheva, S. Sedov, E. McClung de Tapia, Y. Rivera-Uria, C. Salcido-Berkovich, and A. Kuznetsova. 2011. "Fluvial Processes and Paleopedogenesis in the Teotihuacan Valley, México:Responses to Late Quaternary Environmental Changes." *Quaternary International* 233 (1): 40–52. http://dx.doi.org/10.1016/j.quaint.2010.08.005.

Sosis, Richard, and Candace Alcorta. 2003. "Signaling, Solidarity, and the Sacred: The Evolution of Religious Behavior." *Evolutionary Anthropology* 12 (6): 264–74. http://dx.doi.org/10.1002/evan.10120.

Spence, Michael W., Christine D. White, Evelyn Childs Rattray, and Fred J. Longstaffe. 2005. Past Lives in Different Places: The Origins and Relationships of Teotihuacan's Foreign Residents. In *Settlement, Subsistence, and Social Complexity: Essays Honoring the Legacy of Jeffrey R. Parsons*, ed. Richard E. Blanton, 155–197. Cotsen Institue of Archaeology, University of California, Los Angeles.

Spencer, Charles S. 1993. "Human Agency, Biased Transmission, and the Cultural Evolution of Chiefly Authority." *Journal of Anthropological Archaeology* 12 (1): 41–74. http://dx.doi.org/10.1006/jaar.1993.1002.

Spencer, Charles S. 2000. "Prehispanic Water Management and Agricultural Intensification in Mexico and Venezuela: Implications for Contemporary Ecological Planning." In *Imperfect Balance: Landscape Transformations in the Precolumbian Americas*, ed. David L. Lentz, 147–78. New York: Columbia University Press.

Stanish, Charles. 2004. "The Evolution of Chiefdoms: An Economic Anthropological Model." In *Archaeological Perspectives on Political Economies*, ed. Gary M. Feinman and Linda M. Nicholas, 7–24. Salt Lake City: University of Utah.

Stanish, Charles. 2010. "Labor Taxes, Market Systems, and Urbanization in the Prehispanic Andes: A Comparative Perspective." In *Archaeological Approaches to Market Exchange in Ancient Societies*, ed. Christopher P. Garraty and Barbara L. Stark, 185–205. Boulder: University Press of Colorado.

Stanish, Charles, and Kevin J. Haley. 2005. Power, Fairness, and Architecture: Modeling Early Chiefdom Development in the Central Andes. In *Foundations of Power in the Prehispanic Andes*, ed. Kevin J. Vaughn, Dennis Ogburn, and Christina A. Conlee, 53–70. Archaeological Papers of the American Anthropological Association, No. 14. American Anthropological Association, Arlington.

Tarnita, Corina E., Tibor Antal, Hisashi Ohtsuki, and Martin A. Nowak. 2009. "Evolutionary Dynamics in Set Structured Populations." *Proceedings of the National Academy of Sciences of the United States of America* 106 (21): 8601–4. http://dx.doi.org/10.1073/pnas.0903019106.

Taube, Karl A. 2000. *The Writing System of Ancient Teotihuacan*. Ancient America 1, Center for Ancient American Studies, Barnardsville, NC, and Washington, D.C.

Uruñuela, Gabriela, and Patricia Plunket. 2007. "Tradition and Transformation: Village Ritual at Tetimpa as a Template for Early Teotihuacan." In *Commoner Ritual and Ideology in Ancient Mesoamerica*, ed. Nancy Gonlin and Jon C. Lohse, 33–54. Boulder: University Press of Colorado.

Uruñuela, Gabriela, and Patricia Plunket. n.d. De tronco-cónicos a cuexcomates: hacia la exhibición del almacenamiento a finales del Formativo. In *Memorias del "Simposio Internacional Arqueología del Almacenamiento en Tiempos Prehispánicos desde el Norte de México hasta el Altiplano Central*, ed. Severine Bortot, Dominique Michelet, and Veronique Darras. Centro de Estudios Mexicanos y Centroamericanos, México, D.F.
Webster, David. 1996. "Economic Differentiation, Stratification, and the Evolution of Complex Societies: A Teotihuacan Case Example." In *Homenaje a William T. Sanders*. vol. I. ed. Alba Guadalupe Mastache, Jeffery Parsons, Robert Santley, and Carmen Serra Puche Mari, 111–134. D.F., México: Instituto Nacional de Antropolgía e Historia and Arqueología Mexicana.
Wells, E. Christian. 2007. Faenas, Ferias, and Fiestas: Ritual Finance in Ancient and Modern Honduras. In *Mesoamerican Ritual Economy: Archaeological and Ethnological Perspectives*, ed. E. Christian Wells and Karla L. Davis-Salazar, 29–65. University Press of Colorado, Boulder.
West, Michael. 1965. "Transition from Preclassic to Classic at Teotihuacan." *American Antiquity* 31 (2): 193–202. http://dx.doi.org/10.2307/2693984.
Wilken, Gene C. 1968. "Drained-Field Agriculture: An Intensive Farming System in Tlaxcala, Mexico." *Geographical Review* LIX (2): 215–41.
Woodbury, Richard B., and James A. Neely. 1972. Water Control Systems of the Tehuacan Valley. In *The Prehistory of the Tehuacan Valley, Vol. 4, Chronology and Irrigation*, Richard S. MacNeish, General Editor, 81–153. University of Texas Press, Austin.
Zorita, Alonso de. 1963. *Life and Labor in Ancient Mexico: The Brief and Summary Relation of the Lords of New Spain*. Trans. Keen, Benjamin. New Brunswick, NJ: Rutgers University Press.

12

Caste as a Cooperative Economic Entitlement Strategy in Complex Societies of the Indian Subcontinent and Sub-Saharan Africa

Monica L. Smith

The development of complex societies generally is accompanied by an increase in craft specialization, a process that has social, economic, and political correlates. Many labor-intensive traditions of craftmaking such as pottery, metallurgy, and textile manufacturing encompass a long period of apprenticeship that requires the cooperation of group members for the successful transfer of knowledge. Cooperation also is required for the logistical components of manufacture including raw material acquisition, stages of material preparation, the cleaning of work surfaces and installations, and the distribution of finished objects. In many ethnographically and historically documented societies, crafts are learned and carried out by specific subgroups of a population on a hereditary basis, such that the management of skills and the development of cooperation become interwoven with bonds of kinship. In some cases, these social aspects of craft specialization become highly codified, resulting in the development of "castes" in which occupational specialties become overlain with a social hierarchy based on concepts of ritual purity and pollution, along with restrictions on commensality, physical contact, and intermarriage.

Caste systems in the Indian subcontinent and West Africa are among the most robust historical examples of occupational group social hierarchies with moral overtones expressed through endogamy and proscriptions of social contact. However, many other regions of the world have historical and ethnographic evidence for the presence of endogamous, socially restrictive groups that have

TABLE 12.1 Caste occupations in the Indian subcontinent and in West Africa

Indian subcontinent low-caste occupations	Sweeper, disposer of dead cattle, leather worker, rope maker, carrier of messages about death, collector of wood for cremation fires, village watchman, witness to boundary disputes, scavenger, weaver of rough cotton cloth (Karve 1968); minstrel, executioner, menial laborer (Hutton 1961: 143, 206); collector of night-soil, bard, lime worker, clothes washer, shoemaker (Shah 1987:495;498–99); carrier of corpses, basket maker, potter, barber (Dutt 1968 [1931]: 229–231).
West Africa hereditary caste groups	woodworker, leather worker, griot (bard), blacksmith, hairdresser (Babou 2008); general metalworker, weaver, potter, musician (Tamari 1995); repairer of calabashes (Tamari 1991)

been described as caste systems, including Bali (Geertz 1959), Ethiopia (Todd 1977; Weedman 2002, 2006), Japan (Meerman 2003; see also Hutton 1961: 147), Korea (Passin 1956), Myanmar (Hutton 1961), Rwanda (Maquet 1961), Tibet (Holdich 1906: 313), Yap (Marksbury 1982, Throop 2010), and Yemen (Walters 1987). Within the socially restrictive and endogamous group systems of these societies, higher-ranked groups tend to treat lower-ranked groups as economically and morally inferior, even though lower-ranked groups usually provide needed goods and services (such as pottery, iron, leather-working, barbering, basketry, removal of animal carcasses, clothes washing, and burial assistance; see Table 12.1).

This chapter examines how caste emerged to encompass the paradox of essential services and low social status. I propose that the development of "caste" societies, with their strict moral overtones, are the result of specific episodes of dramatic economic decline, characterized by conditions that are severe enough to require significant social retooling but not severe enough to result in complete collapse and population dispersal. The study of group-level cooperative adaptations to economic change illustrates that a decline in the political economy is not merely a phase of loss, but can also be generative of new social traditions. While the case studies of South Asia and West Africa are highlighted with reference to the development of caste systems, the socially creative group response to economic fluctuations in complex societies can be fruitfully modeled for both ancient and historic societies elsewhere in the world.

GROUP RESPONSES TO PERIODS OF DECLINE IN COMPLEX SOCIETIES

The development of social complexity is characterized not by incremental and steady growth but by fluctuations in political and environmental circumstances (including warfare, drought, conquest, the development of hierarchical administrations, and the introduction of new technologies). The most dramatic fluctuations often have been termed as outright "collapse," a phenomenon that has elicited a great deal of interest from the public as well as from social scientists (e.g., Diamond 2005; Tainter 1988; Yoffee and Cowgill 1988). However, scholars are increasingly focused on the more subtle rise-and-fall cycling of political systems (Marcus 1998) and the way in which declines are often accompanied by significant cultural survivals (see Schwartz and Nichols 2006). Events of complete disintegration appear to be relatively rare, and even those circumstances that have been characterized as "collapses" are viewed as situationally variable and contain elements that survive into the postcollapse phase.

All human societies encounter environmental fluctuations for which a wide variety of individual, household, and communal mechanisms have been developed, particularly after the widespread adoption of agriculture (Halstead and O'Shea 1989). But in complex societies that are characterized not only by agriculture and sedentism but also a relatively high population density and an extractive political hierarchy, individuals may experience a reduced number of options for mitigating resource scarcity. Under systems of taxation in kind, produce that might have been managed at the household or community level is instead diverted elsewhere. Systems of food procurement and distribution imposed by political authorities can preclude the ability of individuals and households to mitigate local failures, resulting in a cascade effect: for example, massive state storage facilities designed to support centrally directed projects might also serve as a backup source of supply in the event of crop failure, but if the transportation system also fails, individuals cannot reclaim what might have been their very own produce but is now removed from their direct control. Other potential strategies, such as migration to more favorable locales, may be difficult because those locales are likely to be occupied by other people—and by people whose own surplus similarly has already been extracted and removed.

Other strategies of resource-stress mitigation also become challenged by the need to engage in complex economies that often accompany political complexity. Households might engage in individually brokered relationships of patronage, but the diversity and scale of the economy renders these arrangements of limited utility particularly if large numbers of people are affected and if the slide into decline is relatively rapid. Under such circumstances, group-level actions can emerge as a more effective response to crisis. In an article entitled "When the Going Gets Tough, Think as a Group," D. S. Wilson and his colleagues (2004: 225) suggest that groups can address more complex tasks because "cognitive cooperation often produces substantial gains for everyone at minimal individual cost." The potential for group-level organization with the result of mutual dependence thus provides an example of what Halstead and O'Shea (1989: 4) describe as "high-level" coping mechanisms that are "embedded . . . with radical consequences for the articulation and survival of that society."

Manifestations of embedded group-level coping mechanisms are encoded into previously existing social relationships and material practices. One particularly effective way to reinforce social codes is through the use of ritual, and archaeologists working in diverse regions have used architectural remains to highlight the role of ritual as a component of economic activities in challenging environments (Stanish, chapter 4). In the arid American Southwest, the Hohokam peoples of central Arizona used ritual ballcourt events as a backdrop for trade activities (Abbott 2010); in precontact Hawaii, an increased frequency of temples was associated with the agricultural management of marginal agricultural zones (Kirch et al. 2004). The use of ritual to stabilize volatile political relationships also can be seen for premodern groups (e.g., Drennan [1976: 358–359] for the Olmec; Goodman and Holladay [1986] for the truces associated with the ancient Olympics and other games; Cowdrey [1970] for the Peace of God imposed by bishops on the troubled political landscape of continental Europe in the eleventh century AD). When environmental and/or political volatility reached crisis levels, the archeological record shows that groups increased their ritual activities in response (e.g., di Lernia [2006] for the proliferation of cattle burials along the marginal areas of the Sahara at a time of climate deterioration in the fifth millennium BC; VanDerwarker, Scarry, and Eastman [2007] for the intensity of purification ceremonies in the Contact period of eastern North America). The success of all of these group endeavors was predicated on the capacity of a large proportion of participants to interact according to well-understood rules of action whose intensification was triggered by the presence of environmental, economic, and/or political challenges.

When ritual is incorporated into the structure of group responses to resource stress, participants act in anticipation of a divine as well as human sanction for transgressions. More recent historical cases under conditions of increased social complexity provide additional insights on the intertwined role of ritual and economic roles at a group level of functioning during times of crisis. In a provocative article on the formation of the European guild system, Gary Richardson and Michael McBride (2009) evaluate the intertwined expression of religious and economic commitments expressed during the Black Death. Although there were both religious groups and economic cooperatives in existence starting around AD 1000, the devastating appearance of plague in the mid-fourteenth century led to the formation of "combined cooperatives" that included both crafts and intercessory activities such as praying for deceased members of the guild (Richardson and McBride 2009: 8). To ensure a consistent level of piety that would render those prayers effective, individual members' actions were scrutinized not only in the course of strictly economic transactions but also in their personal lives, such that the guild's regulation of behavior permeated daily activities.

As Richardson and McBride's work suggests, crisis events under conditions of increasing complexity can precipitate a hyperdevelopment of intertwined social and economic interactions. A comparative example from the New World illustrates the way in which group-level responses to political and economic change can be manifested into social hierarchies of occupational specialization. Charles Stanish (2000) has written about the reaction of local elites in the Lake Titicaca basin region to the Inka-Spanish transition of the sixteenth century AD. In this region of the highland Andes of South America, there was a significant shift in the political economy when the labor-tax system of the Inka was superseded by the money-tax system and market-based political economy of the Spanish. In this transition, local elites of the Lupaqa polity intensified a preexisting social class distinction between groups known as the Aymara and the Uru. The Aymara arranged to pay their taxes in money, while simultaneously hiring the group known as the Uru to fulfill their labor-tax requirements. The Uru were viewed negatively and were repeatedly described as "poor"; as early as 1612, they were described by a local chronicler as "despised by all" (Stanish 2000: 330). Although the transition period from the Inka system to the Spanish system would have been relatively brief, with perceptible changes evident within the span of one generation, the result was the creation of a permanent underclass whose low social standing continues even today.

Examples such as the European guild system and the Aymara/Uru interaction indicate the way in which group-level responses to crisis have long-lasting effects. In some instances, ritual prescriptions can become deeply engrained in economic activities; in other instances, expedient economic solutions can become deeply engrained in social structure. We can view caste systems as another type of group-level response to crisis, in which economic activities, social structure, and ritual practices become mutually reinforced. Although caste systems can be characterized as part of a continuum of economic groups including guilds, unions, cooperatives, and gangs, in caste systems all individuals are classified as having both an occupational specialty and a corresponding social rank. Caste groups are defined and maintained not only through within-group understandings and strictures, but also by between-group restrictions and mutual dependencies whose moral overtones are maintained with reference to religious beliefs and ritual practices. In caste systems, individual and group viability is sustained through the obligation of both high-caste and low-caste members to be employed in economically essential tasks, at the cost of accepting social expectations of seclusion, endogamy, and ritual pollution. Today, much of the discussion of caste concerns the desire to mitigate the political uses and social abuses associated with caste discrimination in contemporary societies. In this chapter, however, I wish to examine the origins of caste as a way of understanding group-level responses to downward cycles of political economy.

The word *caste* as it currently exists in the English language comes via the Portuguese from Latin, and means "pure or unpolluted" (Oxford English Dictionary) or, more neutrally, "unmixed" (Dumont 1980: 21); refer to societies "distinguished by relative degrees of ritual purity or pollution and of social status" (Oxford Dictionary of English). The word has been used to describe many different types of social and economic configurations, resulting in considerable confusion in the use of the term.[1] Regardless of how they are defined, however, caste groups are marked by three distinct characteristics: economic specialization, endogamy, and the social envelope of moral overtones.

Economic Specialization: As Akerlof (1976: 611) has observed, "By its very nature the caste system involves trade and the division of labor" (see also Babou

2008: 3; Hoffman 2000: 177; Leach 1960: 6; Tamari 1991: 223). In a caste system, groups are identified with the performance of specific tasks. These tasks can range from specialized skills that require long periods of learning and apprenticeship (ritual services such as singing, healing, and poetry), to the production of goods (such as pottery and metal), to tasks that are viewed as distasteful but necessary (such as disposing of deceased persons, dead animals, and human waste). As economic specialists, caste groups are mutually dependent for the provision of goods and services that encompass both everyday and special-purpose activities. Exchange relationships among economic specialists are themselves socially codified, in which the transfer of products or services often is arranged through long-term mutual association rather than through monetary exchange.

Moral Overtones: The interactions among caste groups, though marked by economic interdependence, also are overlain with moral overtones that include prescriptions on behavior and social relations. The simultaneous social distance and economic interdependence is observed wherever caste groups occur; one succinct encapsulation is offered by Kathryn Weedman in her description of the Gamo people of southwestern Ethiopia as a "submerged artisan group" within the caste system: "Hide-workers hold a low social-political-economic status in Gamo society, yet they produce items from cattle hides used in almost every household, including bedding, chairs, saddles, and bridles" (2002: 732). The social distance and moral overtones of caste groups often are described in terms of ritual purity and pollution, as exhibited through prohibitions against sharing food or engaging in sexual relations with members of other groups (Dumont 1980; Passin 1956: 197; Tamari 1991: 230; Todd 1977; Weedman 2006).

Endogamy: As the biological expression of social segregation, endogamy is a prominent aspect of caste systems (e.g., Todd 1977: 402; Weedman 2006: 193). Endogamy is strongly associated with moral overtones, and prohibition against intermarriage with members of other castes is one of the most frequently mentioned conditions of caste societies. Endogamy is manifested not only in the oral traditions and social practices of caste groups, but also can be discerned through studies that show a low genetic diversity within caste groups indicative of long-term adherence to marriage rules (e.g., Watkins et al. 2008).

We can examine the development of caste and its sociopolitical milieu through reference to two regions of the world in which caste groups are evident in the historical and ethnographic records: the Indian subcontinent and West Africa.

THE INDIAN SUBCONTINENT

The Indian subcontinent today is marked by the presence of deeply ingrained caste groupings as a component of economic, social, and political life. Caste differentiations are based not only on the five principal ritual divisions of people (*brahmans*, *kshatriyas*, *vaisyas*, and *shudras*, along with outcastes, today called *harijans*), but also take into account occupations, resulting in thousands of identifiable caste groups (Hutton 1961: 149; Jaiswal 1998: 15). Caste distinctions are recognized and perpetuated through restrictions on intermarriage, commensality, and other forms of contact; in recent times, caste boundaries also have been expanded and reinforced both through government-sponsored affirmative action programs aimed at redressing historical imbalances, and through programs that bring casteless tribal groups into the contemporary political process (e.g., Guha 1982: 487).

One aspect of caste that is universally recorded in ethnographies and historical treatments is the interdependence of caste groups, an observation that is at the foundation of the most well-known treatments of the Indian caste system (e.g., Dumont 1980; Hutton 1961; Leach 1960). This interdependence has both ritual and practical effects, with a mutual reliance in which

> the unclean occupations are just as necessary to the Hindu's concept of community life as are the clean ones. In other words, castes who remove dead animals are Untouchable on that account but they are also intrinsically essential to the Hindu social system because the Twice Born are ritually prevented from performing this unclean occupation. By the same token, *all* unclean occupations are at once degrading to their practitioners yet essential to the appropriate organization of any orthodox Hindu community. (Gould 1986: 429, emphasis in original; see also Leach 1960: 6–7)

Although in recent years there has been a trend within postmodern scholarship toward claims that the British "invented" (or at least greatly emphasized) caste as a mechanism for subduing its large colonial population starting in the eighteenth century (e.g., Dirks 2001), historical documentation of social divisions linked to occupation dates back to the second millennium BC (Bayly 1999: 4). The idea of four main economic and social groups is first seen in the text known as the *Rig Veda*, whose oral antecedents go back to c. 1500 BC (Kashikar 2000: 31; Roy 1995: 13). As the first "written" versions of the *Rig Veda* do not appear until many hundreds of years afterwards, there is a considerable potential for the interpolation of texts over time and the inclusion of later practices into the text (this circumstance is analogous to the Homeric epics of ancient Greece, which contain both the original kernels of oral tradition and anachronisms added through later copies). The *Rig Veda* contains a passage that describes the human race as symbolically resulting from the division of the body of the cosmic entity Purusha into the four main social groups still recognized today: the brahmans from the head, the rajanya (kshatriya) group from the arms, the vaishya group from the thighs, and the shudras from the feet.

After the sixth century BC, population growth was accompanied by the reemergence of urbanism, the development of new religious traditions including Buddhism, and the growth of political dynasties throughout the subcontinent (Singh 2008). Documents from that era, such as the "Laws of Manu" and the *Jataka* (stories of the Buddha's past lives), record the notion of social divisions as a trope for interactions and political gain (Dumont 1980: 53). The written observations of the Greek traveler Megasthenes from the fourth century BC similarly note the presence of seven social divisions yet "without any association of degrees of purity" (Thapar 2002/2004: 62).

Another external source of information is found in the writings of the fifth century AD Chinese pilgrim Fa-Hian, who observed that the group known as the Chandalas, who dealt with dead bodies, were "outcastes" shunned by other castes, though he makes little other comment about distinctions or social proscriptions among other known groups (Keay 2000: 145). An assessment of historical sources suggests that the caste system's development of rigidity had a long trajectory, and that the ritual divisions seen in the *Rig Veda* and other early texts were not codified into social prescriptions for many centuries. Nor was the development of the caste system uniform throughout the subcontinent: southern India was distinguished not only by a completely distinct language family unrelated to the Indo-European languages of the north, but also a different mode of social divisions, which initially consisted of only a binary grouping equivalent to brahmans and nonbrahmans (Avari 2007: 241).

The textual record of the early subcontinent provides information about economic conditions as well as about social structure. The environment of the subcontinent, while conducive to agriculture and animal husbandry, is marked by variability in agricultural productivity and extreme seasonality in the form of monsoon rains. Rainfall cycles, which are unpredictable, produce years of significant drought as well as years of flooding, both of which adversely affect crop yields even today. These realities are reflected in documents from the earliest period such as prayers related to weather and rainfall. Significant food shortages also are a component of historical texts as early as the first century BC, when the Buddhist text known as the *Dhammapada* mentions the presence of famines (Kumar 1988).

South Asian textual sources also indicate numerous competing political agents over time and space. In the northern subcontinent, the fluorescence of the sixteen "mahajanapada" dynasties starting in the sixth century BC was superceded by the growth of the expansionist Mauryan polity starting in the late fourth century BC. This polity was defeated by rivals in 187 BC, and subsequently replaced by a number of competing polities that fought over territory until the development of robust states starting in the fourth century AD. Throughout the first millennium AD, the subcontinent was marked by the development of regional political dynasties that formed alliances and controlled substantial amounts of territory. Subsequent episodes of political and economic change included Islamic political incursions, particularly in the western subcontinent, starting in the early eleventh century AD. By the sixteenth century, larger political entities such as the Mughals (in the north) and Vijayanagara (in the south) grew large enough to be considered "empires" and continually engaged with other dynasties in the temporal and spatial interstices of territorial control (Keay 2000; Sinopoli 2003).

Increased sociopolitical complexity was accompanied by economic expansion and growth, but also by increased risks. As Breckenridge (1985) notes, empires that make military demands (in the form of people and cash from the hinterlands) and that encourage the expansion of farming into areas previously considered "marginal" actually increase the level of uncertainty in a society, a stress that leads to the development of many storage mechanisms to avert famine and other disasters. In the subcontinent, famine and warfare become increasingly part of the historical record after the end of the first millennium AD, coincident with the growth of political groups.[2] For example, the reign of Muhammad bin Tughluq (1325–1351) in the northern Indian plains was marked by a multiyear famine and a reorganization of the monetary system that resulted in a high level of counterfeiting and economic destabilization (Keay 2000: 268–269). Famines are recorded in the southern subcontinent in AD 1201, 1412, 1424, 1471, 1509, and 1540 (Kotraiah 1995: 13, Srivastava 1968), and again in the northern subcontinent in 1291, 1396, 1399, 1555, 1573, 1577, 1583, 1595–98, 1614, 1630, 1641, 1646, 1650, 1658–60, 1687, 1702–1704, and 1747 (Srivastava 1968).

Given the uncertainties of the environment and the fluctuations in the effectiveness of political authorities to ameliorate hardship, an economic solution that relied on social relationships rather than distant bureaucracies was a more sustainable solution to economic duress. In caste systems, occupational specializations became codified and crystallized into more rigid hierarchies of mutual dependence that encoded socially sanctioned access to essential goods, particularly food. An intense bout of nineteenth- and early twentieth-century scholarship on caste in the subcontinent provided a number of explanations for its origins, including the clash of cultures resulting from ancient migration, geographic isolation, ideas of pollution, belief in reincarnation and magic, hereditary occupations with trade and craft secrets, exclusivity of religious and social privileges, exploitation by the social hierarchy, and "deliberate economic and administrative policies" (summarized in Hutton 1961: 190–191). Even in those early days of scholarship, however, some researchers saw a relationship between the functioning of the caste system and the frequency of famine (Hutton 1961: 186). More recent studies also acknowledge that one of the factors in the subcontinent's development of caste systems are the "persistent oscillations between prosperity and dearth" (Bayly 1999: 29).

Under conditions of political uncertainty, resource fluctuation, and environmental unpredictability in the subcontinent, caste systems were a group-level response to perceptible economic decline as an institutionalized form of mutual dependence. These developments were not dictated by political authorities but emerged as nongovernmental (or extragovernmental) innovations precisely because the state was unable to mitigate economic hardship. This process accelerated during cycles of political instability right up to the modern period; for example, S. Bayly (1999: 5) associates a phase of highly divisive caste-based distinctions as having accompanied the "rapid regional state-building which accompanied the collapse of Mughal rule and the expansion of Western power in the subcontinent" in the early eighteenth century.

WEST AFRICA

West Africa, a large region with a continental scale of environmental diversity, is the home of many historically documented chiefdoms and states. Many cultural groups practice caste divisions that in both historical and modern times segregate people on the basis of social hierarchy, endogamy, specialized occupation, and moral injunctions against contact (Babou 2008; Ezeanya 1967; Hoffman 2000). The nomenclature of these divisions identifies three groups: "freeborn" (sometimes identified as "nobles"), caste people, and slaves (Tamari 1991: 223). Many caste groups are engaged in specialized craftmaking that provides everyday goods and essential ritual services including woodworking, leather working, blacksmithing, music making, and weaving (Babou 2008: 4; Tamari 1991).

In West Africa the terminology of occupational specialization appears, at least in some cases, to have predated the emergence of caste as a system of social segregation. Although indigenous historical texts are limited, some evidence can be found in the Sunjata epic, which narrates the story of the Sosso/Malinke war (the foundational event of the Mali empire during the early thirteenth century AD). In the Sunjata epic, references are made to individuals of blacksmith caste who engage in social life without the restrictions on intermarriage and political authority that afterwards became associated with that group (Tamari 1991: 237). Thus it is possible to see the caste system in portions of West Africa as having considerable longevity (by AD 1300 among the Malinke, and among the Soninke and Wolof by AD 1500 according to Tamari on the basis of historical evidence [1991: 221], while S. K. McIntosh [2001: 17] supports placing this development three centuries earlier on the basis of archaeological evidence). The timing of the emergence of social prescriptions can be suggested by reference to external historical sources, which by the 1590s mention the social segregation of caste groups, including prohibitions on caste individuals entering the homes of higher-status individuals and on intermarriage (Tamari 1991: 233).

As in the case of India, caste divisions and the notion of social segregation make reference to religious actions. One example is found in the West African Osu subgroup of the Igbo as a "cult-slave" designation in which members of this group are regarded as sacrifice offerings of the highest sanctity (Ezeanya 1967). In spite of this unique qualification and proximity to the divine, the Osu also are viewed as polluting and of low social rank, with traditional prohibitions on marriage, sexual contact, commensality, and washing or barbering a non-Osu individual that still reverberate in contemporary society (Ezeanya 1967: 38–39; Okwelume 2010). Osu status is achieved either through birth or through adoption; as S. N. Ezeanya notes, individuals can deliberately seek cult-slave status as a means of avoiding punishment or to escape a repressive family situation (e.g., for women after widowhood). Osu status also entitles individuals "to appropriate certain things as it were by force, and make them his own, and above all, to a share of the food and articles offered in sacrifice to the divinity to which he belonged" (Ezeanya 1967: 38).

The historical trajectory of caste systems reveals that conflict and resource scarcity were implicated in the emergence of socially restrictive hierarchies. In her examination of West African caste systems, Tal Tamari suggests that "castes may have originated in interclan alliances contracted in conditions of extreme inequality" (1991: 239). Although she does not specify the source of the inequality, economic disruption is suggested by the political turmoil associated with warfare and resource stress starting in the early second millennium AD, a time of significant changes in economic conditions. Scott MacEachern (2005) has noted that the relatively stable climate of AD 300–1100 was followed by environmental fluctuations starting in the twelfth century. Political conditions in this era also fluctuated, with both centralized states and smaller entitles interacting in a dynamic and complex fashion that included "decreases as well as increases in degree of social hierarchy and political centralization" (MacEachern 2005: 452). Additional factors of change in the political economy of West Africa would have included the coming of Islam starting in the late first millennium AD in which Islamic states "waged *jihads* or holy wars on neighboring non-Islamic populations" with economic, social, and political consequences (DeCorse 2001: 6). Specific historical events include the Sosso/Malinke war of the thirteenth century and what Tamari (1991: 235) calls the "final collapse" of the Mali empire c. AD 1600, events of regional magnitude that would have had an impact on large numbers of people.

Another obvious stressor of the political economy is the Atlantic slave trade starting in the fifteenth century. As DeCorse notes (2001: 7), 12–15 million persons were transported from Africa during the slave era, an economic action that would have affected local labor relations. The selection of individuals for removal would have reinforced preexisting ideas of social division. At the same time, however, the removal of those perceived as "low status" would have reduced the size of the workforce otherwise needed for labor-intensive and craft-making activities. The enforcement of a rigid standard of caste with its expectations of interdependence would have protected higher-ranking individuals' rights to the labor pool, at the same time that it enabled them to justify the removal of lower-ranking people and transship them to ports on the coast for export as slaves. The reinforcement of caste concepts as a stabilizing force during the extreme political and economic fluctuations of the slave trade era might therefore have been responsible for the "broad continuity in social systems" in regions such as Senegambia from AD 1500 to 1900 cited by Scott MacEachern (2005: 456).

THE DEVELOPMENT OF CASTE AS A GROUP-LEVEL RESPONSE TO ENVIRONMENTAL AND POLITICAL FLUCTUATIONS

The brief examination of caste systems in the Indian subcontinent and in West Africa illustrates the conditions that serve as the parameters for group-level solutions to resource stress. In complex societies, periodic fluctuations in the natural environment are exacerbated by political systems that affect household and community approaches to resource scarcity. When migration is not an option and when political structures are incapable of providing access to resources, group-level responses become the most efficient means of providing economic stability and the access to essential resources such as food, water, and shelter. Three characteristics mark the inception of caste as a system of socially restrictive economic specializations: entitlements, opportunity hoarding, and cheap signaling.

Entitlements: The economist Amartya Sen uses the term "entitlements" to refer to the social sanctions and "rules of legitimacy" (1981: 1) that people utilize to acquire needed goods. These "entitlements" become particularly critical under conditions of stress such as food shortages; writing about poverty, famine, and government intervention, Sen has proposed that famine is often not the simple result of a lack of food, but of a lack of socially sanctioned access to the food resources that do exist. In caste societies, all members of society are classified as a member of particular groups, with few mechanisms by which individuals can "opt out" of an ascribed social status. Caste groups also become mutually implicated, defined not only by their own occupation but by what they do *not* do.

As Gould (1986) observes, high and low castes are mutually interdependent: lower-caste groups undertake "polluting" work that enables high-caste individuals to retain their ritual status, in return for physical sustenance and protection (see also Dumont 1980: 24, citing Dubois 1906; Levine 1974: 169–170). Caste groups also "bundle" their entitlements through integrating the equipment and domain of craft expertise into social realms of performance and ritual. For example, in the Bambara cultural region of West Africa, "only blacksmiths and woodworkers may cut down large, old trees, believed to be the homes of divinities" (Tamari 1991: 225). In India, Dumont (1980: 54–55) notes that outcaste groups who touch cow skins as part of their occupation in leather working also have a monopoly on village musical bands and therefore render themselves essential in the domains of celebration and ritual (for a parallel case in Ethiopia, see Weedman 2006: 194).

The mechanism by which caste groups protect their entitlements is specialization and interdependence, usually through direct interpersonal relationships. In an examination of Senegalese caste interdependence, Cheikh Anta Babou describes the relationship of high-ranking (*géer*) families with low-ranking (*ñeeño*) families: "*Géer* families collectively offered gifts of clothes, food, and sometimes money, to their clients. The *ñeeño* had rights over certain parts of animals slaughtered at their *géer*'s house; they were owed gifts at the occasion of family ceremonies such as marriages, naming ceremonies, funerals, and circumcisions; and *in time of need, it was a moral obligation of their patrons to provide them help*" (2008: 5, emphasis added). Kathryn Weedman (2006: 193) similarly points out the mutual interdependence of Ethiopian caste groups, in which pollution is interwoven with the specialized activities of lower-caste groups: "While the Gamo consider the *tsoma* artisans to be impure; they are necessary to perform rituals that mediate between people and illness, death and infertility. The *tsoma* artisans are mediators between life, death, and social disharmony in Gamo society by serving as circumcisers, midwives, healers, morticians, and messengers."

Mutual interdependence provides economic security to both high-caste and low-caste groups. Higher-caste people have a dependent labor pool when they need it, the importance of which is underscored by higher-caste households engaging not only in sustained patron-client relationships but also through providing residential spaces, thereby bringing lower-caste people into as close a physical proximity as standards of pollution will permit (see Todd 1977: 402). The fact that low-caste groups provide essential services reveals their substantial power in the high-caste/low-caste relationship, in which clientship provides for the "protection of socially weak individuals" (Maquet 1961 cited in Todd 1977: 400; for the affirmation of agency in the caste relationship, see also Weedman 2006: 194–195). Bayly (1999: 30–31) similarly sees the accentuation of caste prerogatives in India after the decline of the Mughal empire in the mid-eighteenth century as one that "equipped both the weak and the strong with a means to maximise assets and protect themselves from loss."

Opportunity hoarding: The concept of "opportunity hoarding," first developed by Charles Tilly (1998: 10), refers to the phenomenon that occurs when members of a group acquire and monopolize access to resources such as goods or services. Opportunity hoarding is facilitated through group members' ability to include and exclude other individuals on the basis of language, kinship, marriage, housing, religion, ceremonial life, and credit, and is frequently buttressed through the "creation of beliefs and practices" that uphold the group's ability to monopolize a particular resource (1998: 154–155). Although Tilly developed this analytic concept in reference to modern ethnic groups who dominate particular professions and service industries, the concept of "opportunity hoarding" can be applied more broadly to the development of socially restrictive economic categories of all kinds.

In caste systems, both high- and low-caste groups engage in opportunity hoarding through the conservation of ritual knowledge related to their economic specialization, and through the exercise of specialized skills. High-caste groups, regarded as ritually pure, are essential for the performance of rites of passage by all members of society, while low-caste groups perform services that are necessary for the retention of purity of high-ranking groups. The result is the creation of what can be characterized as "economic niches" that in particular serve to shelter the entitlements of members of lower castes (e.g., Todd 1977: 410). In West Africa, Tamari notes that social prohibitions dictate what "freemen" (as higher-ranking persons) may not do: "While caste persons are by no means obliged to engage in craft or musical activities, free persons may not engage in metalwork, woodwork (beyond that necessitated by house construction), leatherwork and pottery-making" (1991: 225). Moreover, as Edmund Leach (1960: 6) proposes, the balance of power is equalized because of those economic niches: "Economic rights are allocated by right to closed minority groups of low social status; members of the high-status 'dominant caste', to whom the low-status groups are bound, generally form a numerical majority and must compete among themselves for the services of individual members of the lower 'castes'."

Caste groups engage in a variety of strategies to prevent their entitlement-bearing knowledge from being acquired or co-opted by others. While it may be obvious that high-ranking groups retain the rights to specialized ritual education that is supported by their generally greater access to the wealth and leisure opportunities that such education depends on, the strict guardianship of knowledge is not limited to the higher castes. Within-group enforcement of specialist knowledge can be found through the use of "secret" languages of craft production by low-caste persons (Weedman 2006: 193). Perhaps the most elaborate form of ritual knowledge guardianship is seen in certain West African blacksmith groups, for whom ironworking has a close relationship with mystical power (e.g., Njoku 1991; Tamari 1991: 238).

The combination of entitlements and opportunity hoarding, although exercised in a pattern of mutual dependence, does not, however, result in a pattern of social equality. In the development of caste systems, preexisting traditions of access to knowledge, ritual, and resources are crystallized into a more durable social stratification with strict expectations about conduct and obligations. Charles Tilly's (2003: 31) examination of group-based inequalities serve to place caste in the continuum of social constraints on access in which inequality is the result of "the conjunction of socially organized categories with (a) clique control of value-producing resources, (b) clique deployment of those resources in relations of exploitation and/or opportunity with members of subordinated or excluded categories, backed up by (c) emulation and adaptation." The likelihood that caste emerges as a group-enforced socioeconomic system during times of crisis also conditions the type of emulation, which takes the form of modifications to activities that are already practiced through the form of cheap signaling.

Cheap signaling: Anthropologists have been drawn to the concept of "costly signaling" as an evolutionary explanation for the development of elaborate rituals and social entanglements (e.g., Smith and Bliege Bird 2000; Sosis 2003). In evolutionary biology, costly signaling is a means by which free-riders are deterred and through which the trustworthiness of exchange partners can be assessed by the participants in advance of commitment.

By contrast, the maintenance of moral systems such as the ones that are engendered in caste, including prohibitions on marriage, commensality, and physical contact, require very little additional effort. This low-energy maintenance can be described as "cheap signaling" (e.g., Smith 2007). Low maintenance costs would be key in any kind of new social behavior born of a crisis moment, when there would be little scope for the invention of labor-intensive modes of signaling and group maintenance. In caste systems, markers of distinction are manifested through regular activities that are already part of the individual and household routine such as having meals, acquiring necessary goods, selecting mates, and burying the dead. Actions such as the acknowledgment of a restricted pool of marriage partners, refusing to eat in the company of others, or demarcating spaces of physical separation require virtually no energy expenditure, yet serve to create and enforce social boundaries on a regular basis.

The effectiveness of cheap signaling as a component of caste systems should not be underestimated. As Akerlof (1976: 610) notes, any move by an individual to go against socially held caste prescriptions can be a risky endeavor, as the individual may be outcast and therefore dramatically lower her or his own social and economic status. This form of reinforcement fulfills the conditions for successful group-level action identified by Wilson, Timmel, and Miller (2004: 226) in which "there is little incentive to cheat because cooperation produces large benefits for everyone at trivial individual cost." Hence, the maintenance of caste requires only a low level of energy to perpetuate through everyday transactions, but requires a high level of energy by a large number of persons to eradicate once established—a factor evident in the contemporary treatment of caste individuals in places such as India and West Africa. Although occupational specialties and even endogamy have been greatly relaxed due to factors such as migration and new technologies that provide new occupational opportunities, the moral overtones of caste membership are highly resistant to change.

Contemporary observations illustrate that the social restrictions of caste can ameliorate in good economic times, which provides additional support for the proposition that its emergence had something to do with bad economic times. Improved economies can be manifested through migration, for example; Babou (2008) notes that caste status, types of employment, and social stigma among the Senegalese diaspora in the United States is considerably lessened in the face of new economic opportunities that are seized by members of both high and low castes. Caste boundaries also can be effaced through new technologies; Karve (1968: 112) observes that people of different castes living in towns could take up tailoring as an occupation without loss of status, a move that was accelerated when the sewing machine came to the Indian subcontinent. Improved economic conditions also can provide the opportunity for entire caste groups to upgrade their status by adopting customs associated with higher-ranked castes, such as refraining from alcohol and meat, adopting particular ceremonies, and shortening the mourning period (e.g., Cohn 1958; Mukherjee 1994).

DISCUSSION

Ancient political economies were challenged by many types of change, including technological change (the introduction of large-scale technologies such as irrigation or transportation systems), political change (absorption into a rival state, warfare, or the growth and collapse of alliances), and ideological shifts (such as new religious traditions that affected the flows of economic interchange). Sometimes, changes facilitated population growth and political expansion, but sometimes political agents were unable to manage complex series of changes or even made conditions worse through their actions. Historical and ethnographic information shows that the development of caste is coincident with extreme fluctuations in the political economy of the kind that can be described as decline that stops short of outright collapse. With institutional hierarchies unable to mitigate significant fluctuations in the availability of food (due to natural disaster, warfare, or other causes), group-level responses provided a form of redress that also resulted in a socially sanctioned reinforcement of occupational categories.

In this chapter, it has been proposed that caste develops as a response to sharp declines in economic conditions, in which strictly defined occupational designations become beneficial to laboring groups and in which low social status is accepted in exchange for a guarantee of livelihood. Caste is thus not a unilateral imposition by a ruling authority, but rather the creation of an interdependent system from both the top down and the bottom up as social groups codify mutual interdependence along a hierarchy. The mechanism of group formation, which would require large-scale coordination, might be characterized as a self-organizing system (e.g., Mullane 2009; Radzicki 1990). Groups may self-organize on the basis of both internal and external pressures, resulting in new and long-lasting social configurations. From the individual perspective, caste at the time of formation might have served as a welcome cognitive shortcut at a moment of economic and social crisis, in which individuals and households could assure their survival through a low-energy expenditure of "cheap signaling" and a reduction of activities.

The recognition that caste is not a unique configuration but one that is part of a continuum of socioeconomic groupings provides new opportunities to examine the role of group-level responses to changes in political economy in ancient states as well as in more recent historical cases, even those that are not traditionally described as having had caste systems. Moreover, we can posit the potential for caste structures throughout complex societies as one of many forms of group-developed mutualism that can develop under crisis conditions. Although scholars tend to focus on spectacular moments of "collapse" as the final degradation of a particular society, it may be more productive to focus on the many intermittent political, economic, and environmental challenges that were survivable through group-level innovations that often made use of preexisting ritual and religious precepts.

The recognition that social and moral assignments have a specific point of inception also enables us to look at social change from the perspective of individuals and households as they face widespread conditions of loss. State-level economic and social fluctuations have an effect on ordinary people, who in a single lifetime may have had to accept and adjust to circumstances of significant social degradation as a trade-off for the entitlement of access to food through services such as making pottery and metal, or dealing with the dead. The acceptance of social degradation in hereditary occupations in one's lifetime also could be perceived as a matter of improved reproductive fitness, in which individuals and households were guaranteed a minimum level of survival moving forward into the next generation. Economic practices thus became intimately integrated with religious practices in an attempt to secure both present-day survival and long-term viability for one's descendants.

Although the regions of the Indian subcontinent and West Africa are the most familiar anthropological examples of caste systems, we should look for the emergence of caste and other socially restrictive systems whenever we see a complex society in which there has been a precipitous decline that stops short of complete collapse. There are a number of types of social relations of production that can be discerned from the historical record that can be applicable to archaeological time periods, including master-apprentice relationships, guilds, unions, caste, debt bondage, and slavery. Of these relationships, caste might actually be one of the more straightforward to examine because of the ability to trace endogamy through DNA (e.g., Watkins et al. 2008; see also summary in Boivin 2007: 350–356). Caste and caste-like systems also can be inferred through the relative quality and abundance of wealth and status indicators such as housing structures, daily-use tools, foods, and storage facilities as well as spatial organization.

CONCLUSION

As a group-level response to resource shortfall, caste systems create effective entitlements to basic human needs at times of crisis through the codification of occupational specialization, mutual dependence, and the "cheap signaling" of social boundaries. The strong internal cohesion of both high- and low-caste groups is evident in self-policing among group members in both daily and ritual life through commensal rules and endogamy, and as they guard the secrets of their economic specialization whether they are ritual specialists or craft specialists. Such a phenomenon may be similar to the "groupishness" discussed by Feinman (chapter 2), and may have wide applicability to past complex societies. Indeed, caste systems may be only one of a variety of systems that people develop at the group level as a response to the decline phases (or stress points) of social complexity.

Crisis phases are not merely times of cultural or economic loss, but also can foster the development of new patterns of social interaction that persist well into resurgent phases of political economy. Models that examine the opportunities provided by crisis phases therefore might help social scientists to make sense of other types of social groupings that appear to emerge under conditions that are adverse but not severe enough to cause the collapse of an entire social or political system. One example is the formation of gangs and other crime syndicates through which individuals can engage in remunerative work at times of economic stagnation in which there are few other opportunities (see Venkatesh and Levitt 2000). Like other craft specialists, gang members develop and harbor secret knowledge about logistics and ritual, and use material culture (ornaments, tattoos, color-coded clothing) to identify group members through signals that also are readable by nongroup members. Like castes, gangs are economically specialized (often in illegal activities), provide items or services that are in high demand (drugs, prostitution, low-wage smuggled laborers), and have rules of association that are analogous to commensality and endogamy. Finally, the dissolution of gangs often occurs only with great difficulty; while gangs usually operate outside of the law and individuals can be punished, successfully countering gangs at the group level usually involves the provision of new economic opportunities. Similarly, the formation of other nongovernmental organizations, ranging from assistance groups to insurgencies, might also be characterized as self-organizing systems that emerge under conditions of significant decline and which often are maintained even when economic conditions improve.

In their anthropological examination of risk management, Halstead and O'Shea (1989: 5) note, "The importance of variability and buffering lies not only in the immediate realm of provisioning and economic activities, but extends beyond these to exert a strong influence on culture at large, shaping social organisation and providing the crucial conditions that give rise to social change and transformation." For caste groups, the emergence and presence of entitlements is particularly evident in times of crisis, but their existence can sustain social interactions well after the crisis period is over, such that "entitlements" are continually reinforced through a perception of tradition, mutual obligations, and social expectations. The mutual dependence of high-caste and low-caste groups indicates that instead of perceiving the developing of caste systems as a top-down organizational imposition, we can instead propose that they are the result of a mutual integration of a top-down/bottom-up approach in which all levels of the social and political hierarchies are deliberately engaged (see Janusek and Kolata 2004; Smith 2006). The development of caste and other socioeconomic groups as coherent, integrated social and economic systems indicates the extent to which crisis and fluctuation in complex societies can be a source of generative change with significant potential for long-term effects at the individual, household, and society-wide scale.

ACKNOWLEDGMENTS

My interest in caste as an economic phenomenon was initially sparked by an invitation by John Arthur and Kathryn Weedman to serve as a discussant for a session at the Society for American Archaeology meetings. The members of that session (which included Lynn Jorde, Scott MacEachern, Gunnar Haaland, Randi Haaland, Judy Sterner, Tal Tamari, and Allen Zagarell) provided stimulating discussion and insights for which I thank them. I also would like to thank Charles Stanish for discussions about the role of labor groups in political economies. Finally, I would like to thank David Carballo for the invitation to contribute the chapter for the present volume; many thanks go to him and to Elizabeth Brite and Robert Marshall for comments on the written version.

NOTES

1. One particular debate is whether the word *caste* can be used to refer to any societies except those in the Indian subcontinent, a factor that enters into nearly every discussion

of the phenomenon of caste, and even affects the dictionary definition of the term in which caste is usually made equivalent to Hindu (Indian) society; for lengthy discussions of this issue, see Leach 1960; Pitt-Rivers 1971; Todd 1977. Because the phenomenon of socially segregated occupational groups is a demonstrable world-wide phenomenon, it might in the future be advisable to develop a more neutral terminology such as "***G***roups that are ***E***conomically specialized and ***E***ndogamous with ***M***oral overtones" (or GEEMs, for short).

2. The increased number of recorded famines is probably due in part to the actual increase of famine that is correlated with an increase in political complexity; however, it is probably also due to the advent of more thorough recording systems for taxation and record keeping resulting in the cataloging of local and regional events that previously might have gone unrecorded.

REFERENCES

Abbott, David R. 2010. "The Rise and Demise of Marketplace Exchange among the Prehistoric Hohokam of Arizona." In *Archaeological Approaches to Market Exchange in Ancient Societies*, ed. Christopher P. Garraty and Barbara L. Stark, 61–83. Boulder: University Press of Colorado.

Akerlof, George. 1976. "The Economics of Caste and of the Rat Race and Other Woeful Tales." *Quarterly Journal of Economics* 90 (4): 599–617. http://dx.doi.org/10.2307/1885324.

Avari, Burjor. 2007. *India: The Ancient Past.* London: Routledge.

Babou, Cheikh Anta. 2008. "Migration and Cultural Change: Money, "Caste," Gender, and Social Status among Senegalese Female Hair Braiders in the United States." *Africa Today* 55 (2): 3–22.

Bayly, Susan, ed. 1999. *Caste, Society and Politics in India from the Eighteenth Century to the Modern Age. The New Cambridge History of India IV: 3.* Cambridge: Cambridge University Press. http://dx.doi.org/10.1017/CHOL9780521264341.

Boivin, Nicole. 2007. "Anthropological, Historical, Archaeological and Genetic Perspectives on the Origins of Caste in South Asia." In *The Evolution and History of Human Populations in South Asia*, ed. M. D. Petraglia and B. Allchin, 341–361. Dordrecht, The Netherlands: Springer. http://dx.doi.org/10.1007/1-4020-5562-5_15.

Breckenridge, Carol Appadurai. 1985. "Social Storage and the Extension of Agriculture in South India, 1350–1750." In *Vijayanagara - City and Empire: New Currents of Research*, ed. Anna Libera Dallapiccola, 41–72. Stuttgart: Steiner Verlag Wiesbaden GMBH.

Cohn, Bernard S. 1958. "Changing Traditions of a Low Caste." *Journal of American Folklore* 71 (281): 413–21. http://dx.doi.org/10.2307/538570.

Cowdrey, H. E. J. 1970. "The Peace and the Truce of God in the Eleventh Century." *Past & Present* 46 (1): 42–67. http://dx.doi.org/10.1093/past/46.1.42.

DeCorse, Christopher R. 2001. "Introduction." In *West Africa During the Atlantic Slave Trade*, ed. Christopher R. DeCorse, 1–13. Leicester: Leicester University Press.

Diamond, Jared M. 2005. *Collapse: How Societies Choose to Fail or Succeed.* New York: Viking.

di Lernia, Savino. 2006. "Building monuments, creating identity: Cattle cult as a social response to rapid environmental changes in the Holocene Sahara." *Quaternary International* 151 (1): 50–62. http://dx.doi.org/10.1016/j.quaint.2006.01.014.

Dirks, Nicholas. 2001. *Castes of Mind: Colonialism and the Making of Modern India.* Princeton: Princeton University Press.

Drennan, Robert D. 1976. "Religion and Social Evolution in Formative Mesoamerica." In *The Early Mesoamerican Village*, ed. Kent V. Flannery, 345–368. New York: Academic Press.
Dubois, Abbé J. A. 1906. *Hindu Manners, Customs, and Ceremonies*. Translated by Henry K. Beauchamp. Oxford: Clarendon.
Dumont, Louis. 1980 [1966]. *Homo Hierarchicus: The Caste System and Its Implications, complete revised English edition*. Chicago: University of Chicago Press.
Dutt, Nripendra Kumar. 1968 [1931]. *Origin and Growth of Caste in India*. vol. I. Calcutta: K.L. Mukhopadhyay.
Ezeanya, S. N. 1967. "The "Osu" (Cult-Slave) System in Igboland." *Journal of Religion in Africa. Religion en Afrique* 1 (1): 35–45. http://dx.doi.org/10.1163/157006667X00020.
Geertz, Clifford. 1959. "Form and Variation in Balinese Village Structure." *American Anthropologist* 61 (6): 991–1012. http://dx.doi.org/10.1525/aa.1959.61.6.02a00060.
Goodman, M. D., and A. J. Holladay. 1986. "Religious Scruples in Ancient Warfare." *Classical Quarterly* 36 (1): 151–71. http://dx.doi.org/10.1017/S0009838800010612.
Gould, Harold A. 1986. "The Hindu Jajmani System: A Case of Economic Particularlism." *Journal of Anthropological Research* 42 (3): 269–78.
Guha, Amalendu. 1982. Appendix: The Medieval Economy of Assam. In *The Cambridge Economic History of India Vol 1: c. 1200–c. 1750*, ed. Tapan Raychaudhuri and Irfan Habib, 478–505. Cambridge University Press, Cambridge.
Halstead, Paul, and John O'Shea. 1989. "Introduction: Cultural Responses to Risk and Uncertainty." In *Bad Year Economics*, ed. Paul Halstead and John O'Shea, 1–7. Cambridge: Cambridge University Press. http://dx.doi.org/10.1017/CBO9780511521218.002.
Hoffman, Barbara G. 2000. *Griots at War: Conflicts, Conciliation and Caste in Mande*. Bloomington: Indiana University Press.
Holdich, Thomas H. 1906. *Tibet the Mysterious*. Stokes, New York: Frederick A.
Hutton, J. H. 1961. *Caste in India: Its Nature, Function, and Origins*. 3rd ed. Oxford: Oxford University Press.
Jaiswal, Suvira. 1998. *Caste: Origin, Function and Dimensions of Change*. New Delhi: Manohar.
Janusek, John Wayne, and Alan L. Kolata. 2004. "Top-Down or Bottom-Up: Rural Settlement and Raised Field Agriculture in the Lake Titicaca Basin, Bolivia." *Journal of Anthropological Archaeology* 23 (4): 404–30. http://dx.doi.org/10.1016/j.jaa.2004.08.001.
Karve, I. 1968. *Maharashtra: Land and its People*. Directorate of Government Printing, Bombay.
Kashikar, C. G. 2000. A Historical Survey. In *Ancient Indian Literature: An Anthology*, Volume I: Vedic Sanskrit and Pali, ed. T.R.S. Sharma, 62. Sahitya Akademi, New Delhi.
Keay, John. 2000. *India: A History*. New York: Grove Press.
Kirch, P. V., A. S. Hartshorn, O. A. Chadwick, P. M. Vitousek, D. R. Sherrod, J. Coil, L. Holm, and W. D. Sharp. 2004. "Environment, Agriculture, and Settlement Patterns in a Marginal Polynesian Landscape." *Proceedings of the National Academy of Sciences of the United States of America* 101 (26): 9936–41. http://dx.doi.org/10.1073/pnas.0403470101.
Kotraiah, C. T. M. 1995. *Irrigation Systems under Vijayanagara Empire*. Directorate of Archaeology and Museums, Mysore.
Kumar, Tuk-Tuk. 1988. *History of Rice in India: Mythology, Culture and Agriculture*. Delhi: Gian Publishing House.

Leach, E. R. 1960. "Introduction: What Should We Mean by Caste?" In *Aspects of Caste in South India, Ceylon and Northwest Pakistan*, ed. E. R. Leach, 1–10. Cambridge: Department of Archaeology and Anthropology, University of Cambridge.

Levine, Donald N. 1974. *Greater Ethiopia: The Evolution of a Multiethnic Society*. Chicago: University of Chicago Press.

MacEachern, Scott. 2005. "Two Thousand Years of West African History." In *African Archaeology: A Critical Introduction*, ed. Ann Brower Stahl, 441–466. Oxford: Blackwell.

Maquet, Jacques J. 1961. *The Premise of Inequality in Ruanda*. London: International African Institute.

Marcus, Joyce. 1998. "The Peaks and Valleys of Ancient States: An Extension of the Dynamic Model." In *Archaic States*, ed. Gary M. Feinman and Joyce Marcus, 59–94. Santa Fe, NM: School of American Research.

Marksbury, Richard A. 1982. "Legislating Social Order: An Example from the Yap Islands." *Oceania* 53 (1): 19–28.

McIntosh, Susan Keech. 2001. "Tools for Understanding Transformation and Continuity in Senegambian Society: 1500–1900." In *West Africa during the Atlantic Slave Trade*, ed. Christopher R. DeCorse, 14–37. Leicester: Leicester University Press.

Meerman, Jacob. 2003. "The Mobility of Japan's Burakumin: Militant Advocacy and Government Response." In *Boundaries of Clan and Color: Transnational Comparisons of inter-Group Disparity*, ed. William Darity, Jr., and Ashwini Deshpande, 130–151. London: Routledge. http://dx.doi.org/10.4324/9780203987711.ch8

Mukherjee, Sibsankar. 1994. "The Social Role of a Caste Association." *Indian Economic and Social History Review* 31 (1): 89–100. http://dx.doi.org/10.1177/001946469403100104.

Mullane, Elizabeth Brownell. 2009. Megaliths, Mounds, and Monuments: Applying Self-Organizing Theory to Ancient Human Systems. Unpublished Ph.D. Dissertation, Cotsen Institute of Archaeology, University of California Los Angeles.

Njoku, Onwuka N. 1991. "Magic, Religion and Iron Technology in Precolonial North-Western Igboland." *Journal of Religion in Africa. Religion en Afrique* 21 (3): 194–215. http://dx.doi.org/10.1163/157006691X00023.

Okwelume, Obinna Charles. 2010. Igbo Caste Practices: Persistence and Public Attitudes in the Media. Unpublished Ph.D. Dissertation, University of Birmingham.

Oxford Dictionary of English online "caste" August 25, 2010.

Oxford English Dictionary online "caste" August 27, 2010.

Passin, Herbert. 1956. "The Paekchong of Korea: A Brief Social History." *Monumenta Nipponica* 12 (3/4): 195–240. http://dx.doi.org/10.2307/2382752.

Pitt-Rivers, Julian. 1971. "On the Word "Caste."." In *The Translation of Culture*, ed. T. O. Beidelman, 231–256. London: Tavistock.

Radzicki, Michael J. 1990. "Institutional Dynamics, Deterministic Chaos, and Self-Organizing Systems." *Journal of Economic Issues* 24 (1): 57–102.

Richardson, Gary, and Michael McBride. 2009. "Religion, Longevity, and Cooperation: The Case of the Craft Guild." *Journal of Economic Behavior & Organization* 71 (2): 172–86. http://dx.doi.org/10.1016/j.jebo.2009.03.012.

Roy, Kumkum. 1995. "The Vedic Age." In *Ancient India*, 13–40. Publications Division, Ministry of Information and Broadcasting, Government of India.

Schwartz, Glenn M., and John J. Nichols, eds. 2006. *After Collapse: The Regeneration of Complex Societies*. Tucson: University of Arizona.

Sen, Amartya. 1981. *Poverty and Famines: An Essay on Entitlement and Deprivation*. Oxford: Clarendon Press.

Shah, A. M. 1987. "Untouchability, the Untouchables and Social Change in Gujarat." In *Dimensions of Social Life: Essays in Honor of David G. Mandelbaum*, ed. Paul Hockings, 493–505. Berlin: Mouton de Gruyter. http://dx.doi.org/10.1515/9783110846850.493.
Singh, Upinder. 2008. *A History of Ancient and Early Medieval India*. Delhi: Pearson Longman.
Sinopoli, Carla M. 2003. *The Political Economy of Craft Production*. Cambridge: Cambridge University Press. http://dx.doi.org/10.1017/CBO9780511489648
Smith, Eric Alden, and Rebecca L. Bliege Bird. 2000. "Turtle Hunting and Tombstone Opening: Public Generosity as Costly Signaling." *Evolution and Human Behavior* 21 (4): 245–61. http://dx.doi.org/10.1016/S1090-5138(00)00031-3.
Smith, Monica L. 2006. "The Archaeology of Food Preference." *American Anthropologist* 108 (3): 480–93. http://dx.doi.org/10.1525/aa.2006.108.3.480.
Smith, Monica L. 2007. "Inconspicuous Consumption: Non-Display Goods and Identity Formation." *Journal of Archaeological Method and Theory* 14 (4): 412–38. http://dx.doi.org/10.1007/s10816-007-9040-6.
Sosis, Richard. 2003. "Why Aren't We All Hutterites? Costly Signaling Theory and Religious Behavior." *Human Nature (Hawthorne, N.Y.)* 14 (2): 91–127. http://dx.doi.org/10.1007/s12110-003-1000-6.
Srivastava, Hari Shanker. 1968. *The History of Indian Famines*. Agra: Sri Ram Mehra and Co.
Stanish, Charles. 2000. Negotiating Rank in an Imperial State: Lake Titicaca Basin Elite under Inca and Spanish Control. In *Hierarchies in Action: Cui Bono?* ed. Michael Diehl, 317–39. Center for Archaeological Investigations, Southern Illinois University, Carbondale Occasional Paper No. 27, Carbondale, IL.
Tainter, Joseph A. 1988. *The Collapse of Complex Societies*. Cambridge: Cambridge University Press.
Tamari, Tal. 1991. "The Development of Caste Systems in West Africa." *Journal of African History* 32 (02): 221–50. http://dx.doi.org/10.1017/S0021853700025718.
Tamari, Tal. 1995. "Linguistic Evidence for the History of West African "Castes."." In *Status and Identity in West Africa*, ed. David C. Conrad and Barbara E. Frank, 61–85. Bloomington: Indiana University Press.
Thapar, Romila. 2002/2004. *Early India: From the Origins to AD 1300*. Berkeley: University of California Press.
Throop, C. Jason. 2010. Suffering and Sentiment: Exploring the Vicissitudes of Experience and Pain in Yap. Berkeley: University of California Press.
Tilly, Charles. 1998. *Durable Inequality*. Berkeley: University of California Press.
Tilly, Charles. 2003. "Changing Forms of Inequality." *Sociological Theory* 21 (1): 31–6. http://dx.doi.org/10.1111/1467-9558.00173.
Todd, D. M. 1977. "Caste in Africa?" *Africa: Journal of the International Africa Institute* 47 (4): 398–412. http://dx.doi.org/10.2307/1158345.
VanDerwarker, Amber M., C. Margaret Scarry, and Jane M. Eastman. 2007. "Menus for Families and Feasts: Household and Community Consumption of Plants at Upper Saratown, North Carolina." In *The Archaeology of Food and Identity*, ed. Katheryn C. Twiss, 16–49. Center for Archaeological Investigations, Southern Illinois University Carbondale Occasional Paper no. 34, Carbondale.
Venkatesh, Sudhir Alladi, and Steven D. Levitt. 2000. ""Are We a Family or a Business?" History and Disjuncture in the Urban American Street Gang." *Theory and Society* 29 (4): 427–62. http://dx.doi.org/10.1023/A:1007151703198.
Walters, Delores M. 1987. Perceptions of Social Inequality in the Yemen Arab Republic. Unpublished Ph.D. thesis, Department of Anthropology, New York University.

Watkins, W. S., R. Thara, B. J. Mowry, Y. Zhang, D. J. Witherspoon, W. Tolpinrud, M. J. Bamshad, S. Tirupati, R. Padmavati, H. Smith, et al. 2008. "Genetic variation in South Indian castes: evidence from Y-chromosome, mitochondrial, and autosomal polymorphisms." *BMC Genetics* 9 (1): 86. http://dx.doi.org/10.1186/1471-2156-9-86 AccessedAugust252010.

Weedman, Kathryn J. 2002. "On the Spur of the Moment: Effects of Age and Experience on Hafted Stone Scraper Morphology." *American Antiquity* 67 (4): 731–44. http://dx.doi.org/10.2307/1593801.

Weedman, Kathryn J. 2006. "An Ethnoarchaeological Study of Hafting and Stone Tool Diversity among the Gamo of Ethiopia." *Journal of Archaeological Method and Theory* 13 (3): 189–238. http://dx.doi.org/10.1007/s10816-006-9010-4.

Wilson, David Sloan, John J. Timmel, and Ralph R. Miller. 2004. "Cognitive Cooperation: When the Going Gets Tough, Think as a Group." *Human Nature (Hawthorne, N.Y.)* 15 (3): 225–50. http://dx.doi.org/10.1007/s12110-004-1007-7.

Yoffee, Norman, and George L. Cowgill, eds. 1988. *The Collapse of Ancient States and Civilizations.* Tucson: University of Arizona Press.

13

The Dynamics of Cooperation in Context

Concluding Thoughts

Gary M. Feinman

> Our breathtaking ability to cooperate is one of the main reasons we have managed to survive in every ecosystem on Earth, from scorched sun-baked deserts to the frozen wastes of Antarctica to the dark, crushing ocean depths.
> (Nowak 2011: xiv)

Potential for cooperation is one of our most distinguishing features as a species. And yet, as the chapters in this collection illustrate, the nature of human cooperation is both variable and contingent across space and time. In concluding this volume, my comments are not intended to arbitrate between the diverse perspectives on cooperation offered here, nor do they summarize synthetically the specific analyses and arguments that have been presented.

My aim instead is to place these studies of cooperation in a broader theoretical context, specifically within anthropological archaeology, by stressing how the approaches offered here, though internally diverse, collectively differ from the perspectives and tenets generally advanced by the key conceptual frameworks that have dominated our discipline since its inception. In a sense, I place our discussions of cooperation in a broader historical and paradigmatic context. Since the nature, scales, and degrees of cooperation underpin the variability and full extent of human social arrangements, an equally significant goal is to position our understanding of the diversity of human social formations and how they shift over time on firm theoretical ground, intellectual turf that facilitates and enhances our communication with scholars in the wide range of disciplines that share these thematic foci (e.g., Little 2000; Steckel 2007). The manifestation of human cooperation is not unique to any set of societies, cultures, or eras. Yet the degrees and nature of cooperation in different settings provide insight into why human societies differ and change. Consequently, unraveling the bases of cooperation, the different forms that it takes, and the ways that cooperative socio-economic arrangements are constructed and deconstructed yields channels for systematically and comparatively studying the great variation in human social formations that has occurred across the human career. It forges theoretical paths away from both the pigeon-holing and idiosyncratic approaches that characterize a grand part of the academic enterprise today, as well as the universal or formalist generalizations that often neglect the significant contingencies of historical process.

Describing the workings of high finance in New York in the 1980s, Michael Lewis, in *Liar's Poker* (1989: 281), states: "If there is one thing I learned on Wall Street, it's that when an investment banker starts talking about *principles*, he is usually also defending his interests and that he rarely stakes out the moral high ground unless he believes there is gold under his campsite" [emphasis added].

More than four centuries earlier, Alonso de Zorita (1994: 93) described the investiture of a new ruler of the Aztec empire. The high priest morally cautioned the incoming lord: "Consider the honor your vassals have done you. Now that you are confirmed as ruler you must take great care of them and regard them as your sons; you must see to it that they be not offended and that the greater do not mistreat the lesser . . . You must be very diligent in affairs of war."

When considering the breadth and scope of human institutions, past or present, we should not generalize too facilely or directly from any single account, be that the specific intrigues of Salomon Brothers or a bureaucrat's decades-removed recounting of the incarnation of the Aztec emperor. Yet Lewis's observation and Zorita's writings do capture a fundamental dialectical tension between self-serving and group-/institutional-re-enforcing motivations that characterize many, if not most, human social formations. Whether analyzing the inner workings of a large investment bank or human societal groupings (as discussed in all of the chapters in this volume), individuals, including those in leadership positions, navigate between the urges and desires that guide their behaviors and the aims and challenges that affect the persistence of the groups, networks, and institutions to which they belong.

This dialectic or tension lies at the essence of human cooperation and also helps account for the diversity of social arrangements that characterize the human career. In a sense, the recurrent dichotomy between "voluntaristic" and "coercive" group organizational principles that Roscoe (chapter 3) discusses has turned out to be somewhat of a false one. Most human groupings, large and small, involve both certain imbalances in power as well as sharing and reciprocity, both self-interested behaviors and elements of altruism. The more fundamental questions concern the whys and hows of specific social arrangements in particular contexts and how and why those social arrangements change or endure.

For social scientists, the key queries revolve around this element of human agency and the socioeconomic networks (large and small, plutocratic and democratic) that are underpinned by the cooperative interactions of agents, albeit often in different balances than the rather selfish and autocratic traditions and restricted networks that characterize the contemporary culture of investment banking outlined by Lewis (1989). Since degrees of cooperation underpin human groupings and social networks, states and market systems, as well as urban neighborhoods, castes, and villages, it is important to ask, what makes the approaches expressed in these papers different from previous interpretations and analyses? Why does the explicit consideration of agency, rational choice, and cooperation hold promise, and what are some of the broader implications of these theoretical frames?

For centuries, historians and social scientists have asked, why do societies rise and fall, why do people participate in social groupings when their motivations tend to be selfish and, why did people settle in early cities when the transition to urban life meant more work and shorter life spans for the majority of the population? In a sense, these queries are all variants on Hobbes's (2003: 100–128) dilemma, "What holds society together given the tendency of individuals to pursue their self-interest?" And why does the nature of human groupings have many crosscutting parallels across time and space and yet also highly significant differences and specific features? How can we understand both the diversity of human social formations and why have they coalesced, disaggregated, and recalibrated numerous times across human history?

Comparatively focused anthropologists and archaeologists, given our holistic perspectives (covering broad swathes of cultural-geographic space and deep time horizons) have long been concerned with these questions concerning human groupings. Slowly through innovations as well as borrowing from cognate disciplines, we have framed their analyses in ever more realistic ways that dovetail more closely with what is broadly recognized as characteristic of human behavior and the human career writ large.

What do I mean by "realistic"? After decades of discussions and debates, I suspect that I am not being too controversial when I argue that most, if not all, of the authors in this volume would broadly agree with the following statements regarding human behavior, the groups that we belong to, and how we conceptualize them. In other words, these four (Table 13.1) basic principles dovetail with what has been learned through decades of archaeological and anthropological research, but they also place our fundamental tenets more in line with perspectives on human behavior and groups in other social and historical sciences.

1. Evolutionarily, the human career has roots both in a primate tendency toward dominance as well as status and gender hierarchies and inequalities and in our genus *Homo*'s long dependence for survival on living in groups, or "groupishness." The legacies as well as the marked differences are both

TABLE 13.1 Human behavior: basic principles

1.	The evolutionary legacy of our species has tendencies toward both dominance hierarchies and high degrees of sociality.
2.	Agency is universal but also constrained by structure and resources.
3.	Human groupings may be open and permeable to varying degrees, but they rarely are entirely closed for lengthy periods.
4.	Multiscalar perspectives are essential for understanding human groups.

significant (see also Boesch and Tomasello 1998). As discussed in several of the chapters, especially Blanton-Fargher (chapter 5), as well as my own work (Feinman, chapter 2), these countervailing tendencies can manifest in varying ways with implications for relations of trust, compliance, and authority.

2. Following Sewell (1992), among others, all people have the capability for agency, as they have the ability to learn language. Yet just as people learn a multitude of different languages, individual agency is constrained in various and significant ways by structure (including learning) and available resources. At the same time, "To be an agent means to be capable of exerting some degree of control over the social relations in which one is enmeshed, which in turn implies the ability to transform those social relations to some degree" (Sewell 2005: 143).
3. Most human groupings, be they villages, cities, states, nations, or households, are generally open at least to a degree. Boundaries of greater or lesser efficacy may be established, but they should never be assumed or presumed as a natural condition and rarely, if ever, are they entirely impermeable (e.g., Barth 2000; Wolf 1982: 4–7).
4. To understand human groupings and their histories, we must examine multiple scales; neither households, nor single settlements, nor regions alone are sufficient (e.g., Kowalewski 1995; Manson 2008). One ought not extrapolate from observations at one scale to generalize about others.

And yet, if we accept these basic propositions in principle, we have a bit of a problem when it comes to the ways that we, as anthropological archaeologists, frame and envision preindustrial societies. In archaeology, paradigmatic frameworks have not replaced one another holistically as Thomas Kuhn (1970) envisioned for physics. Rather, when new theoretical frameworks are advanced, they have built and borrowed heavily on earlier perspectives and tenets.

Decades ago, the British prehistorian Glyn Daniel (1976: 5) wrote, "The present state of archaeology cannot be divorced from its past state." Daniel was right about the discipline of archaeology at that time and the practice of archaeology today; I suspect his words also characterize the situation in the fields of anthropology and historical social sciences more generally. Nevertheless, I focus on archaeology and our framing approaches to the preindustrial world since I am most aware about where matters stand in such studies.

TABLE 13.2 Perspectives on the preindustrial past

Theoretical Frame	*Agency*	*Scalar Focus*	*Boundedness*
Culture history	Elite?	Culture	Closed
Cultural evolutionary systems	Elite	Society	Closed
Marxism/ Marx- influenced	Elite	Society Class (rarely)	Potentially open
Sociobiology (narrow Darwinian)	All	Individuals Kin	Not adequately considered
Postprocessual	Elite (situational for commoners)	Society	Mostly closed
Rational choice/ cooperation	All	Explicitly multiscalar	Open

As I see it, the problem for our consideration of preindustrial societies is that our theoretical frameworks, often carrying conceptual baggage from even earlier perspectives, have not kept pace with what we have learned (Table 13.2). Over the past half-century, most archaeologists, particularly those trained or based in North America, have adhered to one of five main theoretical frames or paradigms (culture history, cultural evolutionary/systems, Marxism or Marx-inspired, sociobiological/Darwinist, or postprocessual).

Although admittedly simplified, there is little question that traditional culture history and more functionally oriented, cultural evolutionary approaches assumed closed cultural systems (corresponding to Wolf's [1982: 6] metaphorical billiard balls only occasionally banging into each other) with little allocation of individual agency except for that afforded to elites or a small subset of the populace (see Nassaney and Sassaman 1995: xxi–xxii). Behavioral practices often were judged to be set by those with influence, and these norms were then followed by the rest of the population. Although evolutionary approaches were not necessarily wedded in an essential manner to an overriding focus on elites, such emphasis may have been intellectual positions or notions carried over from prior theoretical emphases on culture history. Marxist-inspired perspectives applied to the preindustrial/archaeological past rarely consider class conflicts, also giving agency principally to those with power.

In contrast, recent postprocessual perspectives do speak of commoner resistance, yet it is largely considered to be ad hoc, for example, at times of known societal breakdown or collapse (e.g., Joyce, Bustamante, and Levine 2001). Through this circumstantial or situational consideration of agency, the opportunity to understand the shift in the rules of cooperation is lost. Strict sociobiological approaches do ascribe agency more broadly, but, reliant on narrow definitions of self-interest and altruism toward kin, they cannot explain larger social formations as noted by several contributors to this volume (Blanton and Fargher, chapter 5; Pluckhahn, chapter 8; Roscoe, chapter 3; Spencer; chapter 9).

Only a sixth paradigmatic alternative, rational choice/cooperation-based frameworks, both align with the tenets regarding human groups (Table 13.1) and provide explicit, testable hypotheses concerning the nature and character of those cooperative arrangements (Blanton and Fargher, chapter 5; Smith, chapter 12). As illustrated by the papers in this collection, refocusing our efforts toward the bases of cooperative arrangements and how they vary and change over time provides productive avenues to probe the dynamic dialectics between commoners and elites, cooperation and competition, agency and structure, history and process, and biology and culture.

Talk of paradigms can lead to eyes glazing over, but allow me a brief illustrative example of why and how reframing our questions and interpretations can enhance our understanding of the past. In the region where I have studied for most of my career, pre-Hispanic Mesoamerica, the mid-twentieth century ushered in a significant theoretical shift. Anchored in the writings of Marx's (1971) Asiatic mode of production, and its influence in the writings of Wittfogel (1957), Childe (1950), and Polanyi (Polanyi Arensberg, and Pearson 1957), renowned scholars such as Wolf, Palerm (Palerm and Wolf 1957), and Sanders (Sanders, Parsons, and Santley 1979) refocused the study of Mesoamerica's past from culture history to questions concerning the region's economy and its relationship to sociopolitical organization. This shift in emphasis fostered six decades of research that has encouraged the systematic study of regional settlement patterns and domestic excavations rather than merely examinations of tombs, temples, and tourist sites, as had been the primary focus earlier.

Yet hewing to the Asiatic mode and Wittfogel's (1957) hydraulic hypothesis, Mesoamericanists, for decades, tended to think of pre-Hispanic governments as despotic, deriving their power from the control of water for irrigation, which then provided a basis for centralized redistribution and a tributary or command economy that administratively controlled and managed production. Synthesizing this stance, Carrasco (2001: 363) wrote: "Ancient Mexico had a politically integrated economy. The government controlled the basic means of production, land, and labor, and accumulated the surplus in the form of tribute. This basic model, which also deemphasized nonagricultural production, marketplace exchange, and economic interdependency, held sway for decades despite the rich documentary accounts of Aztec markets (e.g., Díaz del Castillo 1908–1916) and only limited evidence for large-scale, centralized water control (Offner 1981). Such practices were often explained away since they, by implication, necessitated degrees of agency beyond those who directly held political power.

And yet, a very different empirical picture emerged from six decades of research. In pre-Hispanic Mesoamerica, large-scale water control was relatively rare, and when it did occur it tended to be well after the rise of hierarchical polities (e.g., Baker 1998; Kirkby 1973; Offner 1981). Little evidence for centralized redistribution has ever been recorded. Even the ample storehouses that sustained the Spanish in their conquest of the Andes were little known in Mesoamerica. And most tellingly, nonagricultural production was not only centered in domestic contexts (e.g., Feinman 1999), but from the dawn of sedentary life, many households engaged part-time in such practices, implying high degrees of household-to-household interdependence and a good deal of commoner agency in the allocation of their labor and how they made a living (Feinman and Nicholas 2004, 2010; Hirth 2009). Not only would such dispersed production have been nearly impossible to manage centrally, but marketplace exchange likely was a more significant institution than many presumed, dating well before Aztec times (Feinman and Garraty 2010; Garraty and Stark 2010).

In recent years, major disjunctures have arisen between our long-standing theoretical frameworks, which limit agency to the elite and envision command economies, and the actual empirical findings that are coming to light. New frames, underpinned in rational choice and cooperation, both better fit what we have learned through empirical research and provide workable theoretical paths to understand variation and change across pre-Hispanic Mesoamerica.

But there is another consideration. When we recognize the agency of pre-Hispanic Mesoamerican householders or South Asian caste members (Smith, chapter 12), we remove the qualitative barriers that many in the social sciences have erected for too long between the theoretical paradigms we employ to understand Euro-American society over the last two centuries, and those marshaled to interpret and explain the history of the rest of the world's population (Blanton and Fargher 2008; Little 1991). Neither cooperation nor rationality is bound to any culture. This artificial divide between the recent West and the rest is particularly odd, since many great scholars cannot agree on precisely why and when the West became exceptional (e.g., Abu Lughod 1989; Goldstone 2002; Polanyi 1944; Wallerstein 1974; Wolf 1982). The kinds of approaches discussed by the authors in this book have the opportunity not only to yield a more empirically grounded (yet nevertheless comparative) vantage on the past, but to offer a wider and more reflective mirror that crosses scientific borders to help explain how we arrived at the present and possible ways to approach the future.

ACKNOWLEDGMENTS

An earlier version of this paper was presented at the 2011 Annual Meeting of the American Association for the Advancement of Science. I would like to thank David Carballo for asking me to prepare a concluding thought for this volume and for leading this project with forethought and care. I also very much appreciate his suggestions on earlier version of this text. Linda M. Nicholas offered helpful comments on several earlier drafts of this piece. I deeply appreciate her assistance and her work on our joint field studies cited in this chapter.

REFERENCES

Abu-Lughod, Janet L. 1989. *Before European Hegemony: The World System A.D. 1250–1350.* Oxford: Oxford University Press.

Baker, Jeffrey L. 1998. "The State and Wetland Agriculture in Mesoamerica." *Culture & Agriculture* 20 (2-3): 78–86. http://dx.doi.org/10.1525/cag.1998.20.2-3.78.

Barth, Fredrik. 2000. "Boundaries and Connections." In *Signifying Identities: Anthropological Perspectives on Boundaries and Contested Values*, ed. Andrew P. Cohen, 15–36. London: Routledge.

Blanton, Richard, and Lane Fargher. 2008. *Collective Action in the Formation of Pre-Modern States.* New York: Springer.

Boesch, Christophe, and Michael Tomasello. 1998. "Chimpanzee and Human Cultures." *Current Anthropology* 39 (5): 591–614. http://dx.doi.org/10.1086/204785.

Carrasco, Pedro. 2001. "Economic Organization and Development." In *The Oxford Encyclopedia of Mesoamerican Cultures: The Civilizations of Mexico and Central America*, ed. David Carrasco, 363–66. Oxford: Oxford University Press.

Childe, V. Gordon. 1950. "The Urban Revolution." *Town Planning Review* 21:3–17.

Daniel, Glyn. 1976. *A Hundred and Fifty Years of Archaeology*. Cambridge, MA: Harvard University Press.

Díaz del Castillo, Bernal. 1908–1916. *The True History of the Conquest of New Spain*, 5 volumes. Translated by Alfred P. Maudsley, Hakluyt Society, New York.

Feinman, Gary M. 1999. "Rethinking Our Assumptions: Economic Specialization at the Household Scale in Ancient Ejutla, Oaxaca, Mexico." In *Pottery and People: A Dynamic Interaction*, ed. James M. Skibo and Gary M. Feinman, 81–98. Salt Lake City: University of Utah Press.

Feinman, Gary M., and Christopher P. Garraty. 2010. "Preindustrial Markets and Marketing: Archaeological Perspectives." *Annual Review of Anthropology* 39 (1): 167–91. http://dx.doi.org/10.1146/annurev.anthro.012809.105118.

Feinman, Gary M., and Linda M. Nicholas. 2004. "Unraveling the Prehispanic Highland Mesoamerican Economy: Production, Exchange, Consumption in the Classic Period Valley of Oaxaca." In *Archaeological Perspectives on Political Economies*, ed. Gary M. Feinman and Linda M. Nicholas, 167–188. Salt Lake City: University of Utah Press.

Feinman, Gary M., and Linda M. Nicholas. 2010. "A Multiscalar Perspective on Market Exchange in the Classic Period Valley of Oaxaca." In *Archaeological Approaches to Market Exchange in Ancient Societies*, ed. Christpher P. Garraty and Barbara L. Stark, 85–98. Boulder, CO: University Press of Colorado.

Garraty, Christopher P., and Barbara L. Stark, eds. 2010. *Archaeological Approaches to Market Exchange in Ancient Societies.* Boulder: University Press of Colorado.

Goldstone, Jack A. 2002. "Efflorescence and Economic Growth in World History: Rethinking the "Rise of the West" and the Industrial Revolution." *Journal of World History* 13 (2): 323–89. http://dx.doi.org/10.1353/jwh.2002.0034.

Hirth, Kenneth G., ed. 2009. *Housework: Craft Production and Domestic Economy in Ancient Mesoamerica.* Archeological Papers of the American Anthropological Association, 19.

Hobbes, Thomas. 2003. *Leviathan.* Ed. G.A.J. Rogers and Karl Schuhmann. Bristol: Thoemmes Continuum.

Joyce, Arthur A., Laura Arnaud Bustamante, and Marc Levine. 2001. "Commoner Power: A Case Study from the Classic Period Collapse on the Oaxaca Coast." *Journal of Archaeological Method and Theory* 8 (4): 343–85. http://dx.doi.org/10.1023/A:1013786700137.

Kirkby, Anne V. T. 1973. *The Use of Land and Water Resources in Past and Present Valley of Oaxaca.* Museum of Anthropology, University of Michigan, Memoirs 5.

Kowalewski, Stephen A. 1995. "Large-Scale Ecology in Aboriginal Eastern North America." In *Native American Interactions: Multiscalar Analyses and Interpretations in the Eastern*

Woodlands, ed. Michael S. Nassaney and Kenneth E. Sassaman, 147–173. Knoxville: University of Tennessee Press.

Kuhn, Thomas. 1970. *The Structure of Scientific Revolutions*. Chicago: University of Chicago Press.

Lewis, Michael. 1989. *Liar's Poker: Rising through the Wreckage on Wall Street*. New York: W. W. Norton.

Little, Daniel. 1991. "Rational-Choice Models and Asian Studies." *Journal of Asian Studies* 50 (1): 35–52. http://dx.doi.org/10.2307/2057474.

Little, Daniel. 2000. "Explaining Large-Scale Historical Change." *Philosophy of the Social Sciences* 30 (1): 89–112. http://dx.doi.org/10.1177/004839310003000105.

Manson, Steven M. 2008. "Does Scale Exist? An Epistemological Scale Continuum for Complex Human-Environmental Systems." *Geoforum* 39 (2): 776–88. http://dx.doi.org/10.1016/j.geoforum.2006.09.010.

Marx, Karl. 1971. *A Contribution to the Critique of Political Economy*. London: Lawrence and Wishart.

Nassaney, Michael S., and Kenneth E. Sassaman. 1995. "Introduction: Understanding Native American Interactions." In *Native American Interactions: Multiscalar Analyses and Interpretations in the Eastern Woodlands*, ed. Michael S. Nassaney and Kenneth E. Sassaman, xix–xxxviii. Knoxville, TN: University of Tennessee Press.

Nowak, Martin A. 2011. *SuperCooperators: Altruism, Evolution, and Why We Need Each Other to Succeed*. New York: Free Press.

Offner, Jerome. 1981. "On the Inapplicability of "Oriental Despotism" and the "Asiatic Mode of Production" to the Aztecs of Texcoco." *American Antiquity* 46 (1): 43–61. http://dx.doi.org/10.2307/279985.

Palerm, Angel, and Eric R. Wolf. 1957. "Ecological Potential and Cultural Development in Mesoamerica." *Pan American Union Social Science Monograph* 3:1–37.

Polanyi, Karl. 1944. *The Great Transformation: The Political and Economic Origins of Our Time*. Boston: Beacon Press.

Polanyi, Karl, Conrad Arensberg, and Harry Pearson, eds. 1957. *Trade and Market in the Early Empires*. Glencoe, IL: Free Press.

Sanders, William T., Jeffrey R. Parsons, and Robert S. Santley. 1979. *The Basin of Mexico: Ecological Processes in the Evolution of a Civilization*. New York: Academic Press.

Sewell, William H., Jr. 1992. "A Theory of Structure: Duality, Agency, and Transformation." *American Journal of Sociology* 98 (1): 1–29. http://dx.doi.org/10.1086/229967.

Sewell, William H., Jr. 2005. *Logics of History: Social Theory and Social Transformation*. Chicago: University of Chicago Press.

Steckel, Richard H. 2007. "Big Social Science History." *Social Science History* 31 (1): 1–34. http://dx.doi.org/10.1215/01455532-2006-012.

Wallerstein, Immanuel. 1974. *The Modern World-System: Capitalist Agriculture and the Origins of the European World-Economy in the Sixteenth Century*. New York: Academic Press.

Wittfogel, Karl. 1957. *Oriental Despotism*. New Haven, CT: Yale University Press.

Wolf, Eric R. 1982. *Europe and the People without History*. Berkeley: University of California Press.

Zorita, Alonso de. 1994. *Life and Labor in Ancient Mexico: The Brief and Summary Relation of the Lords of New Spain*. Trans. Benjamin Keen. Norman: University of Oklahoma Press.

Contributors

RICHARD E. BLANTON, Department of Anthropology, Purdue University, West Lafayette, Indiana 47907 (blantonr@purdue.edu)

DAVID M. CARBALLO, Department of Archaeology, Boston University, 675 Commonwealth Avenue, Suite 347, Boston, Massachusetts 02215 (carballo@bu.edu)

BENJAMIN CHABOT-HANOWELL, Department of Anthropology, University of Washington, Box 353100, Seattle, Washington 98195 (hanowell@u.washington.edu)

JELMER W. EERKENS, Department of Anthropology, University of California at Davis, One Shields Avenue, Davis, California 95616 (jweerkens@ucdavis.edu)

LANE F. FARGHER, Departamento de Ecología Humana, Centro de Investigación y de Estudios Avanzados del IPN–Unidad Mérida, Mérida, Yucatan, México, 97130 (lfargher@purdue.edu)

GARY M. FEINMAN, Department of Anthropology, Field Museum, 1400 South Lake Shore Drive, Chicago, Illinois 60605 (gfeinman@fieldmuseum.org)

LISA J. LUCERO, Department of Anthropology, University of Illinois, 109 Davenport Hall, 607 S. Matthews Ave., Urbana, Illinois 61801 (ljlucero@illinois.edu)

THOMAS J. PLUCKHAHN, Department of Anthropology, University of South Florida, 4202 E. Fowler Ave, SOC107, Tampa, Florida 33620 (tpluckhahn@usf.edu)

PAUL ROSCOE, Department of Anthropology, University of Maine, 5773 S. Stevens Hall, Orono, Maine 04469 (Paul_Roscoe@umit.maine.edu)

DEAN J. SAITTA, Department of Anthropology, University of Denver, Denver, Colorado 80208 (dsaitta@du.edu)

MONICA L. SMITH, Department of Anthropology, University of California at Los Angeles, Los Angeles, California 90095 (smith@anthro.ucla.edu)

CHARLES S. SPENCER, Division of Anthropology, American Museum of Natural History, Central Park West at 79th Street, New York, New York 10024 (cspencer@amnh.org)

CHARLES STANISH, Department of Anthropology, University of California at Los Angeles, Los Angeles, California 90095 (stanish@ucla.edu)

Index

Abelam, 63
accumulator models, 44
Aceh, 110
Acropolis, 111
activity areas, intrahousehold, 162–64
Adams County (Miss.), Saragossa Plantation at, 138
African Americans, 134, 138
African diaspora communities, collective action, 132–35
Africanisms, 132–33
Africans, slaves, 132–33
agency, 102, 179, 302, 305; individual, 129–30; race-based, 132–35, 143
aggrandizer models, 19, 44
aggregation, cooperative, 210–11
agora, public functions of, 111–12
agricultural plots, private claims to, 16–17
agriculture, 18, 89, 90, 178, 282; central Mexico, 251, 252; collective vs. private access to, 16–17; work exchange, 245, 246
aguadas, 223–24, 236
Akbar, 110
Alexander, Richard, 62
Alexander the Great, 74
alliances, 37, 45, 48, 285
altruism, 8–9, 15, 37, 60, 62, 198, 199, 300; reciprocal, 95, 154, 200
ancestor worship, 111
Annapolis, 134–35
Anotche, Komfo, 110
apartment compounds, Teotihuacan, 254–55
Arapesh, 63
architecture, 133, 136, 212; ceremonial/ritual, 21, 202, 238, 258–59; integrative, 43–44; Middle and Late Woodland, 180–82, 184–86, 188; religious, 111–12
armory, household assemblages and, 139–40
Asante, 110
Athens, 111–12
Axial Age theory, 108, 111
Aymara, 279
Aztecs, 113, 258, 300; feasting, 248–49

Bakongo iconography, 134
balance-of-power contexts, 74, 75
"balance-of-power" races, 62, 66, 73, 77(n2)
ballcourts, Hohokam, 278
bands, and Indian outcastes, 286
bargaining models, 24; economic, 236–38, 239; Nash, 233–35; noncooperative sequential, 235–36

Barinas (Venezuela), 200; cooperating households in, 210–11; habitation sites in, 202–5; intercommunity cooperation and conflict, 23, 197; interpolity conflicts in, 212–17; settlement hierarchies in, 202–10
Batatuy, 209
beer bottlers, 139
behavioral ecology models, 12, 58
Bena Bena, warfare, 63–64
Berwind, and Colorado Fuel and Iron Company, 137–38
Big Man, 77(n3), 99; community size and population, 68–69(table); emergence of, 71–72
biological markets, 230
biological reductionism, 94, 114
Black Death, 278
blacksmiths, West African, 288
boardinghouses, 140; Colorado Fuel and Iron Company and, 137–38
Boas, Franz, 143
Boott Mills, 140
Boston, domestic reform in, 135–36
bow and arrow: and household organization, 23, 187; and individual hunting, 159–60
British Commonwealth, 75
British Empire, 75, 281
Buddhism, in India, 281–82
bureaucratization, 105, 106–7(table)
burial mounds, Woodland, 180
burial sites, royal, 110

calpolli, 247
calzada networks, pre-Hispanic Venezuela, 205–6, 208, 211, 213–14
capitalism, 130
caste systems, 24; cheap signaling, 288–89; economic specialization, 279–80; entitlement protection, 286–87, 292; environmental changes and, 285–86; India, 280–83; occupational social hierarchies, 275–76; opportunity hoarding, 287–88; political economies and, 289–90; West Africa, 284–85
Catholic Church, secularization, 112
causeways. *See* calzada networks
cenotaphs, 110
ceramics. *See* pottery
ceremonial centers, 253; craft production at, 255, 256(fig.)
ceremonies: Maya lowlands, 238; Middle and early late Woodland, 180, 183, 187–88
Chandalas, 282
cheap signaling, caste systems and, 288–89
Chesapeake, slaves on, 143
chiefdoms, 46, 71, 87, 77(n3), 206, 212, 217; Barinas, 197, 200
circumscription theory, 58, 73
City Beautiful Movement, 136
civic life, 112
clans, 63, 285. *See also* kinship
class, 131; collective action and, 139–42; and material culture, 136–37
Classic period (Maya), 24
Classic period (Mexico), labor collectives, 250–61
class solidarity, 141–42
Coal Field Strike (1913–1914), 137, 140–41
coalitional enforcement hypothesis, 13
coal towns, boarders in, 137–38
coatequitl, cuatequitl, 247, 248
coercive theories, 19–20
cognition, social, 36, 37
cognitive constraints, 49
collaboration, 8, 198; pre-Hispanic Venezuelan households, 210–12
collective action, 9–10, 23, 58, 130, 131, 136–38; class and, 139–42; gender and, 135–38; race and, 132–35; use of term, 14–15
collective action scale, 106–7(table), 117(n14); religions, 110–13
collective-action theory, 19, 60–61, 104
collectives: Mesoamerican labor, 245–61; water management, 251–52
collective unity, 141
collectivities, 10, 48, 61, 179; land tenure, 16–17
colonoware, slave use of, 133–34
Colorado Fuel and Iron Company, working-class women and, 137–38
commoners, 20, 236, 238. *See also* patron-client relationships
common-pool resources, 9, 10, 11, 16, 17–18
common property, 102
communities, 21, 23, 138, 243; cooperative labor, 246–47; growth in size of, 44–45; nation-state as, 65–66; in New Guinea, 63, 67; size of, 38, 39
competition, 3, 4, 8, 153, 177, 237; and cooperation, 239–40
complexity, 35, 38; hierarchical, 40–41
complex societies, 15, 275; decline of, 276–80; evolution of, 93–94; group-level coping mechanisms, 277–78; patron-client relations and, 225–29
compounds, Teotihuacan, 254–55
comunidad, 246
conflict, 18, 175, 253; intercommunity, 23, 189; interpolity, 212–17
conflict (coercive) theories, 57, 58, 59; interpolity, 212–17; political centralization and, 72–73

Mexico, Basin of, irrigation system, 251
middle class, domestic architecture, 136
Middle Sepik River, warfare on, 63–64
Middle Woodland period, 177; competitive feasting, 188–89; households, 180–83
military, 70, 73
militias, Venezuelan chiefdoms, 212
Ming Dynasty, 111
mining permits, 153
Mixtec, 246
Mongol Confederation/Empire, 74, 75
Monticello, 143
Moon Pyramid (Teotihuacan), obsidian production at, 255, 261
moral code, 280; Islamic, 112–13
mortuary treatments, Woodland, 180
mounds, at El Gaván site B12, 213–14; Hohokam, 17; Middle and Late Woodland, 178, 180, 184; pre-Hispanic Venezuelan, 202–5, 207–8, 209–10
Mughals, 110, 282, 287
mullet fishing, ritual cycle and, 89–90
multiactor (-person) models, 20
Murdock, George, 7, 41
mutualism, 8, 290

Nahua, 244, 251, 262; communal labor, 246, 247–48
Nash bargaining solution, 233–35, 239
nation-state, 57; as community, 65–66; social complexity of, 64–65
neighborhoods, 248, 258
Neolithic package, 38–39
network systems, 46–47
neural plasticity, 100
New Guinea, 60, 67; Big Man systems in, 70, 71–72, 99; community size and population, 68–69(table); warfare in, 18, 63–64
New Philadelphia (Ill.), 135
New Pragmatism, 143
New York City, 136–37
New Zealand, 90
9/11, US responses to, 65–66
nonexcludable goods/resources, 9–11, 14, 153, 167

obsidian, 255; access to, 160–62
occupational specialization: Indian caste system, 280–81; Lake Titicaca Basin, 278–79
Odwira, 110
Old God of Fire, 260
opportunity hoarding, castes and, 287–88
organizational complexity, 42, 47–48; and population size, 39–40, 49(fig.)
Osu, 284
Otomís, 251
Ottoman Empire, 112–13
outcaste groups, leatherworking, 286
Owens Valley (Calif.), 23, 152, 155; cooperative behavior in, 164–66; households and activity areas, 162–64; leaders in, 166–67, 168; obsidian in, 160–62; social organization in, 156–57; subsistence, 158–60
ownership, 11; resource, 153, 165

Paiutes, 155, 165–66
palisades, 212
paternalism, corporate, 140
Paterson (N.J.), 135
patron-client relationships, 239–40; in complex societies, 225–29; Maya lowlands, 223–24, 238; surplus distribution in, 229–30; water control systems, 230–33
Philadelphia, 136
pipes, as identifiers, 140
plasticity, behavioral, 100
platform mounds, Woodland, 180, 184
platforms, temple, 21
plazas, 21; as ritual space, 258–59
political centralization, scale and, 68–72
political economies, and caste systems, 289–90
political evolution theory, 57–58, 75–76, 77(n1)
political institutions, hierarchical complexity, 40–41
political intelligence, 104
politics, corporate, 244
polities, 22, 48, 57, 59, 67, 257, 282; conquest warfare and, 73–74; horizontal and vertical dimensions, 72–73; interests and, 76–77; Maya lowland, 230–33; political centralization of, 68–72; pre-Hispanic Venezuelan, 202–12
population: and community size, 68–69(table); and organizational complexity, 39–40, 49(fig.)
population density, and political centralization, 70–71
population growth, 38, 106
population size, 226, 277; and patron-client relationships, 228–29
postcolonial states, 75
postprocessualism, 303
potlatch, 11
pottery, 21, 206; colonoware, 133–34; Kolomoki, 182–83, 186–87; ritual use of, 260–61
power, 87, 167; political centralization of, 68, 70–72
power brokers, 44
power sharing, 45

internment, Japanese American, 135
interpersonal connectivity, 48
IPD. *See* Iterated Prisoner's Dilemma
Irish Americans, 135
irrigation systems, 11, 17, 251, 304
Islam, 110, 285; Ottoman Empire, 112–13
Italy, kin-based cooperation, 98
Iterated Prisoner's Dilemma (IPD), 86

Japanese Americans, internment. *See* internment
Jataka, 281
jihads, 285

kin, kinship, 37, 63, 154, 156–57, 257
kin psychology, 95, 98
kin selection, 98, 199
kivas, 17
knowledge, 143, 287
Kolomoki site (Ga.), 23, 188, 257; household changes in, 178, 180–87
kula exchange, 11
Kumasi, 110
!Kung, 70

Labai, mullet fishing in, 89–90
La Betania, 209
labor, 19, 139, 258–59, 279, 285; collective, 22, 24, 243, 244, 250–57, 260, 261; cooperative, 7, 176, 245–46; Maya lowlands, 236, 238; organization of, 87–88; in patron-client relationships, 228–29, 236; reciprocal exchange of, 244–50; ritualization of, 88–91
labor pools, 287
labor tribute, 246–48
La Calzada, 209
La Laguna, 252; ritual architecture in, 258–59; ritual ceramics in, 260–61
land, 16, 17; corporate management of, 245–46; Maya lowlands access to, 224–25, 236
land management, corporate, 245–46
landscape, 67, 135
land tenure, 261; collective vs. privatized, 16–17; Puebloan, 16–17
language, 102, 302
largest autonomous polity (LAP), 67
Late Formative period (Mexico), labor collectives in, 250–61
Late Graván phase, interpolity conflict, 212, 213
Late Paleolithic societies, 37, 47
Late Woodland sites, 189; households in, 177–78, 180–87
"Laws of Manu," 281
leadership, 38, 44, 167, 217; small-scale societies, 70–71. *See also* rulership
leadership package, 18
learning, social, 36, 37
leatherworking, 286
legitimacy, rules of, 286
Leviathan, 6–7, 64
Locust Grove Plantation, 138
longhouse communities, New Guinea, 67
Louisville (Kent.), Locust Grove Plantation, 138
Lowell (Mass.), 140
Lozi, 110–11
Ludlow Tent Colony, 140–41, 144
Lupaqa polity, 279

Macedonian Kingdom/Empire, 74, 75
Machiavellian intelligence, 100
macrosocial groups, 62
Magdalen Society Asylum (Philadelphia), 136
magic, 89, 90
"mahajanapada" dynasties, 282
maize, 186; competitive feasting, 188–89
Mali empire, 284
Malinke, 284, 285
Malinowski, Bronislaw, on Trobriand Islanders, 89–90
managerial (adaptational) models, 58
managers, material culture of, 139–40
Manambu, 63
Maori, 90
marketplaces, exchange, 305
Marxist theory, 176, 303
Massachusetts Ratifying Convention, 65
material cultures, 131; of cooperation, 20–22; gender and class in, 136–37, 140, 141–42; identity maintenance, 134–35; slave, 132–34
Mauryan polity, 282
Maya, 24, 246
Maya lowlands; patron-client relationships in, 223–24, 226, 239–40; surplus distribution in, 229–30; water control in, 230–33, 236, 237–38
Mead, Margaret, 9; *Cooperation and Competition among Primitive Peoples,* 7–8
Melanesia, 89
meme, 94
memory, social, 22
mental modules, 94, 95, 96, 101, 116(n3)
Mesa Verde region, 17
Mesoamerica, 243, 244; cooperative labor, 245–48, 249–50; ritual networks, 248–49; water control, 304–5. *See also* Maya lowlands; Mexico
mestizos, comunidad, 246
Mexico: collective labor in, 247–48, 249–51; resource management, 251–53; ritual and symbolism, 257–61; social institutions, 253–57; water control, 18, 304

facial expression, 102
faena, 246
fairness, principles of, 89
familial sociology, 95
families, 98, 157, 165, 166–67. *See also* households
famines, 282, 283, 286, 293(n2)
farming. *See* agriculture
feasting, feasts, 21, 151, 248; competitive, 188–89; work, 245, 249, 260
Federalist Papers, 65
field houses, 16–17
fiesta finance, 249
Firth, Raymond, on ritual cycles, 89, 90
fishing, ritual behaviors and, 89–90
flooding, Maya lowlands, 231
food: distribution of, 277, 286; production of, 137–38; sharing of, 17, 98, 117(n6), 157; storage of, 23, 256–57
foodways, slave, 133–34
foragers, 17, 23. *See also* hunter-gatherers
forest management, central Mexico, 252
Formative period (Mexico), labor collectives in, 250–61
Foundations of Human Sociality (Henrich et al.), 7
free-riding, 10, 18, 49, 61, 88, 100, 102, 152, 157; controlling, 153–55, 167–68; second-order, 12–13

game animals, 159, 160
game theory, 7, 11–12, 199; evolutionary, 83, 84; maintenance of cooperation and, 85–88
Gamo people, 280, 286–87
gangs, structure and organization, 291–92
gaze, 102
gender, 131; collective action and, 135–38
Genghis Khan, 74
gifting, reciprocal, 95
Golden Stool, 110
governments, and excludable resources, 153
Granada (Amache), 135
grazing permits, 153
Greenwich Village, material culture in, 136–37, 142
group behavior, 8(fig.), 11, 47, 48; altruism and, 198, 199; size and, 36, 43–44
groupishness, 37, 301–2
group selection, 85, 96
Gulf Coast, Woodland period, 178–87
guilds, 10; European, 278, 279

Harpers Ferry, 139
Hawaii, 278
headmen, Paiute, 165–66
helping, 15
hide-workers, Ethiopian, 280
hierarchy, 37; occupational, 275–76; political institutions, 40–41
Hindu, 110
historical archaeology/anthropology, 23; collective action in, 131–42; emancipatory knowledge of, 143–44; and evolutionary anthropology, 142–43
hoarding, opportunity, 287–88
Hobbes, Thomas, social contract, 6–7, 64
Hohokam, 17, 278
households, 17, 23, 89, 133, 152, 190(n1), 246, 304–5; access to resources, 160–62; activity areas, 162–64; boarders in, 137–38; collaborative, 210–12; hunter-gatherer, 156–57; Middle and Late Woodland, 177–78, 180–87; manager and worker, 139–40; ritual, 111, 258; Teotihuacan, 254–55
housekeeping, cooperative, 136
houses, 133, 206; Middle and Late Woodland, 180–82, 184–86, 188; slave, 133
human behavior, basic principles of, 301–2
human groupings, 300–301, 302
hunter-gatherers, 40, 64; cooperative behavior, 164–66; houses and activity areas, 162–64; interhousehold comparisons, 160–62; subsistence changes, 155–57, 158–60
hunting, 138, 151, 166, 187; cooperative vs. individual, 159–60
hydraulic hypothesis, 304

iconography, resistance and, 134, 135
Igbo, 133, 284
Ilahita, 63
inclusive fitness theory, 199
income, potable water and, 236
identities, 131, 140; maintenance of, 134–35
India, caste systems, 24, 275–76, 280–84, 286, 291
indirect bias, 18
individual networks, 44
individuals, 8, 9, 167, 243; agency, 129–30; and communal labor requirements, 246–47
industrialization, 139–40
inequality, 37, 131–32, 225, 231, 232
in-group vs. out-group, 99
Inka, labor-tax system, 278–79
institutions, 136; integrative, 43–44
integration, 35, 41–42, 48, 74
integrative architecture, village-level, 43–44
integrative strategies, 46–47
intelligence, 100
intentions, 23, 102
interconnection webs, 45, 46
interests: polities and, 76–77; vs. self-interests, 60, 61

Confucianism, 111
Congo-Angola region, 134
Connecticut River Valley, 139
consciousness, shared, 130
consensus building, 45
Constitutional Convention, 65
consumption, household, 23
contingent cooperation approach, 103
cooperation, 3–4, 21, 37, 49, 101, 151–52, 153; causes of, 5–6; and competition, 239–40; definitions of, 5, 179; evolution of, 22, 93, 198–99; maintenance of, 85–88; surplus value of, 237–38; use of term, 14–15
Cooperation and Competition among Primitive Peoples (Mead), 7–8
cooperation networks, size of, 40–41
cooperator dilemmas, 114–15
cooperators, 200–201
coordination, group, 11, 197
corporate-kin groups, 24
corporate mode, 46
corporate/network spectrum, 19, 46–47
corporate politics, Mesoamerican, 244, 257
Coso rock art, 160
cost-benefit considerations, 18, 152
cost signaling, 154
craft production, 10, 19; ritual behavior and, 90, 261; Teotihuacan, 255–56, 261
craft specialization, 225, 275; castes and, 280, 284, 286–87; Teotihuacan, 255–56
Cuicuilco, 258, 260
cult-slave status, 284
cultural anthropology, and political evolution theories, 57–58
cultural change, 101, 103–4
cultural evolution, 15, 19, 24
cultural production, 99–100, 101
cultural reductionism, 114
cultural transmission, 18, 84
cutlery factories, 139
Curbatí complex, 209

Darwinian anthropology, 93, 94, 96–98, 101, 114, 116(nn1, 5)
deceptive behavior, 100
defense, 10, 18, 65, 74, 212; interests in, 62–63; social groups and, 63–64, 66–67
defensibility, economic, 225
defensive group, scale of, 66–67
demographic stress points, 39
determinism, demographic, 38
Dewey, John, New Pragmatism, 143–44
Discourse on the Origin and the Foundations of Inequality among Men (Rousseau), 7
discrimination, 134
division of labor, 19
domestic reform, in Boston, 135–36
domestic space, 133
dominance, 37, 301–2
drift, in patron-client relationships, 228–29
drives, cooperative, 159
drought, Maya lowlands, 238
dual-processualists, 176–77
Ducal Procession, 112

earthworks, at El Gaván site B12, 212
economic bargaining models, and Maya lowlands, 233–38, 239
economic defensibility, of resources, 225, 226
economic rights, castes and, 287–88
economics, 262; maintenance of cooperation, 87–88; nonmarket, 84–85
economic specialization, 16; caste system and, 279–80; gangs and, 291–92
economies, 22, 283, 304
economies of scale, 19, 42, 87, 259
effigy vessels, La Laguna, 260–61
egalitarianism, 37–38, 70, 98
El Cedral polity, 207–9, 211; interpolity conflict and competition, 212–17
El Gaván polity (Venezuela), 197; calzada network, 205–6; interpolity conflict and competition, 212–17; description of, 202–5
El Gaván site B12, 212; burning of, 216–17; cooperating households at, 210–11; mounds at, 213–14; uninhabited space at, 215–16
elites, 87, 108; Titicaca Basin, 278–79; water control, 231, 232–33, 236, 237–38
empires, 74, 75, 283, 284
endogamy, caste, 280
Enlightenment, 6–7, 57, 104
entitlements, 288; caste protection of, 286–87 292
environmental change: caste system and, 285–86; complex societies and, 277–78, 285
ephemeral resource gluts, 225
Ethiopia, caste system in, 280, 286
ethnicity, 74, 141, 251
evolution: of complex societies, 93–94; of cooperation, 198–99
evolutionary anthropology, 93; and historical archaeology, 142–44
evolutionary cooperation, 22
evolutionary game theory, 83, 84
evolutionary psychology, 93, 96, 98
evolutionary theory, 93, 143
exchange, 89, 178, 231, 305
excludable goods/resources, 10–11, 19, 153

praxis, 143
preindustrial societies, 303
pressure groups, 60
prestige bias, 18
prestige goods, Maya exchange of, 231
priests, 110, 113
primary groups, 102
primates, 100, 117(nn9, 10), 301–2
principals, control of, 105, 106–7(table)
Prisoner's Dilemma, 86, 199
private goods, 10
privatization, of resources, 17, 157, 165, 167
privatization model, and Owens Valley archaeology, 158–66
problematization, religion and state, 109–13
problem solving, 100
products, distribution of, 88–89
projectile technologies, 13
prosocial behavior, 84; biological evolution of, 94–96
psychology, 83; evolutionary, 93, 96; kin, 95, 98
public goods, 16, 18, 20, 105, 106–7(table); defined, 9–10
public space, 136; ritual, 24, 111–12, 184, 258–59
Puebla, 251
Puebloans, land tenure, 16–17
punishment, 12, 13, 86–87, 88, 96, 154, 165
Purrón Dam, 251, 261
Purusha, 281

race, 131; collective action and, 132–35
raiding, inter-village, 253, 261
rainfall cycles, India, 282
rational-choice theory, 23, 48, 84, 152, 304
reciprocity, 11, 22, 23, 38, 91(n2), 99, 138, 260, 300; collective labor exchange, 244–50; direct, 199–200; indirect, 95, 200; strong, 12–13
recycling, 14
Red Hand symbol, 135
refugees, from interpolity conflict, 215–16
religion, 117(n11), 133, 154; group-wide cooperation and conformity, 167–68; in India, 281–82; and state structure, 107–13
representational understanding, 23, 100
reputation, 11, 12, 22, 86
reservoirs, Maya lowlands, 231, 238
residential mobility, 155
resistance, 19; and colonoware use, 133–34; in industrial settings, 139, 140; material culture used in, 134–35
resource problems, 16
resources, 23, 48, 70, 153, 232, 277, 291; access to, 160–61; central Mexico, 251–53; economic defensibility of, 225, 226; food, 158–60; privatization of, 17, 157; rulership and, 104–5; surplus distribution, 229–30
resource stress: group response to, 277–78
retribution, 11, 21; and strong reciprocity, 12–13
revenues, 104, 105
reverse dominance hierarchy, 37
rewards, 11, 13, 21, 22
Rig Veda, 281, 282
risk management, 292
ritual(s), 17, 109, 111, 279; castes and, 281, 284, 286–87; central Mexican, 257–61; labor scheduling, 88–90; public, 21, 22, 23, 24, 112, 278; Woodland, 180, 187–88
ritual centers, 109
ritual cycles, 111, 112, 113
ritualization, of domestic labor, 88–91
ritual networks, communal labor, 248–49
rock art, Coso-style, 160
Roman Empire, 112
Rousseau, Jean-Jacques, *Discourse on the Origin and the Foundations of Inequality among Men,* 7
Rubinstein solution, 235–36, 239
rulership, 20, 46, 48, 238; Aztec, 113, 300; divine, 111, 112; lineages, 231–32; Maya lowlands, 223–24; religion and, 109, 110; resources and, 104–5
rules of legitimacy, 286
Russell Cutlery Company, John, 139

saa sa'a, 246
Sacred Law of Islam (shari'a), 112–13
sacred sites, 110
sacrifices, 184, 259, 284
sake production, 135
Saragossa Plantation, 138
scalar hierarchy, 201
scalar issues, 22
scale, 35, 47; of defensive groups, 66–67; and political centralization, 68–72, 73
scrounging, tolerated, 98
secularization, Venice, 112
security, state, 64
seed collecting, 17; Owens Valley, 158–59
self-interest, 60
selfish behavior, 100
self-organization, 4, 16, 19, 21, 244
Senegal, 286
sequential bargaining model, noncooperative, 235–36
settlement hierarchies: El Cedral polity, 208–9; El Gaván polity, 202–5, 210
settlement size, growth in, 44–45; and organizational complexity, 40, 41

shamans, 168
shari'a, 112–13
sharing, 98, 157, 165, 167, 300; altruistic, 95, 154
shark fishing, 90
shrines, 133
slaves, 138; collective agency, 133–35, 143
slave trade, Atlantic, 285
small groups, 98
small-scale societies, 13, 18, 37, 67; political centralization and, 70–71
Smith-Choi simulation, 237
social change, 101, 103–4
social cognition, 36, 37
social complexity, 37, 47, 177; cooperation-competition dynamic, 175–76; decline of, 276–80; group dynamics, 278–79; India, 282–83; nation-state, 64–65; response to crises, 277–78
social contract, 6–7
social evolution, 244
social formation, 299–300
social groups, 61, 188; defensive, 63, 64, 66–67; egalitarian, 156–57
social hierarchy, 18, 206, 279, 285
social instincts, 95, 101
social institutions, 15–16; centralized, 17–18; central Mexican, 253–57
social intelligence, 100, 117(n9)
sociality, 37, 49; warfare and, 61–62
social memory, 22
social mobility, 141
social partners, 230
social production, 99–100, 101
social signaling, 18; caste systems and, 288–89
socioeconomic complexity, 22, 35
socioeconomic networks, 300–301
solidarity, expressions of class, 141–42
Soninke, 284
Sosso/Malinke war, 284, 285
South Carolina Lowcountry, slaves in, 133–34
springs, agricultural systems, 252
STAR, 22–23, 114–15
state, 59, 64; formation of, 101–2; and religion, 107–13; taxpayers and, 104–5
status recognition, earned, 21
stoning, 13
storage, at Woodland sites, 182, 186
Storm God, on effigy vessels, 260
strangers, cooperation among, 102
strikes, 137; Ludlow and, 140–41
subsistence, 182, 238; Owens Valley, 158–60
subtractability, 10
Sufism, 110
Sunjata epic, 284
surplus: cooperation and, 237–38; distribution of, 229–30
Swahili Lamu, 111
swidden agriculture, La Laguna, 252
Swift Creek ceramic tradition, 178
Sycamore site, 185
symbolic systems, state, 111
symbols, and central Mexican ritual, 257–61
symbols in action, 132–33
systems (multivariant) theories, 57, 58, 59

Tahiti, 71
Tamari, 284
Tarascans, 247
taxation, 20, 24, 248, 249, 277, 279; state services, 104–5
teaware, 137, 141
Tehuacán Valley, irrigation systems, 18, 251
temple-plaza-ballcourt configuration, 258–59
Teotihuacan, 250, 251, 257, 258, 260, 261, 262; social institutions, 253–57
Teotihuacan Valley, 252
Tepotzlan, 248
tequitl, tequio, 246–47; organization of, 249–50
Terminal Formative period, 257
terraces, Trincheras, 17
territories, foraging, 17
Tetimpa, 257
theft, 153
Theory of Mind, 22–23, 94, 100–101, 102, 115, 117(n10)
Thomas, Mary, 141
Tikopia, 89
time, as resource, 70, 238
time-space, and offensive warfare, 73–74
Titicaca Basin, Lake, 259; occupational specialization, 278–79
tit-for-tat models, 11–12, 86
Titus, 112
Tlaxcala, 246, 251
trade, 105, 278, 285
tragedy of the commons, 16, 153
Trajan, 112
transgression, 13
transportation, 70
trash, domestic, 137–38
Traulsen-Nowak formula, and El Gaván polity, 201–2, 210, 211–12
tribal societies, 43
Trincheras culture, 17
Trobriand islanders, 11; labor scheduling, 89–90
trust, 86, 101, 154
tsoma artisans, 286–87
Tughluq, Muhammad bin, 283
Tula plaza plan, 258–59

Turner's Falls (Mass.), 139
Tutu, Osei, 110
Tzintzuntzan, 247

Ultimatum Game (UG), 86
United States, 9/11 response, 65–66
unity, 141
upper class, 136
Upper Paleolithic societies, 37
urbanization, 24, 262
Uru, 279
utility, 83
Utopia Plantation, 133

Venetian Republic, 98
Venezuela, 18, 200; intercommunity cooperation and conflict, 23, 197; interpolity conflict, 212–17; polities in, 202–12
Venice, ritual cycles, 112
Vespasian, 112
Vijayanagara, 282
village-level societies, 43
villages, 17, 18, 63, 67, 253; growth in size, 44–45, 46(fig.); integrative architecture, 43–44; Paiute, 165–66; Woodland, 180–87
violence, 18, 259
Virginia, plantation architecture, 133
voluntaristic theories, 57, 58, 59; political centralization and, 72–73
vuelta mano, 245

warfare, 16, 18, 22, 67, 99; conquest, 73–74; El Gaván polity and, 212–17; and human sociality, 61–62; offensive and defensive, 62–64; West Africa, 284, 285
War of Independence, 65
water control/management, 16; centralized, 17–18; central Mexico, 251–52; Maya lowlands, 223–24, 230–33, 236, 237–38; Mesoamerica, 304–5
weaponry, polity use of, 72–73
Weeden Island ceramic tradition, 178
West Africa: caste systems, 24, 275–76, 284–85, 288, 291; foodways, 133–34
witchcraft, 168
Wolof, 284
women, working-class, 137–38
Woodland culture, households, 177–78, 180–87
Woodland Terrace, 186
work exchanges, agricultural, 245–46
work gangs, 247
working class, 135; domestic architecture, 136–37; domestic trash, 137–38; Ludlow Colony, 140–42; material culture, 139–42
work units, 42
World War II, Japanese American internment camps, 135

Xochitecatl, 251, 258; craft production at, 255–56

yam harvest festival, 110
yams, ritual cycle, 89
Yangoru Boiken, 63
Yoruba, 133
YWCA, 136

Zapotecs, 246
zero-sum resources, 10